Dietary Supplements and Human Health

Dietary Supplements and Human Health

Editor

Tsuyoshi Chiba

MDPI • Basel • Beijing • Wuhan • Barcelona • Belgrade • Manchester • Tokyo • Cluj • Tianjin

Editor
Tsuyoshi Chiba
National Institutes of Biomedical
Innovation, Health and Nutrition
Japan

Editorial Office
MDPI
St. Alban-Anlage 66
4052 Basel, Switzerland

This is a reprint of articles from the Special Issue published online in the open access journal *Nutrients* (ISSN 2072-6643) (available at: https://www.mdpi.com/journal/nutrients/special_issues/Dietary_Supplements_Human_Health).

For citation purposes, cite each article independently as indicated on the article page online and as indicated below:

LastName, A.A.; LastName, B.B.; LastName, C.C. Article Title. *Journal Name* **Year**, *Volume Number*, Page Range.

ISBN 978-3-0365-6119-6 (Hbk)
ISBN 978-3-0365-6120-2 (PDF)

Contents

Preface to "Dietary Supplements and Human Health"

Dietary supplements are used for various purposes beyond supplementation of nutrients, such as vitamins and minerals. However, the evidence levels for each ingredient and its function are not enough. In this regard, accumulation of evidence through human studies is important. This Special Issue of Nutrients invites papers on randomized controlled trials conducted to gauge the effects of ingredients, such as cod residual protein, paramylon derived from Euglena gracilis, sage, and Cosmos caudatus. In addition, the efficacy and safety of dietary supplements might differ among adolescents and younger generations in general and older generations, and this is something that must receive further attention. Thus, this Special Issue invites studies on everything that we should know about dietary supplements and on ideas for further investigations on dietary supplements.

Tsuyoshi Chiba

Editor

Article

The Effect of Antioxidant Supplementation in Patients with Tinnitus and Normal Hearing or Hearing Loss: A Randomized, Double-Blind, Placebo Controlled Trial

Anna I. Petridou [1], Eleftheria T. Zagora [2], Petros Petridis [3], George S. Korres [4], Maria Gazouli [5], Ioannis Xenelis [1], Efthymios Kyrodimos [1], Georgia Kontothanasi [2] and Andriana C. Kaliora [6,*]

1 1st ENT Department, School of Medicine, National and Kapodistrian University of Athens, Hippokration Hospital, 11527 Athens, Greece; anna.petridou@outlook.com (A.I.P.); cxeneli@yahoo.com (I.X.); timkirodimos@hotmail.com (E.K.)
2 ENT Department, General Hospital of Nikaia "Agios Panteleimon", Nikaia, 18454 Athens, Greece; zagoraeleftheria@gmail.com (E.T.Z.); georgia.kontothanasi@gmail.com (G.K.)
3 ENT Department, St. Johannes Hospital, 44137 Dortmund, Germany; petridispppeter@hotmail.com
4 2nd ENT Department, School of Medicine, National and Kapodistrian University of Athens, Attikon Hospital, 12462 Chaidari, Greece; gfkorres@gmail.com
5 Department of Biology, School of Medicine, National and Kapodistrian University of Athens, 11527 Athens, Greece; maria.gazouli@gmail.com
6 Department of Dietetics and Nutritional Science, School of Health Science and Education, Harokopio University, 17671 Athens, Greece
* Correspondence: akaliora@hua.gr or rianakaliora@gmail.com; Tel.: +30-210-954-9226

Received: 7 November 2019; Accepted: 3 December 2019; Published: 12 December 2019

Abstract: Tinnitus is the perception of sound in the absence of any external stimulus. Oxidative stress is possibly involved in its pathogenesis and a variety of antioxidant compounds have been studied as potential treatment approaches. The objective of the present study was to assess the effects of antioxidant supplementation in tinnitus patients. This is a randomized, double-blind, placebo-controlled clinical trial. Patients ($N = 70$) were randomly allocated to antioxidant supplementation ($N = 35$) or to placebo ($N = 35$) for a total of 3 months. Demographic, anthropometric, clinical, and nutritional data were collected. Serum total antioxidant capacity (TAC), oxidized LDL (oxLDL), and superoxide dismutase (SOD), tinnitus loudness, frequency, and minimum masking level (MML), and scores in Tinnitus Handicap Inventory questionnaire (THI), Tinnitus Functional Index (TFI), and Visual Analogue Scale (VAS) were evaluated at baseline and follow-up. Tinnitus loudness and MML significantly decreased from baseline to post measure ($p < 0.001$) only in the antioxidant group, the overall change being significantly different between the two groups post-intervention ($p < 0.001$). THI and VAS decreased only in the antioxidant group. Differences in changes in serum TAC, SOD, and oxLDL post-intervention were insignificant. In conclusion, antioxidant therapy seems to reduce the subjective discomfort and tinnitus intensity in tinnitus patients.

Keywords: tinnitus; antioxidant supplementation; oxidative stress; a-lipoic acid; multi-vitamin supplement

1. Introduction

Tinnitus is the perception of sound in the absence of any external stimulus [1]. Its prevalence is estimated at approximately 10%–15% of the adult population and its impact varies from a mild perception to a severe disturbance in everyday life [2]. Causes of tinnitus remain unknown and the pathological mechanisms are not fully understood. Currently, no drug treatment is effective, whereas the treatment approaches described in literature provide different results [2].

Oxidative stress has been proposed to play a critical role in the pathogenesis of tinnitus, since it could lead to cellular changes in hair cells, hair cell apoptosis, cochlear degeneration, changes in supporting structures and stria vascularis, changes in nerve fibers of the acoustic nerve, irregular neural activity in the auditory pathway and dysfunction of the central cortex [3–5]. Hair cell apoptotic pathways linked to oxidative stress have been studied in animal models in conditions of aging, noise trauma and ototoxicity [3]. Oxidative stress activates mitogen-activated protein kinase/c-Jun N-terminal kinase (MAPK/JNK) pathway, which subsequently leads to the release of cytochrome c from mitochondria [3,6]. Cytochrome c causes mitochondrial membrane damage and activates caspase pathway, promoting apoptosis [3,6]. In organ of Corti cultures, low dose oxidative stress has been shown to induce mitochondrial DNA (mtDNA) deletions in hair cells, which make them more vulnerable to further injury [7]. Moreover, oxidative stress may be linked to endothelial damage within cochlear microcirculation [8].

Higher plasma concentrations of oxidative stress biomarkers and lower antioxidant activity have been reported in tinnitus patients compared with healthy subjects [8–12]. However, research data on the efficacy of antioxidant supplementation in tinnitus are limited and conflicting. Moreover, oxidative stress biomarkers have not been assessed in most studies. Gingko Biloba, a rich source of flavonoids, has been shown to reduce the subjective discomfort and intensity of tinnitus [13,14]. Additionally, preliminary outcomes of a brief report have shown that an 18-week supplementation with antioxidants and phospholipids regulated oxidative stress and reduced the subjective discomfort and intensity of tinnitus [15]. However, other studies failed to show any benefit [16,17].

A great deal of the protective mechanisms of antioxidants against cochlea damage has been identified from animal models For example, alpha-lipoic acid (ALA) has been shown to provide protection from noise-induced hearing loss in animal models [18,19]. Moreover, many experimental studies have proven the protective effect of polyphenols against cisplatin-induced ototoxicity [20–22] and cochlear hair cell damage after intense noise exposure [23–25]. In human studies, vitamin E has been shown effective in the treatment of sudden idiopathic hearing loss [26] and ALA in the prevention of noise-induced hearing loss [27].

The aim of the study was to explore the efficacy of antioxidant supplementation which provides vitamins, minerals, and phytochemicals combined with ALA on tinnitus parameters and subjective discomfort. A second aim was to assess whether antioxidant administration had an impact on biomarkers of oxidative stress.

2. Materials and Methods

2.1. Study Design

The study protocol was reviewed and approved by the Scientific Council of the General Hospital of Athens, "Hippocratio" (11888/24-6-2009) and it was conducted according to the principles of the Declaration of Helsinki of 1975 as revised in 2013.

This was a randomized, double-blind, placebo-controlled, parallel group clinical trial (Appendix A). The trial was registered with ClinicalTrials.gov (Identifier: NCT04105426).

The primary endpoint was change in tinnitus loudness. The secondary endpoints were changes in tinnitus frequency and MML, the impact of tinnitus on daily life, hearing thresholds, and serum oxidative stress biomarkers. Tinnitus patients were enrolled based on certain inclusion and exclusion criteria. Both males and females, aged between 25 and 75 years old, with chronic unilateral or bilateral persistent tinnitus of at least 6 months' duration, with normal hearing or up to moderate sensorineural hearing loss were included. Tinnitus maskable with noise of at least 5 decibel and a score of minimum 4 in Tinnitus Handicap Inventory (THI) questionnaire were set as inclusion criteria. Patients whose tinnitus resulted from acute acoustic trauma, sudden deafness or traumatic head or neck injury were excluded from trial enrolment. Patients who were taking ototoxic or potentially tinnitus-inducing medication (e.g., aminoglycosides, chemotherapeutics, loop diuretics, high doses of aspirin or quinine) were also excluded. Moreover, patients with Meniere's disease, otosclerosis and acute or chronic otitis media were excluded. Gastrointestinal disease, active malignant diseases, autoimmune diseases, hemorrhagic diathesis, cardiovascular, renal or hepatic disorders, psychiatric disorders, and unregulated diabetes mellitus, hypertension, or thyroid disease were also set as exclusion criteria. Additionally, alcohol or drug abuse, dietary supplement use, a vegan or macrobiotic diet <2 years prior to screening, pregnancy and lactation were exclusion criteria. Patients who changed their medication, diet or physical activity habits during the trial were also excluded.

One hundred patients registered in medical archives with the symptom of tinnitus were invited for screening. An Ear Nose and Throat (ENT) physician, unrelated to the project, took a complete medical and tinnitus history followed by an ENT review examination including an audiological evaluation. Audiological assessment included conventional pure tone audiometry (PTA), tympanometry and brainstem response (BSR). Moreover, a computer tomography scanning (CT) and magnetic resonance imaging (MRI) were conducted where appropriate, in order to exclude any retrocochlear lesion. Seventy patients were eligible for the study, based on the above criteria. The ENT physician informed the eligible patients regarding the aims, methods, anticipated benefits and potential hazards of the study, and provided them with the information leaflet of the study. Each patient who agreed to take part in the study, signed an informed consent form, a copy of which was given to them.

After the initial screening, participants were randomly allocated to an intervention, either active or placebo. Randomization was conducted by an unrelated to the study person, who prepared a random number list in the computer. Neither the participants nor the investigators were aware of the treatment allocation.

The antioxidant group received one multivitamin-multimineral tablet once a day with their meal and one tablet of alpha-lipoic acid twice a day on an empty stomach, whereas the placebo group received three placebo tablets per day at the same time points. To prevent any acute supplementation effects, participants were asked not to take any tablets on the day of the follow-up measurements. To check upon compliance, each participant was seen monthly by a research coordinator who checked the compliance and tolerance of the supplement. Additionally, blood levels of vitamins and minerals were estimated in all participants before and after the intervention.

The dietary supplements received by the antioxidant group are commercially available. A-lipoic acid supplement contained 300 mg a-lipoic acid per tablet. The ingredients of the multivitamin-multimineral supplement are shown in Appendix B. The nutritional supplementation doses in the multivitamin-multimineral supplement were in line with or above the Recommended Dietary Allowances and Adequate Intakes (RDA) [28] and did not exceed the Tolerable Upper Intake Levels (UL) [29]. These doses are commonly available in commercial multivitamin supplements. Placebo pills were produced by a local manufacturing pharmacy according to good manufacturing practice (GMP) and contained sorbitol. They were manufactured with similar shape and color to the other supplements.

Dietary supplements and placebo pills were packaged in bottles and then in bags of identical appearance and labeled with the participant's number by an investigator who was not involved in the study. The bags were given to participants by an independent investigator. Instructions for the consumption of pills were included in the bags.

The intervention lasted 3 months. Participants were instructed to keep their usual medical treatment, diet, and exercise habits stable during the intervention. Patients were recruited between January 2019 and March 2019. Follow-up visits were completed on May 2019.

2.2. Baseline Assessment

After enrollment, patients underwent a complete baseline assessment which included anthropometric, audiometric, tinnitus psychoacoustic measures, tinnitus discomfort, psychological, physical activity, and dietary assessment as well as blood sample collection.

2.2.1. Audiometric Assessment and Psychoacoustic Measures of Tinnitus

Patients underwent conventional pure tone audiometry and extended high frequency (EHF) audiometry, to determine any hearing loss. Pure tones were delivered to the ear where tinnitus was more intense at the frequencies from 250 to 12,000 Hz. Pure tone audiometric threshold (PTA) is the minimum volume required to hear each tone at the examined frequency. PTA thresholds were then plotted on the audiogram. The degree of hearing loss was determined using the average of values in four consecutive frequencies (500–1000–2000–4000) and was classified as normal hearing, mild hearing loss or moderate hearing loss [30]. High frequency hearing loss was determined based on figures of hearing thresholds at extended high frequencies [31].

Moreover, tinnitus assessment tests were conducted, using psychoacoustic techniques which included frequency pitch and loudness matching, as well as minimum masking level (MML) method [32]. When tinnitus was bilateral, tests were performed to the ear where tinnitus was more intense.

Frequency pitch matching test, which determines the possible frequency of tinnitus, was conducted using the two-forced alternative choice procedure. Patients were given pairs of different tones in the ear without or less intense tinnitus and were asked to choose which tone is closer to the perceived tinnitus. This was continued until a definite match was made [32].

Loudness matching test, which determines the loudness of tinnitus, was conducted using the ascending method. Tones close to or exactly the frequency determined with the pitch matching test were presented to the patients in the ear without or less intense tinnitus. The intensity level started from just below threshold and was increased until there was a match to the perceived tinnitus loudness [32].

The MML method, which determines the least intensity needed to just mask patient's tinnitus, was conducted using the ascending method. Tones close to or exactly the frequency determined with the pitch matching test were presented to the ear with tinnitus. The intensity level started from below threshold and was increased until the patient stopped to perceive his/her tinnitus [32].

2.2.2. Tinnitus Discomfort Assessment

Patients completed the questionnaires Tinnitus Handicap Inventory (THI), Tinnitus Functional Index (TFI), and the Visual Analogue Scale (VAS), which measure the subjective discomfort a patient experiences because of tinnitus.

THI comprises 25 questions which are divided in functional, emotional, and catastrophic subscales [33]. Total scores of THI range from 0 to 100.

Visual Analogue Scale assessed the annoyance patients experienced because of tinnitus during work, sleep, relaxing, and concentration [34]. Each score of VAS ranged from 0 to 10 and the total score was the mean of the scores.

TFI includes eight subscales which concern different aspects of daily life: Intrusiveness (I), sense of control (SC), cognition (C), sleep (SL), audition (A), relaxation (R), quality of life (Q), and emotions (E) [35]. These sub-scales contribute to a subscale score and to an overall score ranging from 0 to 100.

2.2.3. Anthropometric Assessment

All anthropometric measurements were recorded after a ≥12-h fast. Body weight was measured with light clothing and without shoes using a flat scale (Tanita WB-110MA, Tokyo, Japan) and was recorded to the nearest 0.1 kg. Height was measured on a stadiometer (Seca Model 220, Hamburg, Germany) and was recorded to the nearest 0.1 cm. BMI was calculated as weight (in kg) divided by height^2 (in m^2). The waist and hip circumferences were measured using a stretch-resistant tape.

2.2.4. Nutrition and Physical Activity Evaluation

Dietary intake of patients was assessed by a dietitian using the 24-h recall method for 3 nonconsecutive days (2 weekdays and 1 weekend) and a Food Frequency Questionnaire [36]. Nutritional data were then analyzed by Nutritionist Pro nutrient analysis software version 5.2.0 (Axxya Systems, Nutritionist Pro, Stafford, TX, USA). Additionally, adherence to the Mediterranean dietary pattern was assessed by the MedDietScore, resulting to a score ranging from 0 to 55 [37]. Physical activity was assessed by a self-administered long form of the International Physical Activity Questionnaire (IPAQ) [38].

2.2.5. Psychological Assessment

The Center for Epidemiologic Studies-Depression (CES-D) and Hospital Anxiety and Depression (HADS) scales were used to assess the psychological situation of patients. Both scales are self-administered and validated in Greek language [39,40].

2.2.6. Blood Sample Collection and Analyses

Standard blood sampling (20 mL) was performed through a catheter in an antecubital vein after a 12 h overnight fast. Freshly drawn blood samples were used for routine biochemical profiles. Serum and plasma were isolated for further processing.

Vitamin D was measured using the Vitamin D Elisa Kit (Vitamin D Elisa Kit, Cayman Chemical Company, Ann Arbor, MI, USA). Folate was measured using a competitive immunoassay according to the instructions of manufacturers (ADVIA Centaur Folate assay, Siemens Healthcare Diagnostics, NY, USA). Iron was determined using the ADVIA Chemistry Iron_2 method (Siemens Healthcare Diagnostics, NY, USA) [41]. Magnesium was measured using the ADVIA Chemistry Magnesium method which is based on the modified xylidyl blue reaction (Siemens Healthcare Diagnostics, Tarrytown, NY, USA) [42]. Vitamin B12 was determined using a competitive immunoassay according to the instructions of manufacturers (ADVIA Centaur VB12 assay, Siemens Healthcare Diagnostics, Tarrytown, NY, USA). Vitamins A, E, C, B1, B2, and B6 were measured applying high performance liquid chromatography (HPLC) according to the instructions of the manufacturer (ClinRep, HPLC kit, Recipe Chemicals, München, Germany). Zinc was determined using Colorimetric method 5-Br-PAPS-Zinc complex with deproteinization according to the instructions of manufacturers (Wako Zinc Test, FUJIFILM Wako Chemicals Europe, Neuss, Germany). Selenium was measured using a methylene blue kinetic catalytic spectrophotometric method (Sigma-Aldrich Chemie GmbH, Taufkirchen, Germany).

The rest of the samples were stored at −80 °C for subsequent analyses.

2.3. Follow-Up Assessment

Compliance and any side effects were checked with a weekly telephone contact. Adherence to supplementation was assessed by counting the remaining pills in the package of each participant at the end of the intervention. At the end of the intervention, all baseline assessments were repeated apart from the psychological and physical activity assessment.

2.4. Oxidative Stress Biomarkers

Analyses of oxidative stress and antioxidant capacity biomarkers were performed in the Lab of Biology in Medical School of National and Kapodistrian University of Athens.

Serum total antioxidant capacity (TAC) was measured using the Trolox Equivalent Antioxidant Capacity (TEAC) method according to the instructions of the manufacturer (Cayman Antioxidant Assay Kit, Cayman Chemical Company, Ann Arbor, MI, USA). Serum Superoxide Dismutase (SOD) activity was assessed by measuring the dismutation of superoxide radicals generated by xanthine oxidase and hypoxanthine (Cayman Superoxide Dismutase kit, Cayman Chemical Company, Ann Arbor, MI, USA). Oxidized low-density lipoprotein LDL (oxLDL) was measured by a sandwich enzyme-linked immune-sorbent assay (Human OxLDL ELISA kit, Wuhan Fine Biological Technology, Wuhan, China). Samples and standards were run in duplicate. A Biotek PowerWave XS2 ELISA reader (BioTek Instruments, Winooski, VT, USA) was used for all measurements and analysis.

2.5. Sample Size Determination

Power analysis methodology represents a design, with two levels of the between-subject factor of two study groups and two levels of the within-subjects factor of time. A repeated measures ANOVA power analysis was conducted. The effect size for this calculation used the ratio of the standard deviation of the effects for a particular factor or interaction and the standard deviation of within-subject effects. The power analysis was conducted for a single, two-group between-subjects factor, and a single within-subjects factor assessed over two time points. For this design, 68 participants (34 per group) achieves a power of 0.95 for the within-subjects main effect at an effect size of 0.22; and a power of 0.95 for the interaction effect at an effect size of 0.25.

2.6. Statistical Analysis

Continuous variables are presented with mean and standard deviation (SD) and/or with median and interquartile range (IQR). Quantitative variables are presented with absolute and relative frequencies. For the comparison of proportions, chi-square and Fisher's exact tests were used. For the comparison of study variables between the placebo and antioxidant group the Student's *t*-test was computed. Differences in changes of tinnitus parameters, antioxidant parameters, minerals, and vitamins during the follow up period between the two study groups were evaluated using repeated measurements analysis of variance (ANOVA). Variables that had skewed distribution were log-transformed for the analysis of variance. All *p*-values reported are two-tailed. Statistical significance was set at 0.05 and analyses were conducted using SPSS statistical software (version 22.0) (IBM, Armonk, NY, USA).

3. Results

Seventy patients with tinnitus met the criteria for recruitment. Out of the 70 patients, 35 were randomized to the antioxidant group and 35 to the placebo. One patient from the antioxidant group and 2 patients from the placebo group discontinued the intervention due to an unscheduled surgery. Moreover, four patients from the placebo group were lost and unable to contact during the follow-up. Sample consisted of 63 patients (29 in the placebo group and 34 in the antioxidant group). No adverse events were mentioned in either of the two groups. Patient compliance was good. Average missed tablets (days) in antioxidant and placebo group were four and five, respectively.

Demographic, clinical, biochemical, and anthropometric characteristics for both groups are presented in Supplementary Material, Table S1. The mean age was 59.2 years (SD = 13.5 years) for the placebo group and 56.5 years (SD = 12.4 years) for the antioxidant group ($p = 0.416$). Both groups were similar in terms of sex, education, marital status, and smoking habits. Moreover, there were no differences in BMI, waist and hip circumferences and biochemical blood profile. Scores on HADS and CES-D scales along with physical activity levels were also similar for the placebo and antioxidant group (Supplementary Material, Table S1). Table 1 presents the descriptive characteristics of tinnitus and classification of hearing. Tinnitus duration and severity, family history of tinnitus, and hearing loss, as well as the number of previous therapies, the age of tinnitus onset and the presence of normal hearing and hearing loss were similar in the placebo and antioxidant group.

Table 1. Descriptive characteristics of tinnitus and classification of hearing. The results are given as *N* (%) of the total number.

	Placebo Group	Antioxidant Group	*p*
	N (%)	*N* (%)	
Tinnitus duration (years)			
<1	3 (10.3)	3 (8.8)	0.547 ++
1–2	4 (13.8)	9 (26.5)	
2–3	3 (10.3)	5 (14.7)	
3–5	9 (31)	6 (17.6)	
5–10	2 (6.9)	5 (14.7)	
>10	8 (27.6)	6 (17.6)	
Tinnitus onset			
Gradually	14 (48.3)	15 (44.1)	0.741 +
Abruptly	15 (51.7)	19 (55.9)	
Perceived all day	22 (75.9)	24 (70.6)	0.638 +
Site of the tinnitus			
Right ear	3 (10.3)	3 (8.8)	0.603 ++
Left ear	12 (41.4)	9 (26.5)	
Both ears	13 (44.8)	20 (58.8)	
Inside the head	1 (3.4)	2 (5.9)	
History of exposure to noise	12 (41.4)	18 (52.9)	0.360 +
Stable loudness all the days	14 (48.3)	17 (50)	0.891 +
Description of tinnitus			
Whistle	14 (48.3)	18 (52.9)	0.493 ++
Clatter	0 (0)	2 (5.9)	
Cicadas' noise	8 (27.6)	7 (20.6)	
Blow	2 (6.9)	3 (8.8)	
Buzzing	3 (10.3)	2 (5.9)	
Other	2 (6.9)	0 (0)	
Bees' noise	0 (0)	2 (5.9)	
Previous therapies			
0	11 (37.9)	21 (61.8)	0.063 ++
1	15 (51.7)	8 (23.5)	
2	1 (3.4)	4 (11.8)	
3	2 (6.9)	1 (2.9)	
Tinnitus severity			
THI, mean (SD)	40.6 (27.7)	31.6 (19.3)	0.139 ‡
Tinnitus loudness (db), mean (SD)	47.1 (20.5)	45 (15.3)	0.649 ‡
Tinnitus frequency (Hz), mean (SD)	4431 (3078.3)	5562.5 (3027.7)	0.147 ‡
MML(db), mean (SD)	49.3 (22.5)	57.9 (18.9)	0.110 ‡
Age of tinnitus onset (years), mean (SD)	54 (14.3)	48.2 (15.7)	0.138 ‡
Family history of hearing loss	7 (25)	5 (14.7)	0.307 +
Family history of tinnitus	5 (17.9)	6 (17.6)	1.000 ++
Classification of hearing			
Normal hearing in conventional and EHF audiometry	10 (35.7)	16 (47.1)	0.651 ++
Mild sensorineural hearing loss in conventional audiometry	14 (50)	13 (38.2)	
Moderate sensorineural hearing loss in conventional audiometry	4 (14.3)	5 (14.7)	
High-frequency sensorineural hearing loss in EHF audiometry	11 (37.9)	9 (26.5)	

+ Pearson's chi-square test; ++ Fisher's exact test; ‡ Student's *t*-test. THI: Tinnitus Handicap Inventory questionnaire; EHF: extended high frequency.

Additionally, the two groups were similar as far as the dietary habits are concerned. There were no differences in baseline total energy and macronutrient intake, as well as adherence to the Mediterranean diet assessed by MedDietScore and the frequency of consumption of herbal beverages, chocolate, coffee, and wine as presented in Supplementary Material, Table S2.

The anthropometric and biochemical parameters before and after the intervention are presented in Supplementary Material, Tables S3 and S4. BMI, waist and hip circumferences remained unchanged in both groups after intervention. Moreover, biochemical parameters did not change, apart from LDL, which increased significantly in the antioxidant group. No changes were also reported for the total energy and macro- and micronutrient dietary intake after intervention for both groups (Supplementary Material, Table S5).

The changes for tinnitus loudness, tinnitus frequency and Minimum Masking Level (MML) for both groups after intervention are presented in Table 2. Loudness and MML significantly decreased from baseline to post measure ($p < 0.001$) only in the antioxidant group and the overall change was different between the two groups as indicated from the significant interaction effect of the analysis ($p < 0.001$). Tinnitus frequency did not significantly change in any of the two groups.

Results from the tinnitus questionnaires before and after the intervention revealed that scores of THI, VAS, TFI-Relaxation (TFI-R), and TFI-Emotions (TFI-E) had a significant reduction in the antioxidant group, while no change was recorded in the placebo group (Table 3). A significant interaction effect of group with time indicated a significant treatment difference for THI, TFI-R, and TFI-E (Table 3). Figures 1–4 present the median values for tinnitus loudness, tinnitus frequency, MML, and THI accordingly, displayed as box plots for each group.

To assess if tinnitus duration had any effect on the outcomes, placebo and antioxidant groups were divided in two subgroups according to tinnitus duration. There were no differences in tinnitus loudness, tinnitus frequency, MML and THI between patients with tinnitus duration lower and higher than 10 years at baseline and at follow-up in placebo (Table 4) and antioxidant group (Table 5).

As far as serum concentrations of biomarkers of oxidative stress are concerned (Table 6), serum TAC was decreased significantly in both groups, while SOD and oxLDL did not change in either of the two groups. Differences between the groups in changes in serum TAC, SOD and oxLDL postintervention were insignificant.

Table 2. Tinnitus loudness, frequency, and minimum masking level (MML) at baseline and at follow-up.

		Pre		Post		Change		
		Mean (SD)	**Median (IQR)**	**Mean (SD)**	**Median (IQR)**	**Mean (SD)**	p^1	p^2
Tinnitus loudness (db)	Placebo	47.1 (20.5)	50 (30; 60)	40.4 (15.5)	40 (30; 50)	−6.7 (8.8)	0.168	<0.001
	Antioxidant	45 (15.3)	40 (35; 55)	30.8 (11.2)	27.5 (25; 35)	−14.2 (12.7)	<0.001	
Tinnitus frequency(hz)	Placebo	4431 (3078.3)	6000 (1000; 8000)	4240 (2932.6)	4000 (1500; 6000)	−191 (1401.3)	0.303	0.082
	Antioxidant	5562.5 (3027.7)	6000 (4000; 8000)	5222.7 (2982.1)	6000 (2500; 8000)	−339.8(1565.8)	0.216	
MML (db)	Placebo	49.3 (22.5)	52.5 (30; 70)	47.7 (21)	47.5 (32.5; 65)	−1.6 (7.4)	0.989	<0.001
	Antioxidant	57.9 (18.9)	60 (40; 75)	43.4 (16.5)	40 (30; 60)	−14.5 (14.3)	<0.001	

[1] *p*-value for the time effect (using logarithmic transformations); [2] *p*-value from repeated measurements ANOVA. The effects reported include differences between the groups in the degree of change (using logarithmic transformations).

Table 3. Tinnitus questionnaires' scores at baseline and at follow-up.

		Pre		Post		Change		
		Mean (SD)	Median (IQR)	Mean (SD)	Median (IQR)	Mean (SD)	p^1	p^2
THI	Placebo	40.6 (27.7)	30 (16; 74)	42.8 (24.5)	36 (26; 66)	2.2 (11.5)	0.607	0.015
	Antioxidant	31.6 (19.3)	33 (16; 44)	25.5 (18)	21 (11; 36)	−6.1 (11.7)	0.002	
VAS	Placebo	5.13 (2.81)	5 (3.75; 7.25)	4.86 (2.53)	4.5 (3.25; 6.6)	−0.3(2.11)	0.767	0.147
	Antioxidant	3.66 (2.49)	3.38 (1.25;5.75)	2.88 (2.37)	2.38 (0.75; 4.63)	−0.78 (1.4)	0.013	
TFI	Placebo	37.9 (19.7)	32 (22.8; 52)	41 (22.3)	32.8 (27.2; 57.6)	3.1 (10.3)	0.950	0.410
	Antioxidant	30.4 (20)	29.2 (11.4; 47)	28.6 (20.5)	21.2 (13.6; 49.6)	−1.8 (16.1)	0.129	
TFI-I	Placebo	56.9 (26.4)	60 (31.7; 75)	62.7 (26.6)	70 (45; 83.3)	5.7 (17.2)	0.518	0.666
	Antioxidant	48.3 (27.3)	45 (23.3; 71.7)	48.3 (27.8)	43.3 (20; 76.7)	0.1 (20.5)	0.873	
TFI-SC	Placebo	61.9 (27)	56.7 (48.3;86.7)	65.9 (21.8)	60 (56.7; 76.7)	4 (10.9)	0.964	0.376
	Antioxidant	44.2 (29.3)	40 (20; 66.7)	39.9 (27.7)	40 (20; 56.7)	−4.3 (21.3)	0.120	
TFI-C	Placebo	31 (31.4)	30 (0; 45)	22.1 (26.4)	13.3 (0; 30)	−8.9 (29.2)	0.294	0.939
	Antioxidant	31.6 (28.1)	23.3 (8.3; 51.7)	27.7 (26.9)	13.3 (3.3; 60)	−3.8 (23.4)	0.187	
TFI-SL	Placebo	33.6 (32.7)	26.7 (3.3; 61.7)	36.9 (32.7)	23.3 (10; 66.7)	3.3 (6.9)	0.720	0.236
	Antioxidant	23.1 (26.9)	13.3 (0; 45)	23 (30.3)	10 (0; 46)	−0.1 (24.6)	0.128	
TFI-A	Placebo	31.2 (32.4)	23.3 (0; 56.7)	23.5 (31.9)	6.7 (0; 50)	−7.8 (11.7)	0.612	0.836
	Antioxidant	24.2 (25.2)	13.3 (0; 40)	23.9 (23.7)	16.7 (3.3; 36.7)	−0.3 (18)	0.726	
TFI-R	Placebo	49.3 (31.8)	55 (18.3; 73.3)	60.8 (22.7)	66.7 (43.3; 76.7)	11.5(19.1)	0.189	0.037
	Antioxidant	34.7 (29.6)	25 (8.3; 65)	26.5 (26.2)	13.3 (3.3; 50)	−8.2 (22.5)	0.044	
TFI-Q	Placebo	27.2 (29)	20 (0; 50)	24.9 (29.3)	25 (0; 37.5)	−2.4 (11)	0.842	0.897
	Antioxidant	14.3 (16.2)	10 (0; 30)	18.1 (21.1)	10 (0; 25)	3.8 (17.7)	0.957	
TFI-E	Placebo	41.2 (36.7)	30 (10; 68.3)	41 (36.3)	30 (10; 73.3)	−0.2 (10.8)	0.789	0.042
	Antioxidant	29.2 (26)	20 (6.7; 50)	23.3 (22.6)	20 (6.7; 40)	−5.8 (18.1)	0.006	

[1] *p*-value for the time effect (using logarithmic transformations); [2] *p*-value from repeated measurements ANOVA. The effects reported include differences between the groups in the degree of change (using logarithmic transformations). THI: Tinnitus Handicap Inventory, VAS: Visual Analogue Scale; TFI: Tinnitus Functional Index.

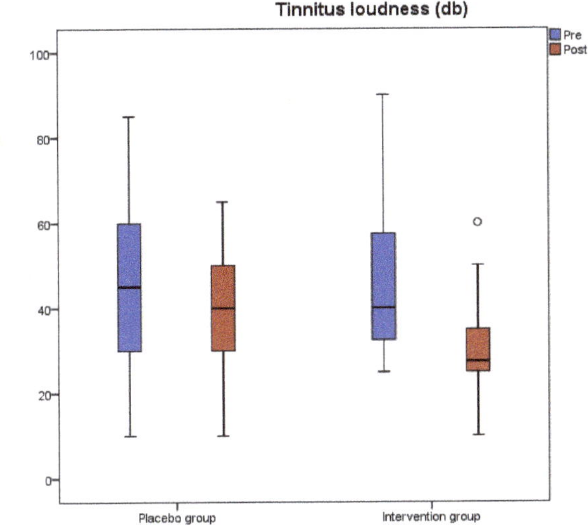

Figure 1. Tinnitus loudness (db) is displayed as box plots for each group at baseline and at follow-up. The line inside the box represents the median and the box portion of the box plot is defined by two lines at the 25th percentile and 75th percentile. Circles denote outliers.

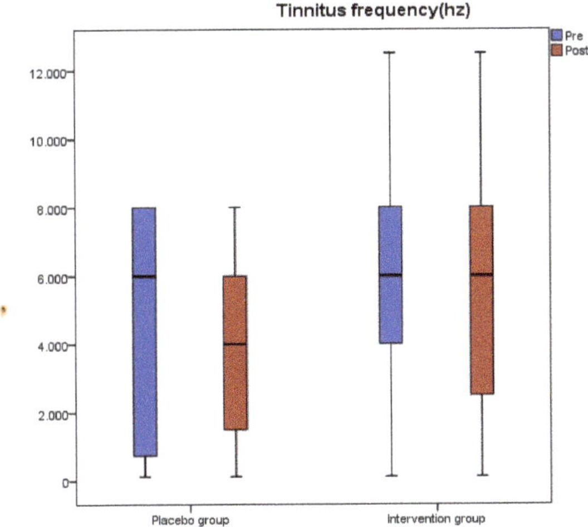

Figure 2. Tinnitus frequency (Hz) is displayed as box plots for each group at baseline and at follow-up. The line inside the box represents the median and the box portion of the box plot is defined by two lines at the 25th percentile and 75th percentile.

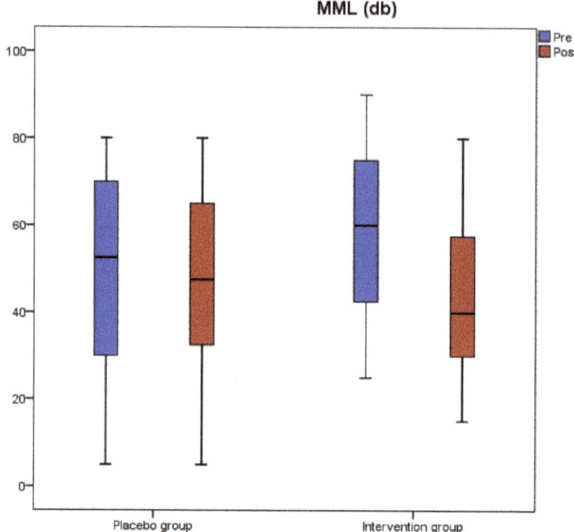

Figure 3. MML (db) is displayed as box plots for each group at baseline and at follow-up. The line inside the box represents the median and the box portion of the box plot is defined by two lines at the 25th percentile and 75th percentile.

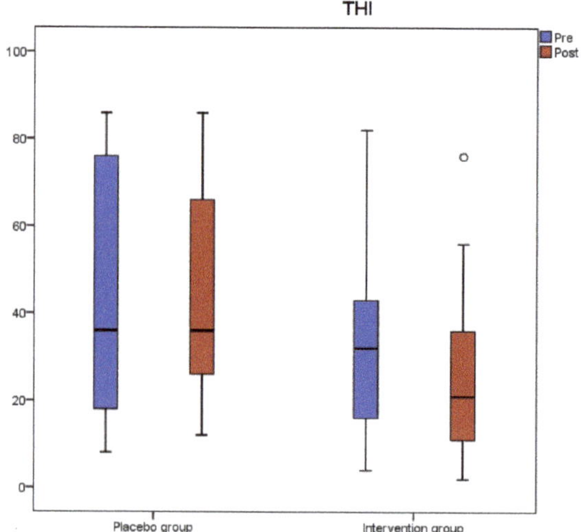

Figure 4. THI is displayed as box plots for each group at baseline and at follow-up. The line inside the box represents the median and the box portion of the box plot is defined by two lines at the 25th percentile and 75th percentile. Circles denote outliers.

Table 4. Tinnitus loudness, frequency, MML, and THI in patients with tinnitus duration of <10 years and >10 years at baseline and at follow-up in placebo group.

	Tinnitus Duration				
	<10 Years		>10 Years		P Mann-Whitney Test
	Mean (SD)	Median (IQR)	Mean (SD)	Median (IQR)	
Tinnitus loudness (db) (pre)	46.9 (22.8)	50 (30; 60)	47.5 (13.9)	47.5 (35; 60)	0.769
Tinnitus loudness (db) (post)	39.7 (16.3)	42.5 (25; 50)	42.1 (14.1)	35 (30; 60)	0.784
Tinnitus frequency (Hz) (pre)	4011.9 (3230)	4000 (750; 8000)	5531.3 (2487.2)	6000 (5000; 7000)	0.345
Tinnitus frequency (Hz) (post)	3875 (3082.2)	4000 (1000; 8000)	5178.6 (2461)	6000 (4000; 6000)	0.373
MML (db) (pre)	47.5 (24.3)	45 (30; 70)	53.8 (17.7)	55 (42.5; 65)	0.557
MML (db) (post)	45.3 (22.7)	45 (25; 65)	53.6 (16.3)	60 (35; 65)	0.426
THI (pre)	42 (30.5)	30 (14; 76)	36.8 (19.3)	31 (22; 48)	0.961
THI (post)	46.1 (27.5)	46 (16; 76)	34.3 (11.7)	30 (26; 50)	0.467

Table 5. Tinnitus loudness, frequency, MML and scores in THI in patients with tinnitus duration of <10 years and >10 years at baseline and at follow-up in antioxidant group.

	Tinnitus Duration				
	<10 Years		>10 Years		P Mann-Whitney Test
	Mean (SD)	Median (IQR)	Mean (SD)	Median (IQR)	
Tinnitus loudness (db) (pre)	43 (13.7)	40 (30; 52.5)	54.2 (20.1)	45 (40; 65)	0.199
Tinnitus loudness (db) (post)	30.6 (11.5)	27.5 (25; 35)	31.7 (10.8)	30 (25; 35)	0.825
Tinnitus frequency (Hz) (pre)	5183 (3187.5)	4500 (2000; 8000)	7333.3 (1032.8)	8000 (6000; 8000)	0.080
Tinnitus frequency (Hz) (post)	4889.4 (3142.9)	5000 (2000; 8000)	6666.7 (1633)	7000 (6000; 8000)	0.152
MML (db) (pre)	56.9 (18.7)	55 (40; 75)	62.5 (21.2)	65 (60; 70)	0.574
MML (db) (post)	43.2 (16.4)	40 (30; 55)	44.2 (18.6)	40 (30; 60)	1.000
THI (pre)	30.4 (17.9)	29 (16; 43)	37.3 (26.3)	37 (20; 44)	0.586
THI (post)	23.5 (15.9)	20 (10; 36)	34 (25.4)	30 (12; 48)	0.371

Table 6. Oxidative stress biomarkers at baseline and at follow-up.

		Pre		Post		Change		
		Mean (SD)	**Median (IQR)**	**Mean (SD)**	**Median (IQR)**	**Mean (SD)**	p^1	p^2
serum TAC (m M)	Placebo	5.5 (1.5)	5.2 (4.5; 7)	3.9 (2.9)	2.4 (1.4; 6.6)	−1.7 (3.5)	0.002	0.420
	Antioxidant	6.3 (1.9)	6.1 (5.3; 7.2)	5.1 (2.7)	5.5 (3.5; 7.2)	−1.1 (3.8)	0.019	
SOD (U/mL)	Placebo	3.2 (2)	2.5 (2.2; 2.9)	3.1 (1.8)	2.6 (2.2; 3.2)	−0.1 (1.4)	0.792	0.154
	Antioxidant	4 (2.5)	3.3 (2.6; 4.4)	3 (1.6)	2.6 (2; 3.3)	−1 (2.3)	0.065	
ox.LDL (ng/mL)	Placebo	12.1 (23.2)	7 (2; 10.9)	10.4 (8.5)	9 (5.1; 11.4)	−1.7 (7.9)	0.062	0.232
	Antioxidant	25 (50.7)	7.7 (3.8; 15.8)	18.7 (37.6)	8.8 (7.8; 9.6)	−6.3 (27.9)	0.399	

[1] *p*-value for the time effect; [2] *p*-value from repeated measurements ANOVA. The effects reported include differences between the groups in the degree of change (using logarithmic transformations).

Figures 5 and 6 present pure-tone thresholds in the frequency range from 250 to 12,000 Hz (dB HL) as box plots for each frequency at baseline and at follow-up in the placebo and antioxidant group accordingly. The degree of change in the PTA thresholds at the frequencies of 250 Hz, 2000 Hz, 4000 Hz, 10,000 Hz, and 12,000 Hz differed significantly between the two groups. Specifically, at the frequencies of 250 Hz, 500 Hz, 1000 Hz, 2000 Hz, and 6000 Hz there was a significant decrease in the auditory threshold only in the antioxidant group whereas in the placebo group there was no significant change post intervention. At the frequency of 10,000 Hz there was a significant increase only in the placebo group while in the antioxidant group there was no significant change after the intervention. At the frequencies of 4000 Hz and 12,000 Hz there was a significant decrease in the antioxidant group and a significant increase in the placebo group.

As a measure of compliance, blood status of vitamins and minerals was evaluated and results are presented in Table 7. Vitamin D, folate, vitamin B2, vitamin B1, and vitamin B6 levels increased significantly in the antioxidant group postintervention, whereas iron, magnesium, zinc, selenium, vitamin B12, vitamin E, and vitamin C levels remained unchanged. The estimated treatment difference was significant for folic acid as indicated from the interaction effect of the analysis ($p = 0.049$). In the placebo group, vitamin and mineral levels did not change post-intervention.

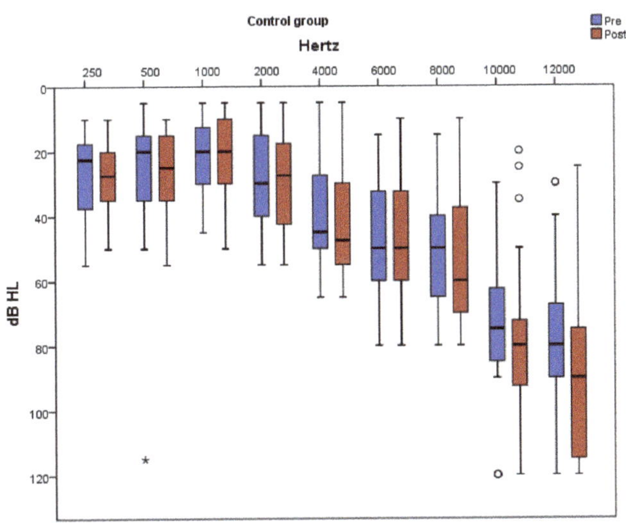

Figure 5. Pure-tone thresholds in the frequency range from 250 to 12,000 Hz (dB HL) are displayed as box plots for each frequency in the placebo group at baseline and at follow-up. The line inside the box represents the median and the box portion of the box plot is defined by two lines at the 25th percentile and 75th percentile. Circles denote outliers and asterisks denote extreme values.

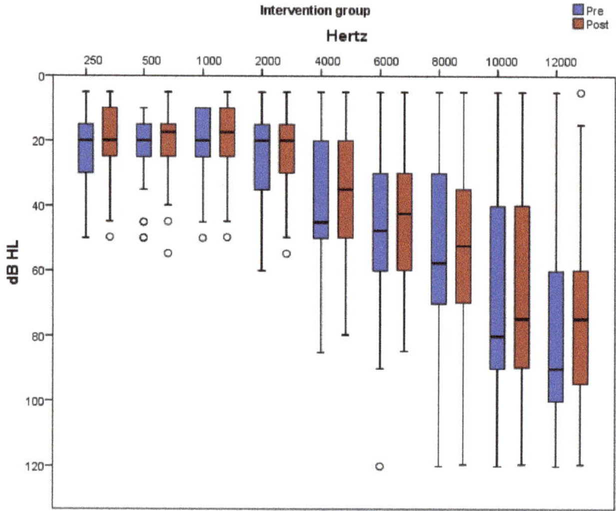

Figure 6. Pure-tone thresholds in the frequency range from 250 to 12,000 Hz (dB HL) are displayed as box plots for each frequency in the antioxidant group at baseline and at follow-up. The line inside the box represents the median and the box portion of the box plot is defined by two lines at the 25th percentile and 75th percentile. Circles denote outliers.

Table 7. Blood status of vitamins and minerals at baseline and at follow-up. Values are expressed as the mean ± SD.

		Pre	Post	Change	p [1]	p [2]
		Mean (SD)	Mean (SD)	Mean (SD)		
Vitamin D (ng/mL)	Placebo	0.32 (0.23)	0.28 (0.13)	−0.05 (0.16)	0.657	0.079
	Antioxidant	0.32 (0.22)	0.4 (0.42)	0.08 (0.36)	0.032	
Magnesium (mg/dL)	Placebo	2.06 (0.11)	2.03 (0.1)	−0.03 (0.16)	0.834	0.355
	Antioxidant	2.01 (0.12)	2.07 (0.16)	0.06 (0.17)	0.066	
Zn (µg/dL)	Placebo	105.8 (14.8)	98.6 (11)	−7.1 (12.7)	0.128	0.148
	Antioxidant	111.6 (15.4)	112 (15.7)	0.5 (23.4)	0.777	
B12 (pg/mL)	Placebo	465.3 (330.6)	358.3 (167.8)	−106.9 (378.2)	0.080	0.095
	Antioxidant	359.8 (130.3)	411.9 (98.5)	52.1 (110.5)	0.657	
Folate (ng/mL)	Placebo	8.16 (4.99)	6.45 (2.14)	−1.71 (1.13)	0.463	0.049
	Antioxidant	8.65 (5.21)	11.31 (5.41)	2.66 (4.29)	0.001	
Fe (µg/dL)	Placebo	128.2 (42.1)	118.5 (62.5)	−9.6 (65.6)	0.963	0.309
	Antioxidant	117.8 (36.2)	104.7 (41)	−13.1 (36)	0.134	
Selenium (µg/L)	Placebo	76.5 (11.8)	68.2 (7)	−8.3 (15.4)	0.080	0.095
	Antioxidant	78.4 (9.5)	80.1 (11.3)	1.7 (13.4)	0.730	
Vitamin E (mg/L)	Placebo	12.09 (0.94)	11.68 (1.04)	−0.41 (0.79)	0.908	0.798
	Antioxidant	11.42 (1.16)	11.92 (1.15)	0.5 (1.56)	0.411	
Vitamin C (µg/L)	Placebo	6.73 (1.65)	7.03 (2.73)	0.31 (3.65)	0.324	0.530
	Antioxidant	7.25 (1.82)	7.73 (2.41)	0.48 (3.22)	0.509	
Vitamin B2 (µg/L)	Placebo	220 (22.4)	223.6 (18)	3.6 (11.1)	0.918	0.326
	Antioxidant	221.7 (25.2)	238.4 (26.6)	16.7 (32.4)	0.039	
Vitamin B1 (µg/L)	Placebo	55 (12.4)	56.8 (14.8)	1.8 (27.2)	0.757	0.468
	Antioxidant	50.4 (11.1)	61.5 (13.5)	11.1 (17.7)	0.023	
Vitamin B6 (µg/L)	Placebo	21.5 (3.6)	22.8 (5.6)	1.3 (6)	0.911	0.463
	Antioxidant	24.2 (17.2)	35.6 (12.9)	11.4 (19.9)	0.050	

[1] p-value for the time effect; [2] p-value from repeated measurements ANOVA. The effects reported include differences between the groups in the degree of change (using logarithmic transformations).

4. Discussion

Herein, the efficacy of antioxidant supplementation with vitamins, minerals and phytochemicals combined with ALA on tinnitus parameters and subjective discomfort has been investigated. Furthermore, the effect of antioxidant supplementation on biomarkers of oxidative stress has been assessed.

Given that tinnitus is a symptom that has multiple dimensions, psychoacoustic measures of tinnitus (loudness, frequency, and MML) together with questionnaires assessing the tinnitus subjective discomfort (THI, TFI, and VAS-annoyance scale) were applied in the present study in order to assess the efficacy of antioxidant supplementation. This was done in accordance with the Consensus for tinnitus patient assessment and treatment outcome measurement [43]. Separate assessment of psychoacoustic features of the tinnitus and tinnitus related distress in everyday life is highly recommended, since they have a weak correlation with each other [44,45]. Our findings showed that tinnitus loudness, MML and scores in THI, VAS, TFI-relaxation, and TFI-emotions decreased significantly only in the antioxidant group, with the overall changes being significantly different between the two groups post-intervention. This means that antioxidant supplementation was effective in improving tinnitus' sensory aspects (tinnitus loudness and MML) and in alleviating patients from tinnitus-related distress compared to placebo. The mean reduction in the THI score of 6 points in the antioxidant group, is considered a clinically relevant change [46]. In accordance with present findings, a significant reduction in the VAS scale and tinnitus loudness was reported after an 18-week supplementation with a mix of phospholipids and vitamins [15].

At the same time, this study showed that antioxidant supplementation led to significant improvements of hearing thresholds across all frequencies, with the overall changes being significantly different between the two groups at the frequencies of 250 Hz, 2000 Hz, 4000 Hz, 10,000 Hz, and 12,000 Hz. Improved thresholds at frequencies between 250 Hz and 8000 Hz is of high importance, since these are the frequencies used for speech recognition and thus tinnitus patients could benefit from a better understanding of speech and language.

Despite the key role of oxidative stress in the pathogenesis of tinnitus, data on the efficacy of antioxidant supplementation on oxidative stress biomarkers in tinnitus patients are scarce. This study assessed oxidative stress by measuring serum oxLDL. OxLDL, measured in blood using immunological methods (i.e., antibodies), is considered by the European Food Safety Authority (EFSA) as a reliable in vivo marker of oxidative damage with appropriate specificity [47]. Moreover, serum TAC and SOD were measured in order to assess total antioxidant status and endogenous antioxidant activity accordingly. No statistically significant differences were observed in changes in serum TAC, SOD, and oxLDL in the antioxidant group compared to placebo group post-intervention. These results are consistent with other studies which have been done in healthy adults or adults with concomitant diseases (i.e., diabetes, CVD) and have shown that antioxidant supplementation had no effect on ox.LDL [48,49], SOD [50–52], or TAC [51,52]. The absence of any effect of antioxidant supplementation on serum ox. LDL and TAC may be due to the low doses of vitamins C (150 mg) and a-tocopherol (150 IU) in the supplement used. As such, the plasma levels of vitamins C and E in the antioxidant group post-intervention remained unchanged. TAC measures total endogenous and food-derived antioxidants and its values are affected by serum concentrations of vitamins E and C [53]. Moreover, it has been shown that a significant reduction in LDL susceptibility to oxidation could be achieved with a minimum dose of 400 IU/d of α-tocopherol supplementation [54].

Antioxidant supplementation studies in tinnitus are limited. Moreover, in most of them a monotherapy treatment approach has been proposed such as zinc [16,55,56], with no benefit in tinnitus. On the contrary, our hypothesis was that an antioxidant combination might be more effective compared with single nutrients, since various antioxidants have a synergistic/complementary activity [57]. A-lipoic acid was used hereby, at a dose of 600 mg daily that has been demonstrated to increase lipoic acid levels in plasma or cells [58,59]. Moreover, the standardized grape seed extract (GSE) contained in the multivitamin-multimineral supplement constitutes a rich source of phenolic compounds including epicatechin, resveratrol and procyanidin oligomers [60]. To the best of our knowledge, this is the first randomized double-blind placebo-controlled study which used a mixture of vitamins, minerals, phytochemicals and ALA in tinnitus.

The beneficial effect observed on tinnitus in this study is most likely attributed to the protective mechanisms of antioxidants contained in the supplements against ROS-induced cochlear hair cell damage. This has been verified from experiments using animal models, in-vitro assays, or auditory cell lines in cases of ototoxicity, age-related or noise-induced hearing loss. A recent study using cochlea explant culture from mice showed that resveratrol, DL-α-lipoic acid and a-tocopherol protected against gentamicin-induced hair cell loss [61]. In animal experiments, vitamins E or C treatment resulted in better auditory sensitivity and less mtDNA deletions with aging [62] and vitamin E or a-lipoic acid attenuated the noise-induced increase of cochlear malondialdehyde [18]. Moreover, a combination of vitamins B1, B2, B6, E, and C was effective in protecting against cisplatin ototoxicity [63] and α-lipoic acid inhibited the kanamycin-induced high expression of phosphorylated p38 and phosphorylated JNK, which mediate cochlear hair cell apoptosis [64]. In studies using auditory cell lines, treatment with D-α-tocopherol or epicatechin reduced the cisplatin-induced increase of ROS and decreased cellular necrosis and apoptosis [65,66].

Apart from their antioxidant mediated effects, compounds in the supplements used may have acted via additional mechanisms. ALA has been widely researched as a neuroprotectant [67] and vitamin B complex play a key role in the functioning of nerve tissue, cellular metabolism, vascular function, and myelin synthesis [68]. Low levels of Vitamin B12 and folate have been associated with decreased endocochlear potential due to the destruction of the microvasculature of the stria vascularis [69]. Moreover, a diet deficient in folate in animals may lead to impaired cochlear homocysteine metabolism and oxidative stress [70].

Despite the interesting results of this study and the advantage of being randomized and double-blinded, it has some limitations, including the enrollment of patients with high levels of oxidative stress (e.g., smokers, diabetics, elderly), the presence of compounds other than antioxidants in the supplement used and the heterogeneity of participants as far as tinnitus duration and severity, hearing loss, family history and age of tinnitus onset are concerned, shown to influence the response to treatment [71,72]. However, these limitations are compensated by the tight control of participants to ensure their compliance with the protocol, as well as the long follow-up and the adequate sample size of the study.

5. Conclusions

Results of this study showed that antioxidant supplementation with vitamins, minerals, phytochemicals and ALA could exhibit favorable effects in tinnitus by reducing the subjective discomfort and tinnitus intensity. However, the effect of this antioxidant supplementation in oxidative stress biomarkers in tinnitus patients needs further investigation.

Supplementary Materials: The following are available online at http://www.mdpi.com/2072-6643/11/12/3037/s1, Table S1: Sample characteristics by study group, Table S2: MedDietScore, antioxidant food consumption frequency and macronutrient intake by study group, Table S3: Anthropometrics at baseline and at follow up. Values are expressed as the mean ± SD, Table S4: Biochemical parameters at baseline and at follow up, Table S5: Macro- and micronutrient intake at baseline and at follow-up.

Author Contributions: Conceptualization, A.I.P. and G.K.; Methodology, I.X. and A.C.K.; Investigation, A.I.P., G.K., and E.T.Z. (Data collection), M.G. (Contribution to laboratory work) and E.T.Z. (Participation in patient recruitment); Resources, A.I.P.; Writing—Original Draft Preparation, A.I.P., E.T.Z., P.P., G.S.K., E.K.; Writing—Review and Editing, A.C.K. and A.I.P.; Visualization, A.C.K.; Supervision, A.C.K. and I.X.; Project Administration, I.X.

Funding: This research received no external funding.

Acknowledgments: All authors have approved the final article. We are grateful to the patients for participating in this study. We are thankful to Lamberts for the kind donation of supplements. We are also thankful to Chara Tzavara for her assistance in statistical analysis.

Conflicts of Interest: The authors declare no conflict of interest.

Appendix A

CONSORT 2010 Flow Diagram

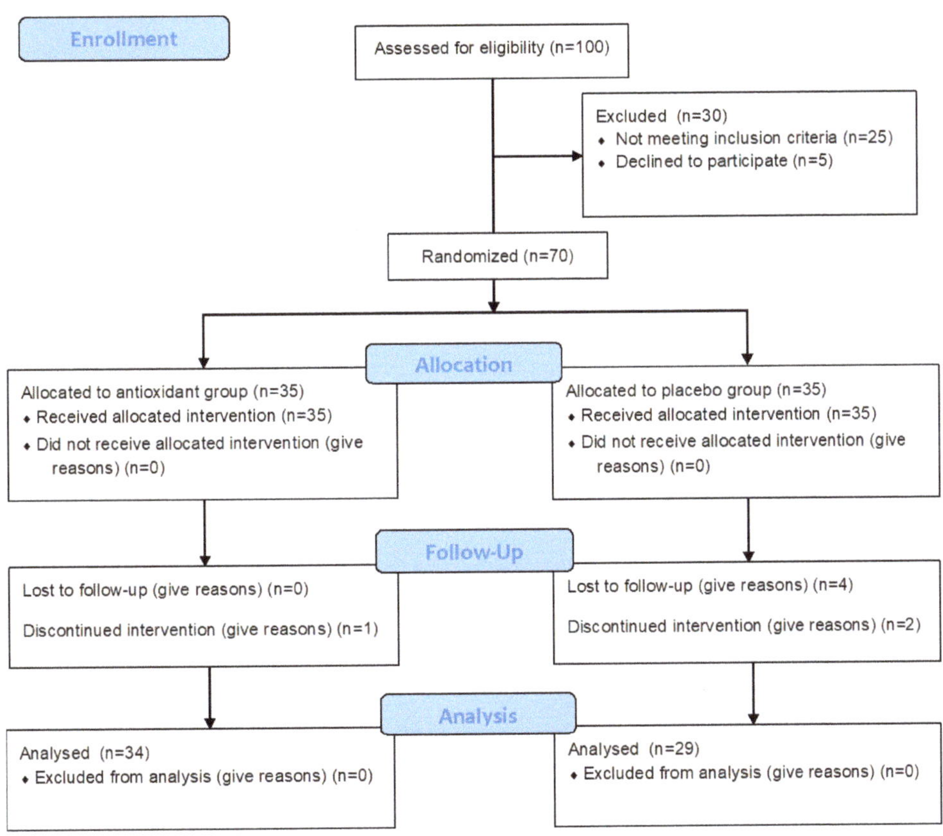

Appendix B

Composition of the Multivitamin-Multimineral Supplement (1 Table)					
Ingredient	Amount per tablet	%RDA	Ingredient	Amount per tablet	%RDA
Vitamin A (acetate)	781 µg (2600 iu)	98	Magnesium (as oxide)	50 mg	13
Vitamin D_3	10 µg (400 iu)	200	Zinc (gluconate)	15 mg	150
Vitamin E (dl-alpha tocopherol acetate)	100 mg (150 iu)	833	Copper (gluconate)	1.2 mg	120
Vitamin C (Ascorbic acid)	150 mg	188	Manganese (gluconate)	4 mg	200
Thiamine (Vitamin B1) (mononitrate)	25 mg	2272	Selenium (as L-Selenomethionine and sodium selenite)	100 µg	200
Riboflavin (Vitamin B2)	25 mg	1786	Chromium (picolinate)	200 µg	500
Niacin (Vitamin B3)	25 mg	156	Molybdenum (as molybdate)	500 µg	1000
Pyridoxine (Vitamin B6) (Pyridoxine Hydrochloride)	10 mg	714	Iodine (potassium iodide)	150 µg	100
Folic acid	200 µg	100	Choline (Bitartrate)	25 mg	
Vitamin B12	10 µg	400	Inositol	25 mg	
Biotin	150 µg	300	PABA	25 mg	
Pantothenic acid (Vitamin B5) (calcium pantothenate)	25 mg	417	Grapeseed extract (GSE) (Provided by 1mg of a 500:1 extract)	500 mg	
Calcium (as phosphate)	62 mg	8			
Iron (ferrous fumarate)	14 mg	100			

References

1. Jastreboff, P.J. Phantom auditory perception (tinnitus): Mechanisms of generation and perception. *Neurosci. Res.* **1990**, *8*, 221–254. [CrossRef]
2. Baguley, D.; McFerran, D.; Hall, D. Tinnitus. *Lancet* **2013**, *382*, 1600–1607. [CrossRef]
3. Ciorba, A.; Bianchini, C.; Pastore, A.; Mazzoli, M. Pathogenesis of Tinnitus: Any Role for Oxidative Stress? *J. Int. Adv. Otol.* **2013**, *9*, 249–254.
4. Poirrier, A.L.; Pincemail, J.; Van Den Ackerveken, P.; Lefebvre, P.P.; Malgrange, B. Oxidative stress in the cochlea: An update. *Curr. Med. Chem.* **2010**, *17*, 3591–3604. [CrossRef] [PubMed]
5. Gonzalez-Gonzalez, S. The role of mitochondrial oxidative stress in hearing loss. *Neurol. Disord. Ther.* **2017**, *1*. [CrossRef]
6. Sha, S.-H.; Chen, F.-Q.; Schacht, J. Activation of cell death pathways in the inner ear of the aging CBA/J mouse. *Hear. Res.* **2009**, *254*, 92–99. [CrossRef] [PubMed]
7. Baker, K.; Staecker, H. Low Dose Oxidative Stress Induces Mitochondrial Damage in Hair Cells. *Anat. Rec.* **2012**, *295*, 1868–1876. [CrossRef]
8. Neri, S.; Signorelli, S.; Pulvirenti, D.; Mauceri, B.; Cilio, D.; Bordonaro, F.; Abate, G.; Interlandi, D.; Misseri, M.; Ignaccolo, L.; et al. Oxidative stress, nitric oxide, endothelial dysfunction and tinnitus. *Free Radic. Res.* **2006**, *40*, 615–618. [CrossRef]
9. Celik, M.; Koyuncu, İ. A Comprehensive Study of Oxidative Stress in Tinnitus Patients. *Indian J. Otolaryngol. Head Neck Surg.* **2018**, *70*, 521–526. [CrossRef]
10. Ekinci, A.; Kamasak, K. Evaluation of serum prolidase enzyme activity and oxidative stress in patients with tinnitus. *Braz. J. Otorhinolaryngol.* **2019**, in press. [CrossRef]
11. Koç, S.; Akyüz, S.; Somuk, B.T.; Soyalic, H.; Yılmaz, B.; Taskin, A.; Bilinc, H.; Aksoy, N. Paraoxonase Activity and Oxidative Status in Patients with Tinnitus. *J. Audiol. Otol.* **2016**, *20*, 17–21. [CrossRef] [PubMed]
12. Pawlak-Osińska, K.; Kaźmierczak, H.; Marzec, M.; Kupczyk, D.; Bilski, R.; Mikołajewska, E.; Mikołajewski, D.; Augustyńska, B. Assessment of the State of the Natural Antioxidant Barrier of a Body in Patients Complaining about the Presence of Tinnitus. *Oxidative Med. Cell. Longev.* **2018**, *2018*, 1439575. [CrossRef] [PubMed]

13. Procházková, K.; Šejna, I.; Skutil, J.; Hahn, A. Ginkgo biloba extract EGb 761®versus pentoxifylline in chronic tinnitus: A randomized, double-blind clinical trial. *Int. J. Clin. Pharm.* **2018**, *40*, 1335–1341. [CrossRef] [PubMed]
14. Morgenstern, C.; Biermann, E. The efficacy of Ginkgo special extract EGb 761 in patients with tinnitus. *Int. J. Clin. Pharmacol. Ther.* **2002**, *40*, 188–197. [CrossRef]
15. Savastano, M.; Brescia, G.; Marioni, G. Antioxidant Therapy in Idiopathic Tinnitus: Preliminary Outcomes. *Arch. Med. Res.* **2007**, *38*, 456–459. [CrossRef]
16. Person, O.C.; Puga, M.E.; da Silva, E.M.; Torloni, M.R. Zinc supplementation for tinnitus. *Cochrane Database Syst. Rev.* **2016**, *2016*. [CrossRef]
17. Polanski, J.F.; Soares, A.D.; de Mendonça Cruz, O.L. Antioxidant therapy in the elderly with tinnitus. *Braz. J. Otorhinolaryngol.* **2016**, *82*, 269–274. [CrossRef]
18. Xiong, M.; Lai, H.; Yang, C.; Huang, W.; Wang, J.; Fu, X.; He, Q. Comparison of the Protective Effects of Radix Astragali, α-Lipoic Acid, and Vitamin E on Acute Acoustic Trauma. *Clin. Med. Insights Ear Nose Throat* **2012**, *5*, 25–31. [CrossRef]
19. Karafakioğlu, Y.S. Effects of α lipoic acid on noise induced oxidative stress in rats. *Saudi J. Biol. Sci.* **2019**, *26*, 989–994. [CrossRef]
20. Erdem, T.; Bayindir, T.; Filiz, A.; Iraz, M.; Selimoglu, E. The effect of resveratrol on the prevention of cisplatin ototoxicity. *Eur. Arch. Otorhinolaryngol.* **2012**, *269*, 2185–2188. [CrossRef]
21. Simsek, G.; Taş, B.M.; Muluk, N.B.; Azman, M.; Kılıç, R. Comparison of the protective efficacy between intratympanic dexamethasone and resveratrol treatments against cisplatin-induced ototoxicity: An experimental study. *Eur. Arch. Otorhinolaryngol.* **2019**, *276*, 3287–3293. [CrossRef]
22. Yumusakhuylu, A.C.; Yazici, M.; Sari, M.; Binnetoglu, A.; Kosemihal, E.; Akdas, F.; Sirvanci, S.; Yuksel, M.; Uneri, C.; Tutkun, A. Protective role of resveratrol against cisplatin induced ototoxicity in guinea pigs. *Int. J. Pediatr. Otorhinolaryngol.* **2012**, *76*, 404–408. [CrossRef]
23. Li, I.-H.; Shih, J.-H.; Jhao, Y.-T.; Chen, H.-C.; Chiu, C.-H.; Chen, C.-F.F.; Huang, Y.-S.; Shiue, C.-Y.; Ma, K.-H. Regulation of Noise-Induced Loss of Serotonin Transporters with Resveratrol in a Rat Model Using 4-[18F]-ADAM/Small-Animal Positron Emission Tomography. *Molecules* **2019**, *24*, 1344. [CrossRef]
24. Seidman, M.D.; Tang, W.; Bai, V.U.; Ahmad, N.; Jiang, H.; Media, J.; Patel, N.; Rubin, C.J.; Standring, R.T. Resveratrol decreases noise-induced cyclooxygenase-2 expression in the rat cochlea. *Otolaryngol. Head Neck Surg.* **2013**, *148*, 827–833. [CrossRef]
25. Xiong, H.; Ou, Y.; Xu, Y.; Huang, Q.; Pang, J.; Lai, L.; Zheng, Y. Resveratrol Promotes Recovery of Hearing following Intense Noise Exposure by Enhancing Cochlear SIRT1 Activity. *AUD* **2017**, *22*, 303–310. [CrossRef]
26. Joachims, H.Z.; Segal, J.; Golz, A.; Netzer, A.; Goldenberg, D. Antioxidants in treatment of idiopathic sudden hearing loss. *Otol. Neurotol.* **2003**, *24*, 572–575. [CrossRef]
27. Quaranta, N.; Dicorato, A.; Matera, V.; D'Elia, A.; Quaranta, A. The effect of alpha-lipoic acid on temporary threshold shift in humans: A preliminary study. *Acta Otorhinolaryngol. Ital.* **2012**, *32*, 380–385.
28. Institute of Medicine (US). *Dietary Reference Intakes: The Essential Guide to Nutrient Requirements*; National Academies Press: Washington, DC, USA, 2006; ISBN 978-0-309-15742-1.
29. Institute of Medicine (US). Using the Tolerable Upper Intake Level for Nutrient Assessment of Groups. In *Dietary Reference Intakes: Applications in Dietary Assessment*; National Academies Press: Washington, DC, USA, 2000; Chapter 6; pp. 113–126.
30. Kapul, A.; Zubova, E.; Torgaev, S.; Drobchik, V. Pure-tone Audiometer. *J. Phys. Conf. Ser.* **2017**, *881*, 012010. [CrossRef]
31. Rodríguez Valiente, A.; Trinidad, A.; García Berrocal, J.R.; Górriz, C.; Ramírez Camacho, R. Extended high-frequency (9–20 kHz) audiometry reference thresholds in 645 healthy subjects. *Int. J. Audiol.* **2014**, *53*, 531–545. [CrossRef]
32. Henry, J.A.; Meikle, M.B. Psychoacoustic Measures of Tinnitus. *J. Am. Acad. Audiol.* **2000**, *11*, 138–155.
33. Newman, C.W.; Jacobson, G.P.; Spitzer, J.B. Development of the Tinnitus Handicap Inventory. *Arch. Otolaryngol. Head Neck Surg.* **1996**, *122*, 143–148. [CrossRef]
34. Figueiredo, R.R.; de Azevedo, A.A.; de Mello Oliveira, P. Correlation analysis of the visual-analogue scale and the Tinnitus Handicap Inventory in tinnitus patients. *Braz. J. Otorhinolaryngol.* **2009**, *75*, 76–79. [CrossRef]

35. Meikle, M.; Henry, J.; Griest, S.; Stewart, B.; Abrams, H.; McArdle, R.; Myers, P.; Newman, C.; Sandridge, S.; Turk, D.; et al. The Tinnitus Functional Index: Development of a New Clinical Measure for Chronic, Intrusive Tinnitus. *Ear Hear.* **2012**, *33*, 153–176. [CrossRef]
36. Willett, W.C.; Sampson, L.; Stampfer, M.J.; Rosner, B.; Bain, C.; Witschi, J.; Hennekens, C.H.; Speizer, F.E. Reproducibility and validity of a semiquantitative food frequency questionnaire. *Am. J. Epidemiol.* **1985**, *122*, 51–65. [CrossRef]
37. Panagiotakos, D.B.; Pitsavos, C.; Stefanadis, C. Dietary patterns: A Mediterranean diet score and its relation to clinical and biological markers of cardiovascular disease risk. *Nutr. Metab. Cardiovasc. Dis.* **2006**, *16*, 559–568. [CrossRef]
38. Craig, C.L.; Marshall, A.L.; Sjöström, M.; Bauman, A.E.; Booth, M.L.; Ainsworth, B.E.; Pratt, M.; Ekelund, U.; Yngve, A.; Sallis, J.F.; et al. International Physical Activity Questionnaire: 12-Country Reliability and Validity. *Med. Sci. Sports Exerc.* **2003**, *35*, 1381–1395. [CrossRef]
39. Fountoulakis, K.; Iacovides, A.; Kleanthous, S.; Samolis, S.; Kaprinis, S.G.; Sitzoglou, K.; St Kaprinis, G.; Bech, P. Reliability, Validity and Psychometric Properties of the Greek Translation of the Center for Epidemiological Studies-Depression (CES-D) Scale. *BMC Psychiatry* **2001**, *1*, 3. [CrossRef]
40. Michopoulos, I.; Douzenis, A.; Kalkavoura, C.; Christodoulou, C.; Michalopoulou, P.; Kalemi, G.; Fineti, K.; Patapis, P.; Protopapas, K.; Lykouras, L. Hospital Anxiety and Depression Scale (HADS): Validation in a Greek general hospital sample. *Ann. Gen. Psychiatry* **2008**, *7*, 4. [CrossRef]
41. Artiss, J.D.; Vinogradov, S.; Zak, B. Spectrophotometric study of several sensitive reagents for serum iron. *Clin. Biochem.* **1981**, *14*, 311–315. [CrossRef]
42. Mann, C.K.; Yoe, J.H. Spectrophotometric Determination of Magnesium with Sodium 1-Azo-2-hydroxy-3-(2,4-dimethylcarboxanilido)-naphthalene-1′-(2-hydroxybenzene-5-sulfonate). *Anal. Chem.* **1956**, *28*, 202–205. [CrossRef]
43. Langguth, B.; Goodey, R.; Azevedo, A.; Bjorne, A.; Cacace, A.; Crocetti, A.; Del Bo, L.; De Ridder, D.; Diges, I.; Elbert, T.; et al. Consensus for tinnitus patient assessment and treatment outcome measurement: Tinnitus Research Initiative meeting, Regensburg, July 2006. *Prog. Brain Res.* **2007**, *166*, 525–536.
44. Hiller, W.; Goebel, G. Factors influencing tinnitus loudness and annoyance. *Arch. Otolaryngol. Head Neck Surg.* **2006**, *132*, 1323–1330. [CrossRef]
45. Landgrebe, M.; Azevedo, A.; Baguley, D.; Bauer, C.; Cacace, A.; Coelho, C.; Dornhoffer, J.; Figueiredo, R.; Flor, H.; Hajak, G.; et al. Methodological aspects of clinical trials in tinnitus: A proposal for an international standard. *J. Psychosom. Res.* **2012**, *73*, 112–121. [CrossRef]
46. Zeman, F.; Koller, M.; Figueiredo, R.; Aazevedo, A.; Rates, M.; Coelho, C.; Kleinjung, T.; de Ridder, D.; Langguth, B.; Landgrebe, M. Tinnitus Handicap Inventory for Evaluating Treatment Effects: Which Changes Are Clinically Relevant? *Otolaryngol. Head Neck Surg.* **2011**, *145*, 282–287. [CrossRef]
47. EFSA Panel on Dietetic Products, Nutrition and Allergies (NDA). Guidance on the scientific requirements for health claims related to antioxidants, oxidative damage and cardiovascular health. *EFSA J.* **2011**, *9*, 2474. [CrossRef]
48. Chang, J.W.; Lee, E.K.; Kim, T.H.; Min, W.K.; Chun, S.; Lee, K.-U.; Kim, S.B.; Park, J.S. Effects of α-Lipoic Acid on the Plasma Levels of Asymmetric Dimethylarginine in Diabetic End-Stage Renal Disease Patients on Hemodialysis: A Pilot Study. *AJN* **2007**, *27*, 70–74. [CrossRef]
49. Kinlay, S.; Behrendt, D.; Fang, J.C.; Delagrange, D.; Morrow, J.; Witztum, J.L.; Rifai, N.; Selwyn, A.P.; Creager, M.A.; Ganz, P. Long-term effect of combined vitamins E and C on coronary and peripheral endothelial function. *J. Am. Coll. Cardiol.* **2004**, *43*, 629–634. [CrossRef]
50. Kim, Y.J.; Ahn, Y.H.; Lim, Y.; Kim, J.Y.; Kim, J.; Kwon, O. Daily Nutritional Dose Supplementation with Antioxidant Nutrients and Phytochemicals Improves DNA and LDL Stability: A Double-Blind, Randomized, and Placebo-Controlled Trial. *Nutrients* **2013**, *5*, 5218–5232. [CrossRef]
51. Kolahi, S.; Mirtaheri, E.; Pourghasem Gargari, B.; Khabbazi, A.; Hajalilou, M.; Asghari-Jafarabadi, M.; Mesgari Abbasi, M. Oral administration of alpha-lipoic acid did not affect lipid peroxidation and antioxidant biomarkers in rheumatoid arthritis patients. *Int. J. Vitam. Nutr. Res.* **2019**, *89*, 13–21. [CrossRef]
52. Mendoza-Núñez, V.M.; García-Martínez, B.I.; Rosado-Pérez, J.; Santiago-Osorio, E.; Pedraza-Chaverri, J.; Hernández-Abad, V.J. The Effect of 600 mg Alpha-lipoic Acid Supplementation on Oxidative Stress, Inflammation, and RAGE in Older Adults with Type 2 Diabetes Mellitus. *Oxidative Med. Cell. Longev.* **2019**, *2019*, 3276958. [CrossRef]

53. Serafini, M.; Rio, D.D. Understanding the association between dietary antioxidants, redox status and disease: Is the Total Antioxidant Capacity the right tool? *Redox Rep.* **2004**, *9*, 145–152. [CrossRef]
54. Jialal, I.; Fuller, C.J.; Huet, B.A. The effect of alpha-tocopherol supplementation on LDL oxidation. A dose-response study. *Arterioscler. Thromb. Vasc. Biol.* **1995**, *15*, 190–198. [CrossRef]
55. Arda, H.N.; Tuncel, U.; Akdogan, O.; Ozluoglu, L.N. The role of zinc in the treatment of tinnitus. *Otol. Neurotol.* **2003**, *24*, 86–89. [CrossRef]
56. Coelho, C.; Witt, S.A.; Ji, H.; Hansen, M.R.; Gantz, B.; Tyler, R. Zinc to treat tinnitus in the elderly: A randomized placebo controlled crossover trial. *Otol. Neurotol.* **2013**, *34*, 1146–1154. [CrossRef]
57. Pisoschi, A.M.; Pop, A. The role of antioxidants in the chemistry of oxidative stress: A review. *Eur. J. Med. Chem.* **2015**, *97*, 55–74. [CrossRef]
58. Shay, K.P.; Moreau, R.F.; Smith, E.J.; Smith, A.R.; Hagen, T.M. Alpha-lipoic acid as a dietary supplement: Molecular mechanisms and therapeutic potential. *Biochim. Biophys. Acta* **2009**, *1790*, 1149–1160. [CrossRef]
59. Teichert, J.; Hermann, R.; Ruus, P.; Preiss, R. Plasma Kinetics, Metabolism, and Urinary Excretion of Alpha-Lipoic Acid following Oral Administration in Healthy Volunteers. *J. Clin. Pharmacol.* **2003**, *43*, 1257–1267. [CrossRef]
60. Yilmaz, Y.; Toledo, R.T. Major flavonoids in grape seeds and skins: Antioxidant capacity of catechin, epicatechin, and gallic acid. *J. Agric. Food Chem.* **2004**, *52*, 255–260. [CrossRef]
61. Noack, V.; Pak, K.; Jalota, R.; Kurabi, A.; Ryan, A.F. An Antioxidant Screen Identifies Candidates for Protection of Cochlear Hair Cells from Gentamicin Toxicity. *Front. Cell. Neurosci.* **2017**, *11*, 242. [CrossRef]
62. Seidman, M.D. Effects of Dietary Restriction and Antioxidants on Presbyacusis. *Laryngoscope* **2000**, *110*, 727–738. [CrossRef]
63. Tokgöz, S.A.; Vuralkan, E.; Sonbay, N.D.; Çalişkan, M.; Saka, C.; Beşalti, Ö.; Akin, İ. Protective effects of vitamins E, B and C and l-carnitine in the prevention of cisplatin-induced ototoxicity in rats. *J. Laryngol. Otol.* **2012**, *126*, 464–469. [CrossRef] [PubMed]
64. Wang, A.; Hou, N.; Bao, D.; Liu, S.; Xu, T. Mechanism of alpha-lipoic acid in attenuating kanamycin-induced ototoxicity. *Neural Regen. Res.* **2012**, *7*, 2793–2800. [PubMed]
65. Kim, S.K.; Im, G.J.; An, Y.S.; Lee, S.H.; Jung, H.H.; Park, S.Y. The effects of the antioxidant α-tocopherol succinate on cisplatin-induced ototoxicity in HEI-OC1 auditory cells. *Int. J. Pediatric Otorhinolaryngol.* **2016**, *86*, 9–14. [CrossRef] [PubMed]
66. Lee, J.S.; Kang, S.U.; Hwang, H.S.; Pyun, J.H.; Choung, Y.H.; Kim, C.H. Epicatechin protects the auditory organ by attenuating cisplatin-induced ototoxicity through inhibition of ERK. *Toxicol. Lett.* **2010**, *199*, 308–316. [CrossRef]
67. Li, D.-W.; Wang, Y.-D.; Zhou, S.-Y.; Sun, W.-P. α-lipoic acid exerts neuroprotective effects on neuronal cells by upregulating the expression of PCNA via the P53 pathway in neurodegenerative conditions. *Mol. Med. Rep.* **2016**, *14*, 4360–4366. [CrossRef]
68. Kennedy, D.O. B Vitamins and the Brain: Mechanisms, Dose and Efficacy—A Review. *Nutrients* **2016**, *8*, 68. [CrossRef]
69. Houston, D.K.; Johnson, M.A.; Nozza, R.J.; Gunter, E.W.; Shea, K.J.; Cutler, G.M.; Edmonds, J.T. Age-related hearing loss, vitamin B-12, and folate in elderly women. *Am. J. Clin. Nutr.* **1999**, *69*, 564–571. [CrossRef]
70. Martínez-Vega, R.; Garrido, F.; Partearroyo, T.; Cediel, R.; Zeisel, S.H.; Martínez-Álvarez, C.; Varela-Moreiras, G.; Varela-Nieto, I.; Pajares, M.A. Folic acid deficiency induces premature hearing loss through mechanisms involving cochlear oxidative stress and impairment of homocysteine metabolism. *FASEB J.* **2014**, *29*, 418–432. [CrossRef]
71. De Ridder, D.; Vanneste, S.; Adriaenssens, I.; Lee, A.P.K.; Plazier, M.; Menovsky, T.; van der Loo, E.; Van de Heyning, P.; Møller, A. Microvascular decompression for tinnitus: Significant improvement for tinnitus intensity without improvement for distress. A 4-year limit. *Neurosurgery* **2010**, *66*, 656–660. [CrossRef]
72. Lopez-Escamez, J.A.; Bibas, T.; Cima, R.F.F.; Van de Heyning, P.; Knipper, M.; Mazurek, B.; Cederroth, C.R. Genetics of tinnitus: An emerging area for molecular diagnosis and drug development. *Front. Neurosci.* **2016**, *10*, 377. [CrossRef]

Article

Effect of Cod Residual Protein Supplementation on Markers of Glucose Regulation in Lean Adults: A Randomized Double-Blind Study

Iselin Vildmyren [1,2,*], Alfred Halstensen [2,3], Adrian McCann [4], Øivind Midttun [4], Per Magne Ueland [4], Åge Oterhals [5] and Oddrun Anita Gudbrandsen [1]

1 Dietary Protein Research Group, Department of Clinical Medicine, University of Bergen, 5021 Bergen, Norway; nkjgu@uib.no

2 K. Halstensen AS, P.O. Box 103, 5399 Bekkjarvik, Norway; alfred.halstensen@uib.no

3 Department of Clinical Science, University of Bergen, 5021 Bergen, Norway

4 Bevital AS, Jonas Lies veg 87, 5021 Bergen, Norway; adrian.mccann@bevital.no (A.M.); nkjbm@uib.no (Ø.M.); per.ueland@ikb.uib.no (P.M.U.)

5 Nofima, P.B. 1425 Oasen, 5844 Bergen, Norway; aage.oterhals@nofima.no

* Correspondence: iselin.vildmyren@uib.no; Tel.: +47-5597-5553

Received: 16 April 2020; Accepted: 13 May 2020; Published: 16 May 2020

Abstract: Large quantities of protein-rich cod residuals, which are currently discarded, could be utilized for human consumption. Although fish fillet intake is related to beneficial health effects, little is known about the potential health effects of consuming cod residual protein powder. Fifty lean adults were randomized to consume capsules with 8.1 g/day of cod residual protein (Cod-RP) or placebo capsules (Control group) for eight weeks, in this randomized, double-blind study. The intervention was completed by 40 participants. Fasting glucose and insulin concentrations were unaffected by Cod-RP supplementation, whereas plasma concentrations of α-hydroxybutyrate, β-hydroxybutyrate and acetoacetate all were decreased compared with the Control group. Trimethylamine N-oxide concentration in plasma and urine were increased in the Cod-RP group compared with the Control group. To conclude, the reduction in these potential early markers of impaired glucose metabolism following Cod-RP supplementation may indicate beneficial glucoregulatory effects of cod residual proteins. Trimethylamine N-oxide appears to be an appropriate biomarker of cod residual protein intake in lean adults.

Keywords: dietary supplements; fish protein; marine protein; hydroxybutyrate; TMAO; biomarkers

---

1. Introduction

Fish intake is associated with beneficial effects on risk factors for type 2 diabetes and cardiovascular disease [1–7], and these effects have mainly been attributed to the content of long-chain n-3 polyunsaturated fatty acids (n-3 PUFA) in fish [8]. Although underlying mechanisms are still unclear, several studies have shown that lean fish may also improve insulin sensitivity and glucose regulation, despite containing very low amounts of long-chain n-3 PUFA [9–15]. Small alterations in glucose regulation may be difficult to identify in short-term dietary intervention studies among healthy individuals. Therefore, other markers of glucose regulation that may be more sensitive to detecting small regulatory changes, other than fasting glucose and insulin concentrations, should be explored. Studies have shown that metabolites of fatty acid β-oxidation and ketogenic amino acid catabolism, such as α-hydroxybutyrate (α-HB), β-hydroxybutyrate (β-HB) and acetoacetate (AcAc), may reflect early changes in glucose regulation [16–19].

People in the western world mostly consume the fish fillet, and not the whole fish or the residuals. The residuals from fillet production consist of head, backbone, skin, entrails and trimmings, which contain up to 70% protein when dried [20]. Currently, large quantities of residuals from cod are discarded, and the rest is mostly utilized for non-food purposes, such as feed ingredients for agriculture and aquaculture. To our knowledge, only two clinical studies on cod residual protein intake have been conducted so far, and findings suggest that cod residual protein supplementation may beneficially affect glucose regulation and lipid metabolism [14,21]. Thus, further studies investigating potential health effects of cod residual protein for human consumption should be undertaken.

The assessment of dietary compliance is vital in nutrition intervention studies, and biomarkers reflecting dietary intake would be especially useful for evaluating intervention efficacy [22]. Higher concentrations of creatine, 1-methylhistidine (1-MeHis), 3-methylhistidine (3-MeHis) [23,24] and trimethylamine N-oxide (TMAO) [23,25–31] have been associated with cod fillet intake in both rats and humans. These observations highlight the potential use of these compounds as biomarkers of cod protein intake from residuals in supplement studies.

The current study aimed to investigate the effects of cod residual protein supplementation on circulating markers related to glucose regulation, and potential biomarkers of cod residual protein intake. We hypothesized that eight weeks of 8.1 g cod residual protein (Cod-RP) per day would decrease concentrations of circulating markers of glucose regulation, and increase concentrations of potential plasma and urine biomarkers of cod residual protein intake in lean adults. Cod-RP supplementation caused reduced fasting plasma concentrations of α-hydroxybutyrate, β-hydroxybutyrate and acetoacetate, compared with the Control group. Increased trimethylamine N-oxide concentrations in plasma and urine suggest it could function as a biomarker of cod residual protein intake in lean adults.

2. Materials and Methods

2.1. Participants and Ethics

The study population consisted of healthy, lean adults. The eligibility criteria included BMI ≥ 18.5 kg/m^2 with body fat percentage 12–35% for women and 5–25% for men, stable body weight (<5 kg variation in 3 months), fasting blood glucose < 7.0 millimoles/L, and age 20–55 years. Exclusion criteria included disorders affecting kidney, heart, intestinal function, or insulin secretion, use of prescription medications for high cholesterol, seafood intake > 200 g/week, allergies towards seafood, gluten or milk, and use of dietary supplements. Pregnant and lactating women were not included in the study.

The participants provided written informed consent prior to study enrolment. The Regional Ethics Committee of Western Norway approved the study (approval no.: 2015/75). The study was conducted in accordance with the Declaration of Helsinki and is registered at clinicaltrials.gov (NCT03538834).

2.2. Study Design, Intervention and Protocol

This double-blind, randomized intervention study was conducted at the University of Bergen (Bergen, Norway). In brief, 50 participants were stratified according to gender and age, and randomized to one of two groups, consuming 27 capsules with a total of 8.1 g of cod residual protein (Cod-RP group), or 27 placebo capsules (Control group) daily for eight weeks. Participants were instructed to maintain their dietary and physical activity routines unchanged during the eight week intervention period. The participants attended two study visits: at baseline and after eight weeks. The study visits were performed after an overnight fast (>10 h) without consuming drink (except for water), food, medications or tobacco. Participants were instructed to not consume alcohol or engage in intensive physical activity for 24 h before the study visits.

The participants' height was measured using a stadiometer (Telescopic Measuring Rod MZ10023-3, ADE, Hamburg, Germany), and body composition was measured using a bioelectrical impedance analyzer (InBody 720, Seoul, Korea). Participants provided urine samples in a fasted state. Fasting blood samples for whole blood and isolation of plasma were collected in Vacuette K2EDTA (Greiner Bio-One GmbH, Kremsmünster, Austria) and Vacutainer SST II Advanced Plus (Becton, Dickinson and Company, Franklin Lakes, NJ, USA) for isolation of serum. Postprandial blood samples were collected every 30 min for 2 h after the participants consumed a drink containing 75 g glucose (Dextrose from SunVita AS, Nyborg, Norway) and 35 g whey protein (Proteinfabrikken AS, Stokke, Norway) mixed with 200 mL of water, as a modified oral glucose tolerance test (OGTT).

2.3. Production and Analyses of Intervention Capsules

Northeast Atlantic cod (Gadus morhua) residuals from fillet production, consisting of head, backbone, skin, entrails and trimmings were processed on-board the factory trawler Granit (Halstensen Granit AS, Bekkjarvik, Norway). The cod residuals were ground, heat treated at 90 °C, pressed and dried on-board the trawler, and the solid phase and liquid phase was dried and mixed at Seagarden AS (Karmøy, Norway). All capsules for the study were produced at Pharmatech AS (Fredrikstad, Norway). A detailed description of the production of intervention capsules is described elsewhere [21] and the contents of the capsules are presented in Table 1.

Table 1. Capsule content.

Per Capsule	Cod-RP	Control
Cod residual powder * (mg)	474	0
Microcrystalline cellulose (mg)	42	454
Magnesium stearate (mg)	5.3	4.6
Silica (mg)	5.3	4.6
Total capsule weight (mg)	527	463
Energy † (kcal)	1.59	0.02

* Crude protein content 64%, total fat content 8.6%. † Estimated true metabolizable energy. Cod-RP, cod residual protein powder.

Trimethylamine N-oxide (TMAO) content in the cod residual powder and placebo capsules was determined using the micro-diffusion assay described by Conway and Byrne [32] and anserine was measured by High Performance Liquid Chromatographyusing the Waters Pico-Tag method [33]. The contents of TMAO, anserine, crude protein, total fat and microorganisms were measured by Nofima BioLab (Bergen, Norway). Contents of the heavy metals cadmium and mercury were analyzed by Eurofins AS (Ålesund, Norway).

2.4. Analyses of Biological Samples

Glucose concentration in serum, HbA1c (glycated haemoglobin) concentration in whole blood and concentrations of creatinine and albumin in urine were analyzed on the Cobas c 111 system (Roche Diagnostics GmbH, Mannheim, Germany) using the GLUC2, A1C-3 (with A1CD2 hemolyzing reagent) and CREP2 kits from Roche Diagnostics for the c111 system. Serum insulin concentration was analyzed using the EIA-2935 ELISA kit (DRG Instruments GmbH, Marburg, Germany).

Plasma concentrations of α-HB, β-HB, AcAc, and plasma and urine concentration of 1-methylhistidine (1-MeHis, π-methyl-histidine), 3-methylhistidine (3-MeHis, τ-methyl-histidine), creatine, trimethylamine N-oxide (TMAO) and creatinine were measured by Bevital AS (http://www.bevital.no) using gas chromatography and liquid chromatography with tandem mass spectrometry [34,35]. α-HB, β-HB, AcAc, 1-MeHis, 3-MeHis and TMAO were analyzed by adding the analytes and isotope-labelled internal standards to existing assays [34,35].

2.5. Outcomes

The primary outcome was changes in circulating markers related to glucose regulation in lean adults after eight weeks of cod residual protein supplementation. The secondary outcome was changes in potential urine and plasma biomarkers of cod residual protein intake.

2.6. Sample Size

Power analysis was not feasible for the original study, which was designed as a pilot study and is the first to investigate effects of cod residuals on circulating markers of lipid metabolism in lean adults [21]. Therefore, no estimation of sample size for the current measurements was conducted prior to the study.

2.7. Statistical Analyses

Normality of data was assessed by the Shapiro–Wilk test, histograms and QQ plots. Variables that were not normally distributed underwent log-transformation before parametric statistical tests were performed. The Paired samples *T*-test was used to detect baseline to eight week changes within groups, and the Independent samples *T*-test was used to compare baseline to eight weekchanges between the two experimental groups. The level of statistical significance was set to $p < 0.05$. SPSS Statistics version 25 (IBM Corp. IBM SPSS for Windows, Armonk, NY, USA) was used for statistical analyses.

3. Results

3.1. Participant Characteristics

Of the fifty participants initially included, forty participants (18 women and 22 men) completed the eight weeks intervention (Figure 1). Four participants withdrew from the study, and six participants were excluded from further analyses due to non-compliance (n = 5) or disease (n = 1). Non-compliance was defined as not taking the study supplements, consuming > 200 g/week of seafood, and changes in habitual dietary intake or physical activity. Three participants were excluded from the OGTT due to difficulties drawing blood. Age, body weight, BMI, body fat percentage, body muscle percentage, concentrations of fasting glucose, insulin, α-HB, β-HB, AcAc and cigarette or snus use were similar between the two groups at baseline (Table 2). The concentrations of whole blood HbA1c, plasma creatinine and urine albumin (relative to creatinine) were within normal range in all participants at baseline: for HbA1c, <42 mmol/mol International Federation of Clinical Chemistry and Laboratory Medicine (IFCC); for creatinine, 45–90 μmol/L in women and 60–105 μmol/L in men; and for albumin, 0–2.5 mg/mmol. Concentrations of whole blood HbA1c, plasma creatinine and urine albumin were unchanged after eight weeks intervention (data not presented). Estimated dietary energy and macronutrient intake did not differ between the groups from baseline to eight weeks, as previously presented [21].

Table 2. Participant characteristics at baseline.

	Cod-RP (n = 19)	Control (n = 21)	p
	Mean ± SD	Mean ± SD	
Women/Men	7/12	11/10	0.36
Age (years)	28.0 ± 6.9	30.5 ± 7.2	0.28
Body weight (kg)	77.0 ± 16.0	73.1 ± 11.5	0.39
BMI (kg/m^2)	24.8 ± 2.8	23.8 ± 2.3	0.25
Body fat (%)	19.7 ± 6.8	19.4 ± 6.7	0.89
Body muscle (%)	45.4 ± 4.7	45.4 ± 4.4	0.95
Whole blood HbA1c (mmol/mol)	32.4 ± 2.2	31.5 ± 2.0	0.23
Plasma creatinine (μmol/L)	80.1 ± 13.1	77.2 ± 11.1	0.46
Urine albumin (mg/mmol creatinine)	0.7 ± 0.5	1.0 ± 0.9	0.23
Serum glucose (mmol/L)	5.1 ± 0.4	4.9 ± 0.3	0.21
Serum insulin (pmol/L)	68.1 ± 32.8	56.3 ± 25.9	0.55
Plasma α-HB (μmol/L)	44.2 ± 19.3	35.1 ± 17.0	0.80
Plasma β-HB (μmol/L)	143 ± 156	102 ± 108	0.34
Plasma AcAc (μmol/L)	76.5 ± 67.0	65.1 ± 52.1	0.86
Cigarette/snus * (n)	1	2	1.00

Values are presented as mean ± standard deviation. Groups were compared at baseline using independent samples *T*-test for continuous data and Pearson's Chi-square test for categorical data. * Snus is a Scandinavian tobacco substance which is placed under the upper lip. Cod-RP, cod residual protein powder; BMI, body mass index; HbA1c, glycated haemoglobin; α-HB, α-hydroxybutyrate; β-HB, β-hydroxybutyrate; AcAc, acetoacetate.

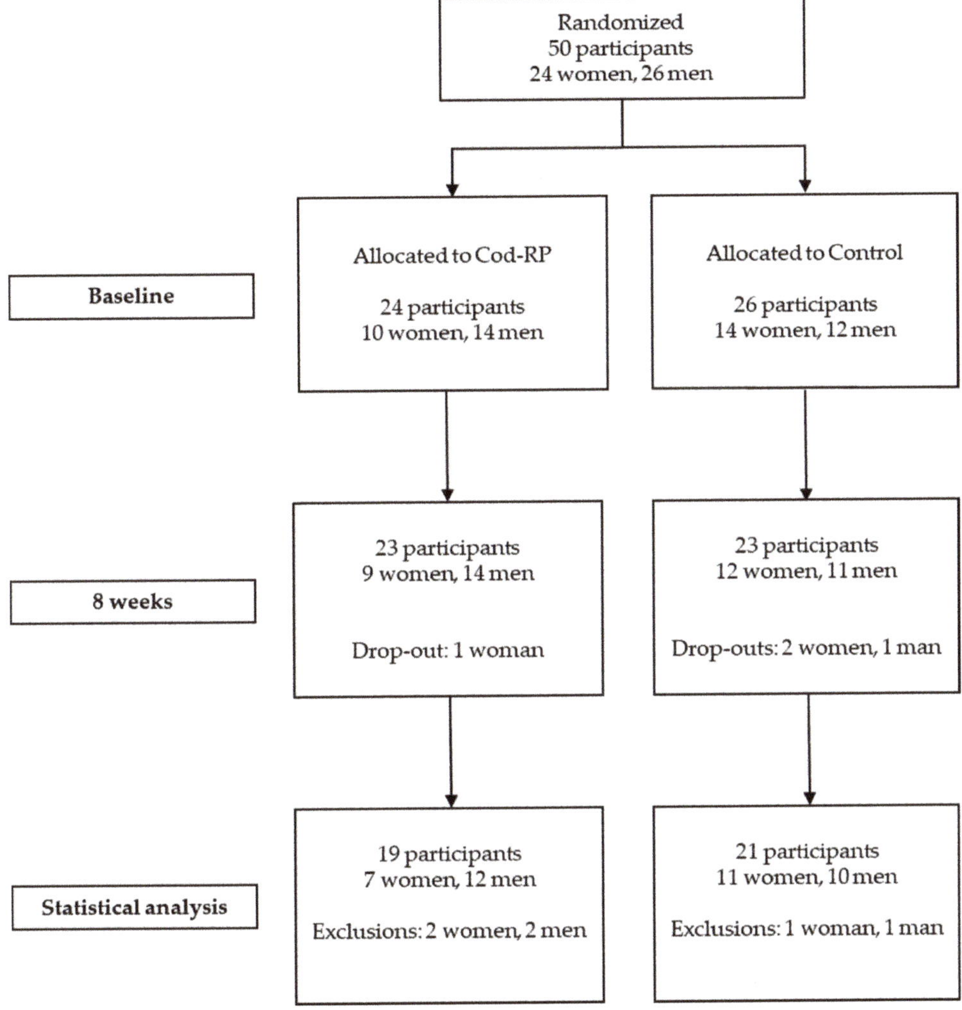

Figure 1. Flow of participants through the study. Participants could withdraw at any time and participants were excluded from analysis if they did not comply with the study protocol. Cod-RP, cod residual protein powder.

3.2. The Contents of Trimethylamine N-oxide, Anserine, Microorganisms and Heavy Metals from the Capsules

The estimated intakes of TMAO and anserine from the Cod-RP capsules were 14.4 and 7.8 mg/day, respectively (average of two measurements of the capsule composition, with deviations < 5% between parallels). TMAO and anserine were not detected in the Control capsules. The contents of microorganisms in the cod residual powder were at an acceptably low level for human consumption. It was estimated that the participants' weekly intake of cadmium and mercury was equivalent to 0.08 and 0.13 µg/kg body weight, respectively (participants' average body weight of 75 kg). The tolerable weekly intake for cadmium and mercury is 2.5 and 1.3 µg/kg body weight, respectively [36,37]. Thus, the estimated intake of the heavy metals cadmium and mercury from cod residual protein powder was within the tolerable weekly intake.

3.3. Circulating Markers Related to Glucose Regulation

Fasting serum concentrations of glucose and insulin were unchanged in both groups from baseline to eight weeks (Figure 2). Comparison of serum glucose after OGTT at baseline, and following eight weeks of supplementation, revealed that only glucose concentration at 60 min decreased in the Cod-RP group compared with the Control group ($p = 0.0078$). Changes in postprandial insulin concentrations, from baseline to eight weeks, did not differ between groups at any time points. Fasting plasma concentrations of α-HB, β-HB and AcAc were decreased in the Cod-RP group when compared with the Control group after eight weeks of supplementation (Figure 3).

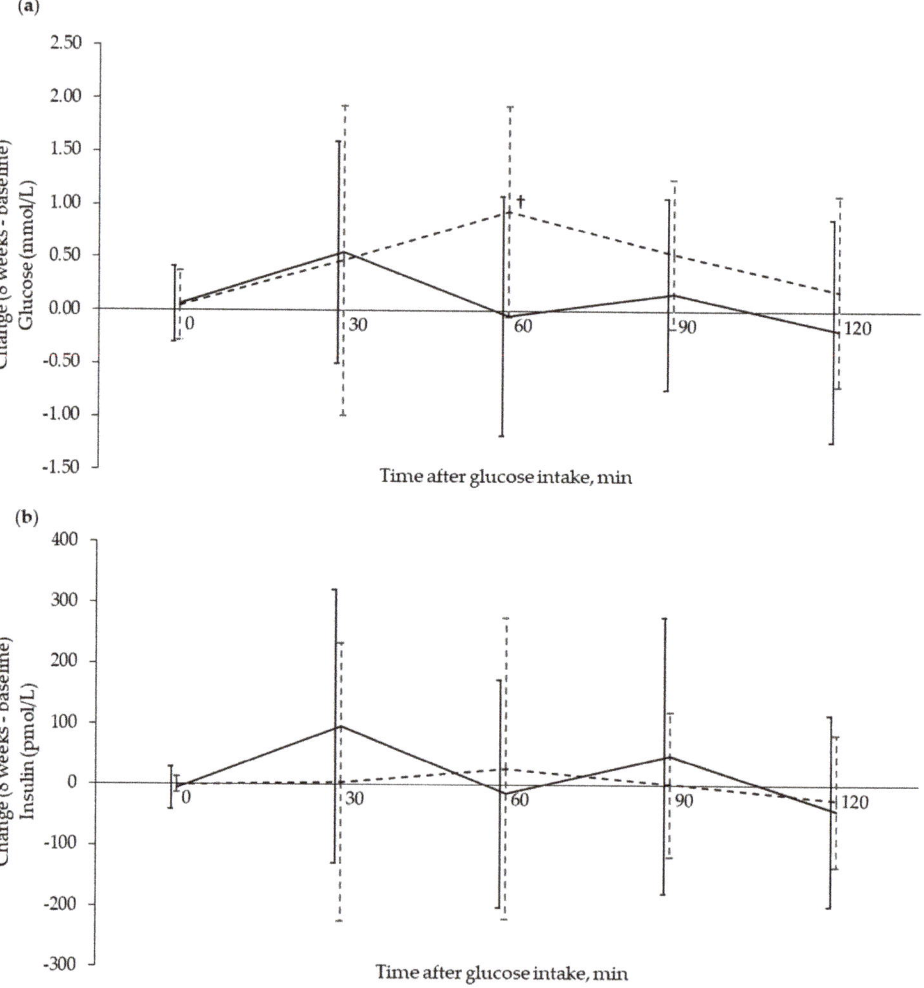

Figure 2. Glucose and insulin response during an oral glucose tolerance test (OGTT) shown as change from baseline to 8 weeks after supplementation. Glucose response (**a**) and insulin response (**b**) are expressed as change (8 weeks–baseline) in the Cod-RP (—) group and the Control group (- -), at 0 (fasting), 30, 60, 90 and 120 min after oral glucose intake. The results are presented as mean with standard deviation for 18 participants in the Cod-RP group and 19 participants in the Control group. † Between-group changes were compared using the independent samples T-test, where $p < 0.05$ was considered significant.

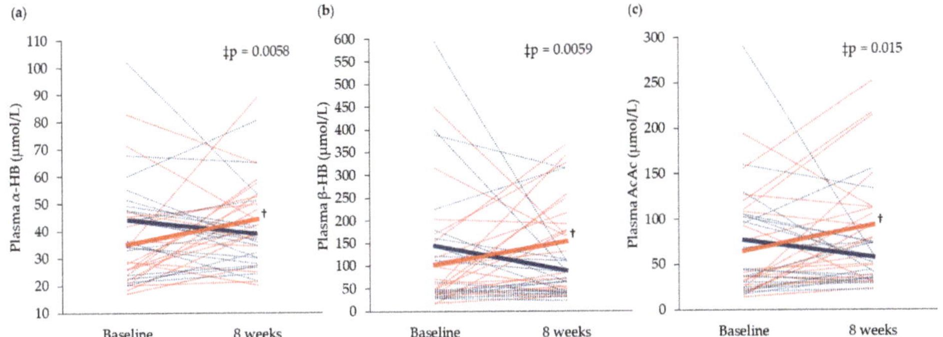

Figure 3. Fasting circulating concentrations of (**a**) α-Hydroxybutyrate (α-HB), (**b**) β-Hydroxybutyrate (β-HB) and (**c**) acetoacetate (AcAc), at baseline and after 8 weeks of supplementation. Concentrations in the Cod-RP group are indicated by blue lines and concentrations in the Control group are indicated by red lines. The means of each group are indicated by bold lines in their respective colors. The results are presented for 19 participants in the Cod-RP group and 21 participants in the Control group. † Within-group changes were tested using the paired samples *T*-test, where $p < 0.05$ was considered significant. ‡ Between-group changes were compared using the independent samples *T*-test.

3.4. Plasma and Urine Biomarkers Related to Cod Residual Protein Intake

Plasma and urine concentrations of TMAO were increased, whereas concentrations of creatine, 1-MeHis and 3-MeHis were unchanged in both plasma and urine in the Cod-RP group, when compared with the Control group, after eight weeks intervention (Table 3).

Table 3. Plasma and urine biomarkers related to cod residual protein intake.

	Baseline	8 Weeks	p †	p ‡
	Mean ± SD	**Mean ± SD**		
Plasma (µmol/L)				
TMAO				0.048
Cod-RP group	4.3 ± 2.2	5.8 ± 3.3	0.032	
Control group	5.2 ± 6.2	4.6 ± 6.3	0.36	
Creatine				0.11
Cod-RP group	24.9 ± 15.3	26.8 ± 15.4	0.36	
Control group	32.2 ± 18.2	30.5 ± 21.8	0.17	
1-MeHis				0.17
Cod-RP group	13.0 ± 11.5	7.8 ± 7.9	0.16	
Control group	9.2 ± 7.2	10.7 ± 9.1	0.68	
3-MeHis				0.24
Cod-RP group	5.1 ± 1.3	4.9 ± 0.9	0.29	
Control group	4.5 ± 1.0	4.7 ± 1.1	0.29	
Urine (µmol/mmol creatinine)				
TMAO				0.026
Cod-RP group	44.3 ± 22.6	63.2 ± 31.5	0.016	
Control group	53.3 ± 68.3	42.0 ± 43.5	0.30	
Creatine				0.18
Cod-RP group	2.7 ± 4.5	6.6 ± 14.9	0.29	
Control group	7.0 ± 11.8	5.6 ± 11.1	0.37	
1-MeHis				0.49
Cod-RP group	90.9 ± 81.4	54.1 ± 54.0	0.16	
Control group	69.0 ± 62.7	72.1 ± 57.9	0.72	

Table 3. *Cont.*

	Baseline	8 Weeks	p †	p ‡
	Mean ± SD	Mean ± SD		
3-MeHis				0.16
Cod-RP group	29.5 ± 9.4	28.2 ± 6.5	0.77	
Control group	25.1 ± 5.6	26.4 ± 6.7	0.41	

Values are presented as mean ± standard deviation for 19 participants in the Cod-RP group and 21 participants in the Control group. The level of significance was set to < 0.05. † Within-group changes were tested using paired samples *T*-test. ‡ Between-group changes were compared using the independent samples *T*-test. Plasma and urine concentrations of TMAO, creatine, 1-MeHis and 3-MiHis were similar between groups at baseline. Cod-RP, cod residual protein powder; TMAO, Trimethylamine N-oxide; 1-MeHis, 1-Methylhistidine; 3-MeHis, 3-Methylhistidine.

4. Discussion

In the current study, we show that eight weeks of cod residual supplementation decreased fasting plasma concentrations of α-HB, β-HB and AcAc, when compared to the Control group. We also observed an increase in TMAO concentration in both plasma and urine in the Cod-RP group, and propose TMAO as a biomarker of cod residual protein intake. To our knowledge, this study is the first to explore the effects of a cod residual protein supplement on glucose regulation and potential biomarkers of cod residual protein intake in lean adults.

The significant reduction in circulating α-HB concentration in the Cod-RP group compared with the Control group suggests that cod residual supplementation may have beneficial effects on glucose regulation, as the concentration of this metabolite is shown to increase under conditions of impaired glucose tolerance [16,17,19]. The current study also demonstrates decreased glucose concentration 60 min after glucose intake in the Cod-RP group, from baseline to eight weeks, compared with the Control group, however, postprandial measurements 30, 90 and 120 min did not differ between the two groups. Previous studies have observed that intake of proteins from cod fillet or cod residuals may improve postprandial glucose regulation in overweight adults and in obese rats, with no effects on fasting serum concentrations of glucose or insulin [12–14,38]. In line with this, two clinical studies, investigating the effects of 750 g cod fillet per week in normal-weight adults for four weeks [39], or overweight adults for eight weeks [40], did not find beneficial effects on fasting glucose and insulin concentrations. In the present study cohort, fasting non-esterified fatty acid (NEFA) concentration decreased after Cod-RP supplementation when compared with Control [21], which could be tied to the decrease in α-HB [16,17,19]. α-HB is synthesized in the liver from α-ketobutyrate, which is a substrate for pyruvate dehydrogenase. The pyruvate dehydrogenase activity is indirectly inhibited by increased NEFA concentration [41,42], resulting in more α-ketobutyrate available for α-HB production. The mechanisms underlying the observations in the present study are currently unknown, but based on the pattern of changes, we speculate that the decreased α-HB in the Cod-RP group may be associated with the decrease in circulating NEFA. This is further supported by the decreased concentrations of the ketone bodies β-HB and AcAc in the Cod-RP group, since these are products of NEFA oxidation, and high circulating concentrations of ketone bodies have been associated with impaired glucose tolerance [18]. The decreases in α-HB, β-HB and AcAc suggest that Cod-RP supplementation increased glucose utilization, despite not affecting fasting concentrations of serum glucose and insulin in this healthy cohort.

The increased TMAO concentrations in both plasma and urine in the Cod-RP group is in agreement with studies demonstrating increased urine or plasma/serum TMAO concentrations after cod fillet intake [23–25,27]. Indeed, higher urinary TMAO concentrations have been observed in populations with high fish intake, when compared to populations with low fish intake [30,31]. TMAO concentration in plasma and urine have also been shown to increase after intake of foods other than fish, if containing precursors of TMAO [25,26,43], but not to the same extent as after fish intake [25–27,44]. Circulating TMAO concentration may increase when kidney function is decreased [45], however, all participants in the current study had normal kidney function, as indicated by normal urine albumin concentration (relative to creatinine) and normal plasma creatinine concentration. Therefore, the increased plasma TMAO concentrations in the Cod-RP group were most likely a result of TMAO intake from the Cod-RP supplements, and not a consequence of impaired kidney function. Following Cod-RP supplementation in the present study, the increased TMAO concentration was more pronounced in urine than in plasma. This is in line with a study demonstrating increased concentrations of TMAO in serum, and particularly in urine, among overweight adults after consumption of 750 g/week of cod fillet for eight weeks [24]. Therefore, this work further suggests that TMAO measured in urine and plasma may be a reliable biomarker of cod intake, and should be measured in future studies assessing compliance to cod residual protein intake.

We found no changes in plasma and urine concentrations of 1-MeHis, 3-MeHis and creatine after Cod-RP supplementation. The concentrations of these biomarkers have been shown to increase after intake of meat [46–49], which was part of the participants' habitual dietary intake. This could explain why no differences in the concentrations of 1-MeHis, 3-MeHis and creatine were observed after Cod-RP supplementation, when compared with Control supplementation. However, cod fillet consumption (750 g/week for eight weeks) increased both serum and urine creatine, and 1-MeHis concentrations, compared with meat intake in overweight adults [24]. These contradictory findings may be attributed to a higher intake of creatine and 1-MeHis from consuming 750 g/week of cod fillet, compared with only 57 g/week of cod residual protein in the current study.

The present study has some limitations. The decreases in circulating concentrations of metabolites related to glucose regulation in the current intervention were observed among lean adults, and may not be translatable to individuals at risk of developing type 2 diabetes or cardiovascular disease. Further exploration of changes in concentrations of α-HB, β-HB and AcAc after cod residual protein intake in individuals with impaired glucose tolerance is therefore suggested.

5. Conclusions

To conclude, eight weeks of supplementation with cod residual proteins lowered plasma concentrations of the early markers of impaired glucose regulation α-HB, β-HB and AcAc, when compared to the Control group, thus suggesting that cod residual proteins may beneficially affect glucose regulation. TMAO concentrations in plasma, and particularly in urine, were increased after cod residual protein supplementation, indicating that TMAO could be a relevant biomarker of cod residual protein intake.

Author Contributions: Conceptualization, I.V., A.H., Å.O. and O.A.G.; Methodology, validation, formal analysis, investigation, I.V., A.M., Ø.M., P.M.U. and O.A.G.; Data curation and writing—original draft preparation, I.V.; Writing—review and editing, I.V., A.H., A.M., Ø.M., P.M.U., Å.O. and O.A.G. All authors have read and agreed to the published version of the manuscript.

Funding: This research was funded by The Research Council of Norway (project no. 252540); K. Halstensen AS and the Regional Research Fund Western Norway.

Acknowledgments: We thank Ingmar Høgøy from Blue Protein (Storebø, Norway) for his significant contribution in the design and pre-production of the intervention capsules and Jørgen Borthen (Norwegian Seafood Center) for organizing and facilitating applications for funding and implementation of the study. The authors wish to thank all participants who have contributed to the study.

Conflicts of Interest: Iselin Vildmyren was employed as an industrial Ph.D.-Candidate at K. Halstensen AS in cooperation with the Research Council of Norway in affiliation with the University of Bergen when the current study was conducted. Alfred Halstensen is shareholder in K. Halstensen AS. The other authors declare no conflict of interest. The funders had no role in the design of the study; in the collection, analyses, or interpretation of data; in the writing of the manuscript, or in the decision to publish the results.

References

1. Kromhout, D.; Bosschieter, E.B.; de Lezenne Coulander, C. The inverse relation between fish consumption and 20-year mortality from coronary heart disease. *N. Engl. J. Med.* **1985**, *312*, 1205–1209. [CrossRef] [PubMed]

2. Nkondjock, A.; Receveur, O. Fish-seafood consumption, obesity, and risk of type 2 diabetes: An ecological study. *Diabetes Metab.* **2003**, *29*, 635–642. [CrossRef]
3. Virtanen, J.K.; Mozaffarian, D.; Chiuve, S.E.; Rimm, E.B. Fish consumption and risk of major chronic disease in men. *Am. J. Clin. Nutr.* **2008**, *88*, 1618–1625. [CrossRef] [PubMed]
4. Zheng, J.; Huang, T.; Yu, Y.; Hu, X.; Yang, B.; Li, D. Fish consumption and chd mortality: An updated meta-analysis of seventeen cohort studies. *Public Health Nutr.* **2012**, *15*, 725–737. [CrossRef] [PubMed]
5. Zhou, Y.; Tian, C.; Jia, C. Association of fish and n-3 fatty acid intake with the risk of type 2 diabetes: A meta-analysis of prospective studies. *Br. J. Nutr.* **2012**, *108*, 408–417. [CrossRef] [PubMed]
6. Chowdhury, R.; Stevens, S.; Gorman, D.; Pan, A.; Warnakula, S.; Chowdhury, S.; Ward, H.; Johnson, L.; Crowe, F.; Hu, F.B.; et al. Association between fish consumption, long chain omega 3 fatty acids, and risk of cerebrovascular disease: Systematic review and meta-analysis. *Bmj* **2012**, *345*, e6698. [CrossRef]
7. Alhassan, A.; Young, J.; Lean, M.E.J.; Lara, J. Consumption of fish and vascular risk factors: A systematic review and meta-analysis of intervention studies. *Atherosclerosis* **2017**, *266*, 87–94. [CrossRef]
8. Mozaffarian, D.; Wu, J.H. Omega-3 fatty acids and cardiovascular disease: Effects on risk factors, molecular pathways, and clinical events. *J. Am. Coll. Cardiol.* **2011**, *58*, 2047–2067. [CrossRef]
9. Feskens, E.J.; Bowles, C.H.; Kromhout, D. Inverse association between fish intake and risk of glucose intolerance in normoglycemic elderly men and women. *Diabetes Care* **1991**, *14*, 935–941. [CrossRef]
10. Lavigne, C.; Tremblay, F.; Asselin, G.; Jacques, H.; Marette, A. Prevention of skeletal muscle insulin resistance by dietary cod protein in high fat-fed rats. *Am. J. Physiol. Endocrinol. Metab.* **2001**, *281*, E62–E71. [CrossRef]
11. Ouellet, V.; Marois, J.; Weisnagel, S.J.; Jacques, H. Dietary cod protein improves insulin sensitivity in insulin-resistant men and women: A randomized controlled trial. *Diabetes Care* **2007**, *30*, 2816–2821. [CrossRef] [PubMed]
12. Vikoren, L.A.; Nygard, O.K.; Lied, E.; Rostrup, E.; Gudbrandsen, O.A. A randomised study on the effects of fish protein supplement on glucose tolerance, lipids and body composition in overweight adults. *Br. J. Nutr.* **2013**, *109*, 648–657. [CrossRef] [PubMed]
13. Drotningsvik, A.; Mjos, S.A.; Hogoy, I.; Remman, T.; Gudbrandsen, O.A. A low dietary intake of cod protein is sufficient to increase growth, improve serum and tissue fatty acid compositions, and lower serum postprandial glucose and fasting non-esterified fatty acid concentrations in obese zucker fa/fa rats. *Eur. J. Nutr.* **2015**, *54*, 1151–1160. [CrossRef] [PubMed]
14. Vildmyren, I.; Cao, H.J.V.; Haug, L.B.; Valand, I.U.; Eng, O.; Oterhals, A.; Austgulen, M.H.; Halstensen, A.; Mellgren, G.; Gudbrandsen, O.A. Daily intake of protein from cod residual material lowers serum concentrations of nonesterified fatty acids in overweight healthy adults: A randomized double-blind pilot study. *Mar. Drugs* **2018**, *16*, 197. [CrossRef]
15. Lavigne, C.; Marette, A.; Jacques, H. Cod and soy proteins compared with casein improve glucose tolerance and insulin sensitivity in rats. *Am. J. Physiol. Endocrinol. Metab.* **2000**, *278*, E491–E500. [CrossRef] [PubMed]
16. Gall, W.E.; Beebe, K.; Lawton, K.A.; Adam, K.P.; Mitchell, M.W.; Nakhle, P.J.; Ryals, J.A.; Milburn, M.V.; Nannipieri, M.; Camastra, S.; et al. Alpha-hydroxybutyrate is an early biomarker of insulin resistance and glucose intolerance in a nondiabetic population. *PLoS ONE* **2010**, *5*, e10883. [CrossRef]
17. Ferrannini, E.; Natali, A.; Camastra, S.; Nannipieri, M.; Mari, A.; Adam, K.P.; Milburn, M.V.; Kastenmuller, G.; Adamski, J.; Tuomi, T.; et al. Early metabolic markers of the development of dysglycemia and type 2 diabetes and their physiological significance. *Diabetes* **2013**, *62*, 1730–1737. [CrossRef]
18. Mahendran, Y.; Vangipurapu, J.; Cederberg, H.; Stancakova, A.; Pihlajamaki, J.; Soininen, P.; Kangas, A.J.; Paananen, J.; Civelek, M.; Saleem, N.K.; et al. Association of ketone body levels with hyperglycemia and type 2 diabetes in 9,398 finnish men. *Diabetes* **2013**, *62*, 3618–3626. [CrossRef]
19. Cobb, J.; Eckhart, A.; Motsinger-Reif, A.; Carr, B.; Groop, L.; Ferrannini, E. Alpha-hydroxybutyric acid is a selective metabolite biomarker of impaired glucose tolerance. *Diabetes Care* **2016**, *39*, 988–995. [CrossRef]
20. Richardsen, R.; Nystøyl, R.; Strandheim, G.; Marthinussen, A. Report: Analysis of marine residual raw material. *SINTEF* **2016**.
21. Vildmyren, I.; Halstensen, A.; Oterhals, A.; Gudbrandsen, O.A. Cod protein powder lowered serum nonesterified fatty acids and increased total bile acid concentrations in healthy, lean, physically active adults: A randomized double-blind study. *Food Nutr. Res.* **2019**, *63*. [CrossRef]

22. Subar, A.F.; Kipnis, V.; Troiano, R.P.; Midthune, D.; Schoeller, D.A.; Bingham, S.; Sharbaugh, C.O.; Trabulsi, J.; Runswick, S.; Ballard-Barbash, R.; et al. Using intake biomarkers to evaluate the extent of dietary misreporting in a large sample of adults: The open study. *Am. J. Epidemiol.* **2003**, *158*, 1–13. [CrossRef] [PubMed]
23. Drotningsvik, A.; Midttun, O.; McCann, A.; Ueland, P.M.; Hogoy, I.; Gudbrandsen, O.A. Dietary intake of cod protein beneficially affects concentrations of urinary markers of kidney function and results in lower urinary loss of amino acids in obese zucker fa/fa rats. *Br. J. Nutr.* **2018**, *120*, 740–750. [CrossRef] [PubMed]
24. Hagen, I.V.; Helland, A.; Bratlie, M.; Midttun, O.; McCann, A.; Sveier, H.; Rosenlund, G.; Mellgren, G.; Ueland, P.M.; Gudbrandsen, O.A. Tmao, creatine and 1-methylhistidine in serum and urine are potential biomarkers of cod and salmon intake: A randomised clinical trial in adults with overweight or obesity. *Eur. J. Nutr.* **2019**, 1–11. [CrossRef] [PubMed]
25. Cho, C.E.; Taesuwan, S.; Malysheva, O.V.; Bender, E.; Tulchinsky, N.F.; Yan, J.; Sutter, J.L.; Caudill, M.A. Trimethylamine-n-oxide (tmao) response to animal source foods varies among healthy young men and is influenced by their gut microbiota composition: A randomized controlled trial. *Mol. Nutr. Food Res.* **2017**, *61*, 1600324. [CrossRef]
26. Cheung, W.; Keski-Rahkonen, P.; Assi, N.; Ferrari, P.; Freisling, H.; Rinaldi, S.; Slimani, N.; Zamora-Ros, R.; Rundle, M.; Frost, G.; et al. A metabolomic study of biomarkers of meat and fish intake. *Am. J. Clin. Nutr.* **2017**, *105*, 600–608. [CrossRef]
27. Schmedes, M.; Balderas, C.; Aadland, E.K.; Jacques, H.; Lavigne, C.; Graff, I.E.; Eng, O.; Holthe, A.; Mellgren, G.; Young, J.F.; et al. The effect of lean-seafood and non-seafood diets on fasting and postprandial serum metabolites and lipid species: Results from a randomized crossover intervention study in healthy adults. *Nutrients* **2018**, *10*, 598. [CrossRef]
28. Sjolin, J.; Hjort, G.; Friman, G.; Hambraeus, L. Urinary excretion of 1-methylhistidine: A qualitative indicator of exogenous 3-methylhistidine and intake of meats from various sources. *Metabolism* **1987**, *36*, 1175–1184. [CrossRef]
29. Drotningsvik, A.; Pampanin, D.M.; Slizyte, R.; Carvajal, A.; Hogoy, I.; Remman, T.; Gudbrandsen, O.A. Hydrolyzed proteins from herring and salmon rest raw material contain peptide motifs with angiotensin-i converting enzyme inhibitors and resulted in lower urine concentrations of protein, cystatin c and glucose when fed to obese zucker fa/fa rats. *Nutr. Res.* **2018**, *52*, 14–21. [CrossRef]
30. Lenz, E.M.; Bright, J.; Wilson, I.D.; Hughes, A.; Morrisson, J.; Lindberg, H.; Lockton, A. Metabonomics, dietary influences and cultural differences: A 1h nmr-based study of urine samples obtained from healthy british and swedish subjects. *J. Pharm. Biomed. Anal.* **2004**, *36*, 841–849. [CrossRef]
31. Dumas, M.E.; Maibaum, E.C.; Teague, C.; Ueshima, H.; Zhou, B.; Lindon, J.C.; Nicholson, J.K.; Stamler, J.; Elliott, P.; Chan, Q.; et al. Assessment of analytical reproducibility of 1h nmr spectroscopy based metabonomics for large-scale epidemiological research: The intermap study. *Anal. Chem.* **2006**, *78*, 2199–2208. [CrossRef] [PubMed]
32. Conway, E.J. An absorption apparatus for the micro-determination of certain volatile substances: The determination of urea and ammonia in body fluids. *Biochem. J.* **1933**, *27*, 430–434. [PubMed]
33. Bidlingmeyer, B.A.; Cohen, S.A.; Tarvin, T.L.; Frost, B. A new, rapid, high-sensitivity analysis of amino acids in food type samples. *J. Assoc. Off. Anal. Chem.* **1987**, *70*, 241–247. [CrossRef] [PubMed]
34. Midttun, O.; McCann, A.; Aarseth, O.; Krokeide, M.; Kvalheim, G.; Meyer, K.; Ueland, P.M. Combined measurement of 6 fat-soluble vitamins and 26 water-soluble functional vitamin markers and amino acids in 50 mul of serum or plasma by high-throughput mass spectrometry. *Anal. Chem.* **2016**, *88*, 10427–10436. [CrossRef]
35. Midttun, O.; Kvalheim, G.; Ueland, P.M. High-throughput, low-volume, multianalyte quantification of plasma metabolites related to one-carbon metabolism using hplc-ms/ms. *Anal. Bioanal. Chem.* **2013**, *405*, 2009–2017. [CrossRef]
36. Chain, E.P.o.C.i.t.F. Statement on tolerable weekly intake for cadmium. *EFSA J.* **2011**, *9*, 1975.
37. Chain, E.P.o.C.i.t.F. Scientific opinion on the risk for public health related to the presence of mercury and methylmercury in food. *EFSA J.* **2012**, *10*, 2985.
38. Hovland, I.H.; Leikanger, I.S.; Stokkeland, O.; Waage, K.H.; Mjos, S.A.; Brokstad, K.A.; McCann, A.; Ueland, P.M.; Slizyte, R.; Carvajal, A.; et al. Effects of low doses of fish and milk proteins on glucose regulation and markers of insulin sensitivity in overweight adults: A randomised, double blind study. *Eur. J. Nutr.* **2019**, 1–17. [CrossRef]

39. Hagen, I.V.; Helland, A.; Bratlie, M.; Brokstad, K.A.; Rosenlund, G.; Sveier, H.; Mellgren, G.; Gudbrandsen, O.A. High intake of fatty fish, but not of lean fish, affects serum concentrations of tag and hdl-cholesterol in healthy, normal-weight adults: A randomised trial. *Br. J. Nutr.* **2016**, *116*, 648–657. [CrossRef]
40. Helland, A.; Bratlie, M.; Hagen, I.V.; Mjos, S.A.; Sornes, S.; Halstensen, A.I.; Brokstad, K.A.; Sveier, H.; Rosenlund, G.; Mellgren, G.; et al. High intake of fatty fish, but not of lean fish, improved postprandial glucose regulation and increased the n-3 pufa content in the leucocyte membrane in healthy overweight adults: A randomised trial. *Br. J. Nutr.* **2017**, *117*, 1368–1378. [CrossRef]
41. Zhang, S.; Hulver, M.W.; McMillan, R.P.; Cline, M.A.; Gilbert, E.R. The pivotal role of pyruvate dehydrogenase kinases in metabolic flexibility. *Nutr. Metab.* **2014**, *11*, 10. [CrossRef] [PubMed]
42. Wu, P.; Inskeep, K.; Bowker-Kinley, M.M.; Popov, K.M.; Harris, R.A. Mechanism responsible for inactivation of skeletal muscle pyruvate dehydrogenase complex in starvation and diabetes. *Diabetes* **1999**, *48*, 1593–1599. [CrossRef] [PubMed]
43. Koeth, R.A.; Wang, Z.; Levison, B.S.; Buffa, J.A.; Org, E.; Sheehy, B.T.; Britt, E.B.; Fu, X.; Wu, Y.; Li, L.; et al. Intestinal microbiota metabolism of l-carnitine, a nutrient in red meat, promotes atherosclerosis. *Nat. Med.* **2013**, *19*, 576–585. [CrossRef] [PubMed]
44. Zhang, A.Q.; Mitchell, S.C.; Smith, R.L. Dietary precursors of trimethylamine in man: A pilot study. *Food Chem. Toxicol.* **1999**, *37*, 515–520. [CrossRef]
45. Zeisel, S.H.; Warrier, M. Trimethylamine n-oxide, the microbiome, and heart and kidney disease. *Annu. Rev. Nutr.* **2017**, *37*, 157–181. [CrossRef] [PubMed]
46. Myint, T.; Fraser, G.E.; Lindsted, K.D.; Knutsen, S.F.; Hubbard, R.W.; Bennett, H.W. Urinary 1-methylhistidine is a marker of meat consumption in black and in white california seventh-day adventists. *Am. J. Epidemiol.* **2000**, *152*, 752–755. [CrossRef] [PubMed]
47. Dragsted, L.O. Biomarkers of meat intake and the application of nutrigenomics. *Meat Sci.* **2010**, *84*, 301–307. [CrossRef]
48. Cross, A.J.; Major, J.M.; Sinha, R. Urinary biomarkers of meat consumption. *Cancer Epidemiol. Biomarkers Prev.* **2011**, *20*, 1107–1111. [CrossRef]
49. Altorf-van der Kuil, W.; Brink, E.J.; Boetje, M.; Siebelink, E.; Bijlsma, S.; Engberink, M.F.; van 't Veer, P.; Tome, D.; Bakker, S.J.; van Baak, M.A.; et al. Identification of biomarkers for intake of protein from meat, dairy products and grains: A controlled dietary intervention study. *Br. J. Nutr.* **2013**, *110*, 810–822. [CrossRef]

Article

Reliability of Self-Administered Questionnaire on Dietary Supplement Consumption in Malaysian Adolescents

Mohamed S. Zulfarina [1], Razinah Sharif [2], Ahmad M. Sharkawi [1], Tg Mohd Ikhwan Tg Abu Bakar Sidik [1], Sabarul-Afian Mokhtar [1,3], Ahmad Nazrun Shuid [1,4] and Isa Naina-Mohamed [1,*]

1. Pharmacoepidemiology and Drug Safety Unit, Department of Pharmacology, Faculty of Medicine, Universiti Kebangsaan Malaysia Medical Centre, Jalan Yaacob Latif, Bandar Tun Razak, Cheras 56000, Kuala Lumpur, Malaysia; P94838@siswa.ukm.edu.my (M.S.Z.); shark_811@yahoo.com (A.M.S.); tgikhwancrc@gmail.com (T.M.I.T.A.B.S.); drsam2020@gmail.com (S.-A.M.); anazrun@yahoo.com (A.N.S.)
2. Centre for Healthy Ageing and Wellness, Faculty of Health Sciences, Universiti Kebangsaan Malaysia, Jalan Raja Muda Abd Aziz 50300, Kuala Lumpur, Malaysia; razinah@ukm.edu.my
3. ProVice-Chancellor Office, Health Campus, Jalan Yaacob Latif, Bandar Tun Razak, Cheras 56000, Kuala Lumpur, Malaysia
4. Department of Pharmacology, Faculty of Medicine, Universiti Teknologi MARA, Sungai Buloh Campus, Jalan Hospital, Sungai Buloh 47000, Malaysia

* Correspondence: isanaina@ppukm.ukm.edu.my; Tel.: +60-3-91459545

Received: 17 August 2020; Accepted: 16 September 2020; Published: 17 September 2020

Abstract: The repeatability of most questionnaires utilized in previous studies related to the consumption of dietary supplements (DS) among youth has not been well documented. Thus, a simple and easy-to-administer questionnaire to capture the habitual use of DS in the past one year known as the dietary supplement questionnaire (DiSQ) was developed and supported with external reliability evaluation. Analyses were done based on a convenience sample of 46 secondary school students. To elicit information regarding the intake of DS, the questionnaire was partitioned into two domains. The first domain was used to identify vitamin/mineral (VM) supplements, while the second domain was utilized to identify non-vitamin/non-mineral (NVNM) supplements. Cohen's kappa coefficient (k) was used to evaluate the test–retest reliability of the questionnaire. Questionnaire administration to the respondents was done twice whereby a retest was given two weeks after the first test. Between test and retest, the reliability of individual items ranged from moderate to almost perfect for the VM (k = 0.53–1.00) and NVNM (k = 0.63–1.00) domains. None of the items had "fair" or "poor" agreement. Various correlation coefficients can be obtained for the DiSQ but are generally reliable over time for assessing information on the consumption of supplements among the adolescent population.

Keywords: test–retest; nutritional supplement; vitamin; mineral; herbal; natural; botanical; nutraceutical

---

1. Introduction

There is no global consensus in terminology on how dietary supplements or supplements in different countries are defined [1,2]. Based on the major regulatory bodies, dietary supplement (DS) is the product or foodstuff that contains dietary ingredients or substances [3] or a concentrated source of nutrients or other substances designed to be taken in small quantities [4,5] either individually or in combination, marketed in the form of pill, tablet, capsule and other similar forms, all of which are intended to supplement the normal diet [3–5].

The global market of DS is a multibillion-dollar commerce, with the Asia Pacific being the second largest market accounting for 31% of the total market share [6]. Studies established that there are substantial national variations regarding the prevalence of dietary supplement consumption in children and adolescents; 20% in Australia [7], 32% to 37% in the United States [8,9], 35% in Italy [10], 20% to 22% in Japan [11,12] and 33% in South Korea [13]. Specifically, the prevalence of vitamin/mineral supplements (54%) and food supplements (40%) intakes are more pronounced among Malaysian adolescents [14]. Regional differences suggest that the guidelines set by the international and national regulations play one of the key roles in the widespread consumption of supplements [15]. For example, countries of the Association of Southeast Asian Nations (ASEAN) have relatively loose regimes that may lead to easy market access compared to those of other international alliances or regions including the European Union (EU) and Japan [1,2].

Given the widespread and increasing use of DS, the contribution of supplement intake towards one's health is an important exposure in any health epidemiologic study and overlooking this may lead to an underestimation of the true relationship between the consumption of DS and health condition [16–18]. Epidemiological study represents an essential research method in understanding the association between risk factors and disease outcomes with the administration of questionnaires being the most commonly utilized scientific tool for acquiring such information. Indeed, the empirical utility of this approach in epidemiological studies to identify risk factors is apparent [19].

To date, many studies have documented the use of supplements and their associated determinants among adolescents and young adults. Most studies over the past 10 years have been conducted in the United States [8,9], Europe [10,20–22] and Australia [7] with limited studies conducted in the Asian countries [12–14,23]. However, the external reliability of most questionnaires utilized in earlier studies has not been well documented. To the best of the authors' knowledge, only one study has demonstrated the test–retest reliability of questionnaires on the use of DS in adolescents [24]. However, the application of dietary methods including Food Frequency Questionnaire (FFQ) used in the previous study required a country-specific nutrient database for DS.

Owing to the scarcity of studies on the reproducibility of questionnaires on the intake of supplements among youth as well as the limitation of local nutrient database, a self-report questionnaire was constructed to offer a simple and easy-to-administer tool provided with an external reliability evaluation. Providing external reliability is essential to ensure that the information provided by the questionnaire is reliable over time. This control process is vital in producing accurate data interpretation and its subsequent use by policy-makers and practitioners to design, implement as well as evaluate policies, practices and intervention strategies in health. Thus, this present study aimed at describing and addressing the degree of consistency of a self-reporting questionnaire on the consumption of DS in a representative group of adolescents aged 15 to 19 years old using test–retest reliability.

2. Methods

A total of 50 national secondary school adolescents (aged 15–19) from the city center of Kuala Lumpur, Malaysia were selected on a convenience basis. The questionnaire was administered to each respondent twice. The elapsed time between the first questionnaire administration and the next was two weeks (14 to 15 days) [25]. The participants took approximately five to fifteen minutes to complete the questionnaire. Oral instruction was also given during each survey, and the respondents were encouraged to seek aid on any items in the questionnaire that they did not fully comprehend. This procedure would allow for the clarification of any ambiguities.

During the re-administration, four participants were not present to complete the second copy of the questionnaire and thus treated as missing data. For this reason, the remaining 46 participants who completed the questionnaire twice were used in the final analysis.

2.1. Questionnaire

For ecological validity, the items for dietary supplement questionnaire (DiSQ) were developed and structured on the basis of existing literature and surveys [8,17,18,26]. This step was carefully reviewed by the panel of experts on pharmacology and nutrition. The questionnaire was partitioned into three domains. The first two domains were used to elicit information regarding the intake of DS while the third domain (additional information) was used to inquire about the intake of medicines over the past one year. Specifically, the first domain was used to identify vitamin/mineral (VM) supplements, whereas the second domain was used to identify non-vitamin/non-mineral (NVNM) supplements, which are also known as natural products or nutraceuticals such as herbals/botanicals, amino acids or omega-3 fatty acid.

A closed-ended format was used to collect information on the consumption of supplements and medicines over the past year (yes, no). An open-ended format was introduced for the next question in which the respondents were required to list the names of the products they consumed. For each supplement and medicine listed, the respondents were asked to provide information that included a closed-ended format on frequency of consumption (at least once a day, a few times in a week/more than once a week, once a week, once in a few weeks/one to three times a month), status of consumption or whether or not the subject is still taking the supplement (current, former), and duration of consumption (short-term refers to less than six months, long-term refers to more than six months).

The respondents may incorrectly identify the type of supplement they consume, which suggests that the respondents' perception of dietary supplements may not match that of the researcher [27]. To amend this, a card that listed the examples of VM supplements for the first domain and NVNM supplements or natural products for the second domain was introduced (Appendix A).

2.2. Sample Size

The minimum sample size required was determined using the tabulated contingency table by [28], which was calculated based on the formula by [29]. The minimum sample size required was derived based on the pre-specified power of 80%, alpha of 0.05 two-sided and the expected Cohen's kappa (k) coefficient (k_2 is 0.42) for every item when no agreement was assumed for the test–retest at the first place (k_1 is 0.0). Thus, a minimum sample of 43 respondents was required based on the assumption that the proportions in each category were proportionate to one another.

2.3. Ethics

This initial work is a part of the larger population-based study where the details of the methodology have been published previously [30]. The participants involved in this reliability study were not included in the analysis mentioned above. Formal approval was obtained from the Universiti Kebangsaan Malaysia Medical Centre (UKMMC) research committee (UKM 1.5.3.5/244), Ministry of Education (MOE), the Department of State Education (JPN), and school authorities. Participation in this study was completely voluntary and parents/guardians who preferred their children not to participate in the study were required to sign and return the negative consent form.

2.4. Statistics

All statistical analyses were conducted using the Stata Statistical software version 15 (Stata Corporation LP, College Station, TX, USA). The percentage of agreement and Cohen's kappa (k) coefficient were assessed for each item and its sub-components. Percent agreement was calculated by dividing the number of agreement scores with the total number of scores. Unweighted k statistic was used to evaluate nominal variables, whereas weighted k statistic with quadratic weighting was used to evaluate ordinal variables. Four respondents with missing data were excluded from all computations.

The strength of agreement was interpreted with due caution according to the reference values described by Landis and Koch in which Cohen's k coefficient of less than 0.20 was considered as poor agreement, 0.21–0.40 as fair agreement, 0.41–0.60 as moderate agreement, 0.61–0.80 as substantial agreement, and 0.81–1.00 as almost perfect agreement [31].

3. Results

From the initial sample of 50 respondents, only 46 (92%) individuals completed the questionnaire twice. Among the 46 respondents, 50% were male students and 50% were female students (mean age 16.9 ± 0.8). The majority of the volunteers were of Malay race (85%). The characteristics of adolescents who completed the survey are shown in Table 1. Table 2 shows the prevalence of DS users in the first questionnaire administration and the follow-up after two weeks. The prevalence of the VM supplement use was 39% for the baseline administration and 37% for the re-administration. On both occasions, vitamin C was the most popular subtype of the VM supplements among the adolescents (baseline: 72%; follow-up: 71%). Regarding the NVNM supplement intake, the prevalence for the baseline administration was 46% with 48% for the re-administration, while the most consumed nutraceuticals on both occasions were herbs and botanicals (baseline: 62%; follow-up: 50%). Additionally, the second most consumed nutraceutical products in the baseline administration and re-administration, respectively, were honey and bee products (baseline: 48%; follow-up: 36%).

Table 1. Characteristics of study participants.

	Mean ± SD/n (%)
All	46
Age	16.9 ± 0.8
Gender	
Male	23 (50.0)
Female	23 (50.0)
Ethnicity	
Malay	39 (84.8)
Non-Malay *	7 (15.2)

* Other races including Chinese and Indian were grouped under one category due to low representation.

Table 2. Descriptive data on the baseline administration and re-administration.

	Baseline Prevalence, n (%)	Follow-Up Prevalence, n (%)
Vitamin/mineral		
Number of users	18 (39.1)	17 (37.0)
Number of supplements		
1 type	13 (72.2)	13 (76.5)
2 types	5 (27.8)	4 (23.5)
>3 types	0 (0.0)	0 (0.0)
Type of DS		
Single vitamin	14 (77.8)	13 (76.5)
Single mineral	3 (16.7)	5 (29.4)
Combination of vitamin(s) and mineral(s) *	7 (38.9)	4 (23.5)
Non-vitamin/non-mineral		
Number of users	21 (45.7)	22 (47.8)
Number of supplements		
1 type	8 (38.1)	10 (45.5)
2 types	9 (42.9)	8 (36.4)
>3 types	4 (19.1)	4 (18.2)
Type of DS		
Honey and bee products	10 (47.6)	8 (36.4)
Fish oil/omega-3	7 (33.3)	7 (31.8)
Meat essence	5 (23.8)	5 (22.7)
Herbs and botanicals	13 (61.9)	11 (50.0)
Protein formula	2 (9.5)	2 (9.1)
Others ‡	2 (9.5)	3 (13.6)
Unspecified #	-	1 (4.5)

DS = dietary supplement * combined preparations containing vitamin and mineral such as multivitamin, multi-mineral, multivitamin multi-mineral, or any type of combination ‡ including functional food # treated as missing data.

The reliability coefficient for the VM supplement use is presented in Table 3. The result of the main item, which was the first listed item in the table, displayed an "almost perfect" test–retest reliability (k = 0.95). For the sub-item, the item on the supplement's name presented the highest agreement (k = 0.68–1.00), while the item on the duration of consumption showed the lowest agreement (k = 0.53–1.00) between test and retest. None of the items had kappa values of less than 0.40 ("fair" or "poor" agreement). The percent agreement for the VM supplement use ranged from 87.0% to 100.0%.

Table 3. Cohen's kappa (k) coefficient for the vitamin/mineral (VM) supplement consumption.

Items	Response Alternatives	n	Percentage Agreement (%)	Kappa Coefficient
Used or taken any vitamins/minerals in the past one year	Yes/no	46	97.8	0.95
Name of supplement	*	46	93.5–100.0	0.68–1.00
Still take	Yes/no	46	91.3–100.0	0.67–1.00
Frequency of intake	At least once a day/more than once a week/once a week/one to three times a month	46	87.0–97.8	0.64–0.80
Took continuously for more than 6 months	Yes/no	46	87.0–100.0	0.53–1.00

* Open-ended answer.

Table 4 shows the reliability coefficient for the NVNM supplement consumption. The kappa value of the main item (k = 0.78) denoted a "substantial" test–retest reliability. For the sub-items, the item on the frequency of intake presented the highest agreement (k = 0.79–0.97), while the item on the status of consumption presented the lowest agreement (k = 0.63–0.88) between the first and second surveys. All items had kappa values of more than 0.60 (indicating a "substantial" to an "almost perfect" agreement). The percent agreement for the NVNM supplement use ranged from 77.3% to 100.0%.

Table 4. Cohen's kappa (k) coefficient for the non-vitamin/non-mineral (NVNM) supplement consumption.

Items	Response Alternatives	n	Percentage Agreement (%)	Kappa Coefficient
Used or taken any NVNM in the past one year	Yes/no	46	89.1	0.78
Name of supplement	*	45	84.4–97.8	0.77–0.87
Still take	Yes/no	43	79.1–97.8	0.63–0.88
Frequency of intake	At least once a day/more than once a week/once a week/one to three times a month	46	77.3–95.7	0.79–0.97
Took continuously for more than 6 months	Yes/no	46	81.8–100.0	0.69–1.00

* Open-ended answer.

Table 5 shows the reliability coefficient for the medicine intake. The kappa statistics of the main item indicated an "almost perfect" test–retest reliability (k = 0.93). For the sub-item, the items listed on the name of medicine and status of consumption presented the highest agreement (k = 0.48–0.94), while the item on the duration of consumption presented the lowest agreement (k = 0.22–0.69) between test and retest. All items had kappa values of more than 0.40 (indicating a "moderate" to an "almost perfect" agreement) except for the item on the duration of consumption. The percent agreement for the medicine intake ranged from 89.1% to 100%.

Table 5. Cohen's kappa (k) coefficient for medicine consumption.

Items	Response	n	Percentage Agreement (%)	Kappa Coefficient
Used or taken any medicines in the past one year	Yes/no	46	97.8	0.93
Name of medicine	*	46	95.7–100.0	0.48–0.94
Still take	Yes/no	43	95.7–100.0	0.48–0.94
Frequency of intake	At least once a day/more than once a week/once a week/one to three times a month	46	91.1–100.0	0.42–0.82
Took continuously for more than 6 months?	Yes/no	46	89.1–100.0	0.22–0.69

* Open-ended answer.

4. Discussion

Across the three domains, the agreement values of k for the two-week interval indicated that all individual items have moderate to almost perfect repeatability. This lies in the preferable measure of agreement except for the item on the duration of the medicine consumption. Thus, it can be deduced that the test–retest reliability of the current self-reported supplement consumption is varied but generally reliable over time.

It was difficult to compare the current results directly with those of previous studies due to the variations in the context of questions and type of response alternatives used in the questionnaire. However, the repeatability of responses was in part comparable to that obtained in another study [24]. The percent agreement for the overall dietary supplement use was 91.7% with $k = 0.62$ for the reported category, while for the corrected category, the percent agreement was 89.8% with $k = 0.57$.

A report suggested that a brief questionnaire can precisely capture the data on the use of the frequently consumed supplements but may not perform well for the less frequently consumed supplements [32]. On the contrary, this present study demonstrated that the reliability estimate for the NVNM supplement consumption was equally comparable to the VM supplement consumption. Concerning the main item (the first question asked in the questionnaire), the reliability estimate regarding the NVNM supplement consumption was lower than that of the VM supplement consumption and medicine use. This result could be partly supported by the fact that supplements are more relevant to the older generation than to the younger generation (prevalence rate) [33]. The prevalence of supplement users may have an influence (albeit small) on the percent agreement estimate between the two questionnaires. The k value would probably be higher in a population with a high prevalence of users. Moreover, unlike the VM products, most NVNM products are consumed intermittently (behavioral differences). Thus, adolescents may face difficulty in recalling the supplements they consumed infrequently.

As shown in this study, the values of Cohen's k statistic for the sub-items 2 to 5 were lower compared to that of the first item under the same domain. It is interesting to note that most teenagers took the supplements as requested by their parents [14]. It was assumed that this reason might also be applied to the NVNM supplement consumption. Although the underlying causes were unclear, adolescents may be aware but did not pay attention to what supplements were given to them since it was not by their own free will or self-awareness [23]. As a result, they were unable to recall specific details such as the type of supplement they consumed because recalling an involuntary action precisely may require more complex cognitive demands [34].

The questions on medicine use were also included as additional information for which the low agreement between test and retest was in line with the previous findings concerning adult respondents [26]. This study postulated that the agreement between test and retest would likely reflect the dependence on time and situation during the period covered [35]. In most cases, medicines are consumed at an interval or when pain is present [26]. This may imply that the agreement between the two responses resulted from the measure of the consistency of the medication use rather than the measure of reliability of the instrument [32].

The open-ended format was chosen for the item on the supplement's name as it allowed for a wide coverage of various types of DS consumed to be recorded. An example of DS includes the *Habbatus sauda*, which is commonly known as black cumin that is native to Southwest Asia. Another example is long jack, locally known as Tongkat Ali, which is indigenous to Southeast Asia and culturally used as folk medicine. By having this format, diverse types of supplement can be documented. Moreover, the DiSQ was developed to cover a supplement intake for one year, and consequently, the previous supplement intake could be missed. Nevertheless, it can still be highlighted that the DiSQ is able to capture the current habitual information on both daily/regular and intermittent/occasional supplement users.

There were some limitations discovered based on the findings of this study. First, the sampling was based on a convenience sample. Second, the questionnaire lacked the questions on an actual dosage level. This was made on account of the insignificant actual dosage level considered in many epidemiologic studies as long as the intakes between two instruments were similar [32]. In addition, the determination of total intake, which includes nutrients consumed from DS, requires a country-specific nutrient database; thus, the inquiry on dosage level was not considered in this study. On the contrary, countries with food database integrated with nutrients from DS have the advantage in measuring total intake that covers nutrients from DS.

This work represents an initial step in a larger-scale study to determine the lifestyle and health outcomes among adolescents specifically or the younger generation in general. Consequently, this current study will serve as a baseline test from which the application to other age groups or populations of interest (adult or elderly) is possible after establishing the test–retest reliability. It is recommended that in the subsequent results evaluation, the items with "fair" kappa statistic should be considered with due caution, modified, or possibly removed from the questionnaire. Further research that appraises additional validity and reliability of the questionnaire is warranted.

5. Conclusions

In accordance with the reference criteria adopted, this study has demonstrated that the concordance or correlation coefficient values of the questionnaire are varied but generally satisfactory and reliable over time. This study will serve as a starter for other new studies related to this topic and DiSQ can be a valuable tool in future studies that aim to investigate the association between various risk factors such as supplement consumption, demographic characteristics, and lifestyle particularly in the adolescent population.

Author Contributions: Conceptualization, M.S.Z. and I.N.-M.; methodology, M.S.Z., R.S., S.-A.M., I.N.-M., and A.N.S.; formal analysis, M.S.Z., I.N.-M., and T.M.I.T.A.B.S.; investigation, M.S.Z., A.M.S., and T.M.I.T.A.B.S.; resources, S.-A.M., I.N.-M., and A.N.S.; data curation, M.S.Z., A.M.S., and R.S.; writing—original draft preparation, M.S.Z.; writing—review and editing, R.S., A.N.S., and I.N.-M.; supervision, R.S. and I.N.-M.; project administration, I.N.-M.; funding acquisition, S.-A.M., A.N.S., and I.N.-M. All authors have read and agreed to the published version of the manuscript.

Funding: This study was financially supported by a university grant of UKM, Dana Impak Perdana (Premier Impact Grant) DIP-2013-002.

Acknowledgments: The authors wish to express their gratitude to the participants and teachers from the participating schools and Nurul Huda Razalli from the Dietetics Program, Universiti Kebangsaan Malaysia (UKM).

Conflicts of Interest: The authors declare no conflict of interest.

Appendix A

Table A1. Examples of VM and NVNM supplements.

Vitamin/Mineral	Non-Vitamin/Non-Mineral
Vitamin C	Fish oil/omega-3/omega-6
B vitamins or B complex	Spirulina
Folic acid/vitamin B6	Evening Primrose Oil
Calcium	Honey/bee products
Vitamin D	Co-enzyme Q10
Zinc	Chicken/Fish essence
Iron	Sea Cucumber
Multivitamin and/or Multimineral	Fiber
Vitamin A	Protein/amino-acids

References

1. Dwyer, J.T.; Coates, P.M.; Smith, M.J. Dietary Supplements: Regulatory Challenges and Research Resources. *Nutrients* **2018**, *10*, 41. [CrossRef] [PubMed]
2. Binns, C.W.; Lee, M.K.; Lee, A. Problems and Prospects: Public Health Regulation of Dietary Supplements. *Annu. Rev. Public Health* **2018**, *39*, 403–420. [CrossRef] [PubMed]
3. Dietary Supplement Health and Education Act (DSHEA) of 1994. Public Law 103-417. 103rd Congress. Available online: https://ods.od.nih.gov/About/DSHEA_Wording.aspx (accessed on 2 January 2020).
4. Codex Alimentarius. *Codex Guidelines for Vitamin and Mineral Food Supplements (CAC/GL 55-2005)*; Codex Alimentarius, 2005. Available online: http://www.fao.org/fao-who-codexalimentarius/codex-texts/guidlines/en (accessed on 2 January 2020).
5. European Parliament and Council 2002. Directive 2002/46/EC of the European Parliament and of the Council of 10 June 2002 on the Approximation of the Laws of the Member States Relating to Food Supplements. Available online: http://www.fao.org/faolex/results/details/en/c/LEX-FAOC037787/ (accessed on 2 January 2020).
6. Dietary Supplements Market Size Analysis Report by Ingredient (Botanicals, Vitamins), by Form, by Application (Immunity, Cardiac Health), by End User, by Distribution Channel, and Segment Forecasts, 2019–2025. California: Grand View Research. May 2019. Available online: https://www.grandviewresearch.com/industry-analysis/dietary-supplements-market (accessed on 2 January 2020).
7. O'Brien, S.K.; Malacova, E.; Sherriff, J.; Black, L.J. The Prevalence and Predictors of Dietary Supplement Use in the Australian Population. *Nutrients* **2017**, *9*, 1154. [CrossRef] [PubMed]
8. Dwyer, J.; Nahin, R.L.; Rogers, G.T.; Barnes, P.M.; Jacques, P.M.; Sempos, C.T.; Bailey, R. Prevalence and Predictors of Children's Dietary Supplement Use: The 2007 National Health Interview Survey. *Am. J. Clin. Nutr.* **2013**, *97*, 1331–1337. [CrossRef]
9. Qato, D.M.; Alexander, G.C.; Guadamuz, J.S.; Lindau, S.T. Prevalence of Dietary Supplement Use in US Children and Adolescents, 2003–2014. *JAMA Pediatr.* **2018**, *172*, 780. [CrossRef]
10. Del Balzo, V.; Vitiello, V.; Germani, A.; Donini, L.M.; Poggiogalle, E.; Pinto, A. A Cross-Sectional Survey on Dietary Supplements Consumption among Italian Teen-Agers. *PLoS ONE* **2014**, *9*, e100508. [CrossRef]
11. Kobayashi, E.; Nishijima, C.; Sato, Y.; Umegaki, K.; Chiba, T. The Prevalence of Dietary Supplement Use Among Elementary, Junior High, and High School Students: A Nationwide Survey in Japan. *Nutrients* **2018**, *10*, 1176. [CrossRef]
12. Mori, N.; Kubota, M.; Hamada, S.; Nagai, A. Prevalence and characterization of supplement use among healthy children and adolescents in an urban Japanese city. *Health* **2011**, *3*, 135–140. [CrossRef]
13. Yoon, J.Y.; Park, H.A.; Kang, J.H.; Kim, K.W.; Hur, Y.I.; Park, J.J.; Lee, R.; Lee, H.H. Prevalence of Dietary Supplement Use in Korean Children and Adolescents: Insights from Korea National Health and Nutrition Examination Survey 2007–2009. *J. Korean Med Sci.* **2012**, *27*, 512–517. [CrossRef]
14. Sien, Y.P.; Sahril, N.; Mutalip, M.H.A.; Zaki, N.A.M.; Ghaffar, S.A. Determinants of Dietary Supplements Use among Adolescents in Malaysia. *Asia Pac. J. Public Health* **2014**, *26* (Suppl. 5), 36S–43S. [CrossRef]
15. Tiwari, K. Supplement (mis)use in adolescents. *Curr. Opin. Pediatr.* **2020**, *32*, 471–475. [CrossRef] [PubMed]

16. Lentjes, M.A.H. The balance between food and dietary supplements in the general population. *Proc. Nutr. Soc.* **2018**, *78*, 97–109. [CrossRef] [PubMed]
17. Patterson, R.E.; Kristal, A.R.; Levy, L.; McLerran, D.; White, E. Validity of methods used to assess vitamin and mineral supplement use. *Am. J. Epidemiol.* **1998**, *148*, 643–649. [CrossRef] [PubMed]
18. Skeie, G.; Braaten, T.; Hjartåker, A.; Lentjes, M.A.H.; Amiano, P.; Jakszyn, P.; Pala, V.; Palanca, A.; Niekerk, E.M.; Verhagen, H.; et al. Use of dietary supplements in the European Prospective Investigation into Cancer and Nutrition calibration study. *Eur. J. Clin. Nutr.* **2009**, *63* (Suppl. 4), S226–S238. [CrossRef]
19. Robinson, M.; Stokes, K.; Bilzon, J.; Standage, M.; Brown, P.; Thompson, D. Test-retest reliability of the Military Pre-training Questionnaire. *Occup. Med.* **2010**, *60*, 476–483. [CrossRef]
20. Gajda, K.; Zielinska, M.; Ciecierska, A.; Hamulka, J. Determinants of the use of dietary supplements among secondary and high school students. *Rocz. Państwowego Zakładu Hig.* **2016**, *67*, 383–390.
21. Sirico, F.; Miressi, S.; Castaldo, C.; Spera, R.; Montagnani, S.; Di Meglio, F.; Nurzyńska, D. Habits and beliefs related to food supplements: Results of a survey among Italian students of different education fields and levels. *PLoS ONE* **2018**, *13*, e0191424. [CrossRef]
22. Kotnik, K.Z.; Jurak, G.; Starc, G.; Golja, P. Faster, Stronger, Healthier: Adolescent-Stated Reasons for Dietary Supplementation. *J. Nutr. Educ. Behav.* **2017**, *49*, 817–826.e1. [CrossRef]
23. Nishijima, C.; Kobayashi, E.; Sato, Y.; Chiba, T. A Nationwide Survey of the Attitudes toward the Use of Dietary Supplements among Japanese High-School Students. *Nutrients* **2019**, *11*, 1469. [CrossRef] [PubMed]
24. Ishihara, J.; Sobue, T.; Yamamoto, S.; Sasaki, S.; Akabane, M.; Tsugane, S. Validity and reproducibility of a self-administered questionnaire to determine dietary supplement users among Japanese. *Eur. J. Clin. Nutr.* **2001**, *55*, 360–365. [CrossRef]
25. Keszei, A.P.; Novak, M.; Streiner, D.L. Introduction to health measurement scales. *J. Psychosom. Res.* **2010**, *68*, 319–323. [CrossRef]
26. Giammarioli, S.; Boniglia, C.; Carratù, B.; Ciarrocchi, M.; Chiarotti, F.; Sanzini, E. Reliability of a Self-Administered Postal Questionnaire on the Use of Food Supplements in an Italian Adult Population. *Int. J. Vitam. Nutr. Res.* **2010**, *80*, 394–407. [CrossRef] [PubMed]
27. Yetley, E.A. Multivitamin and multimineral dietary supplements: Definitions, characterization, bioavailability, and drug interactions. *Am. J. Clin. Nutr.* **2007**, *85*, 269S–276S. [CrossRef]
28. Bujang, M.A.; Baharum, N. Guidelines of the minimum sample size requirements for Kappa agreement test. *Epidemiol. Biostat. Public Health* **2017**, *14*. [CrossRef]
29. Flack, V.F.; Afifi, A.A.; Lachenbruch, P.A.; Schouten, H.J.A. Sample size determinations for the two rater kappa statistic. *Psychometrika* **1988**, *53*, 321–325. [CrossRef]
30. Zulfarina, M.S.; Sharif, R.; Syarifah-Noratiqah, S.-B.; Sharkawi, A.M.; Aqilah-Sm, Z.-S.; Mokhtar, S.-A.; Nazrun, S.A.; Naina-Mohamed, I. Modifiable factors associated with bone health in Malaysian adolescents utilising calcaneus quantitative ultrasound. *PLoS ONE* **2018**, *13*, e0202321. [CrossRef] [PubMed]
31. Landis, J.R.; Koch, G.G. The Measurement of Observer Agreement for Categorical Data. *Biometrics* **1977**, *33*, 159. [CrossRef] [PubMed]
32. Murphy, S.P.; Wilkens, L.R.; Hankin, J.H.; Foote, J.A.; Monroe, K.R.; Henderson, B.E.; Kolonel, L.N. Comparison of two instruments for quantifying intake of vitamin and mineral supplements: A brief questionnaire versus three 24-hour recalls. *Am. J. Epidemiol.* **2002**, *156*, 669–675. [CrossRef]
33. Zaki, N.A.M.; Rasidi, M.N.; Awaluddin, S.M.; Hiong, T.G.; Ismail, H.; Nor, N.S.M. Prevalence and Characteristic of Dietary Supplement Users in Malaysia: Data From the Malaysian Adult Nutrition Survey (MANS) 2014. *Glob. J. Health Sci.* **2018**, *10*, 127. [CrossRef]
34. Gaspelin, N.; Ruthruff, E.; Pashler, H. Divided attention: An undesirable difficulty in memory retention. *Mem. Cogn.* **2013**, *41*, 978–988. [CrossRef]
35. Bae, J.; Joung, H.; Kim, J.-Y.; Kwon, K.N.; Kim, Y.T.; Park, S.-W. Test-Retest Reliability of a Questionnaire for the Korea Youth Risk Behavior Web-Based Survey. *J. Prev. Med. Public Health* **2010**, *43*, 403. [CrossRef] [PubMed]

Article

Effect of Food Containing Paramylon Derived from *Euglena gracilis* EOD-1 on Fatigue in Healthy Adults: A Randomized, Double-Blind, Placebo-Controlled, Parallel-Group Trial

Takanori Kawano [1,*], Junko Naito [1], Machiko Nishioka [1], Norihisa Nishida [1], Madoka Takahashi [1], Shinichi Kashiwagi [2], Tomohiro Sugino [2] and Yasuyoshi Watanabe [3,*]

1 Kobelco Eco-Solutions Co., Ltd., Kobe, Hyogo 651-2241, Japan; j.naito@kobelco-eco.co.jp (J.N.); m.nishioka@kobelco-eco.co.jp (M.N.); n.nishida@kobelco-eco.co.jp (N.N.); md.takahashi@kobelco-eco.co.jp (M.T.)

2 Soiken. Inc., Toyonaka, Osaka 560-0082, Japan; kashiwagi_shinichi@soiken.com (S.K.); sugino@soiken.com (T.S.)

3 RIKEN Center for Biosystems Dynamics Research, Kobe, Hyogo 650-0047, Japan

* Correspondence: t.kawano@kobelco-eco.co.jp (T.K.); yywata@riken.jp (Y.W.); Tel.: +81-78-992-6957 (T.K.); +81-78-304-7100 (Y.W.)

Received: 11 September 2020; Accepted: 8 October 2020; Published: 12 October 2020

Abstract: *Euglena gracilis* EOD-1, a kind of microalgae, is known to contain a high proportion of paramylon, a type of β-1,3-glucan. Paramylon derived from *E. gracilis* EOD-1 is presumed to suppress cellular oxidative injury and expected to reduce fatigue and fatigue sensation. Therefore, we aimed to examine whether food containing paramylon derived from *E. gracilis* EOD-1 (EOD-1PM) ingestion reduced fatigue and fatigue sensation in healthy adults. We conducted a randomized, double-blind, placebo-controlled, parallel-group comparison study in 66 healthy men and women who ingested a placebo or EOD-1PM daily for 4 weeks (daily life fatigue). Furthermore, at the examination days of 0 and 4 weeks, tolerance to fatigue load was evaluated using mental tasks (task-induced fatigue). We evaluated fatigue sensation using the Visual Analogue Scale, the work efficiency of the advanced trail making test and measured serum antioxidant markers. The EOD-1PM group showed significantly lower levels of physical and mental fatigue sensations and higher levels of work efficiency as well as serum biological antioxidant potential levels than the placebo group. These results indicate that EOD-1PM ingestion reduced fatigue and fatigue sensation, which may be due to an increase in antioxidant potential and maintenance of selective attention during work.

Keywords: *Euglena gracilis* EOD-1; β-glucan; paramylon; fatigue; selective attention; antioxidant

---

1. Introduction

Fatigue is a physiological alarm that reminds an organism of the need to rest to prevent exhaustion from overworking. Therefore, fatigue is a considerably familiar phenomenon in daily activities, such as work, exercise, creativity, and mental activity, and is essential for health maintenance.

Overall, fatigue is the difficulty in initiating or sustaining, voluntary activities [1]. In 2011, the Japanese Society of Fatigue Science defined fatigue and fatigue sensation as follows: Fatigue is a decline in the ability and efficiency of mental and/or physical activities that is caused by excessive mental or physical activities or disease. Fatigue is often accompanied by a peculiar sense of discomfort, a desire to rest, and a decline in motivation, which is referred to as fatigue sensation [2,3]. In other words, fatigue and fatigue sensation are distinguished. The subjective feeling of fatigue as a phenomenon that occurs in organisms can be said to be fatigue sensation. Furthermore, fatigue is classified into physical fatigue and mental fatigue. Physical fatigue, also known as peripheral fatigue, is caused by various muscle activities, while mental fatigue can be described as a state in which attention tasks that require self-motivation cannot be initiated or sustained in the absence of apparent cognitive impairment and deterioration of motor ability [1,4,5]. Moreover, overall fatigue, which is a combination of physical fatigue and mental fatigue, indicates the presence of overall fatigue sensation.

Fatigue can be quantified by various approaches (e.g., dysregulation of autonomic nervous system activity, hypothalamic-pituitary-adrenal axis hypofunction, immune abnormality, and reduction in work efficiency). In other words, fatigue is said to be due to reduced homeostasis (i.e., poor neural, endocrine, and immune interactions) caused by excessive activity and stress [1,3,6–11]. Fatigue has also been reported to be associated with oxidative stress [12] and reduced repair energy [13,14]. These associations are considered to be a few of the mechanisms of fatigue related to exercise and mental stress [15,16]. When an organism cannot process reactive oxygen species that are produced due to excessive activity, cell injury is induced and the immune cells that detect it generate cytokines, which serve as fatigue transmitters and inducers of fatigue sensation [9,17].

Euglena gracilis is a kind of microalgae that produces a β-1,3-glucan called paramylon. *E. gracilis* EOD-1 containing a high proportion (70–80%) of paramylon is known to proliferate depending on culture conditions [18,19]. Moreover, *E. gracilis* EOD-1, which has been suggested to have various functionalities, has recently drawn attention as a health food. Regarding its functionality, *E. gracilis* EOD-1 has been reported to improve health-related quality-of-life and immune function in humans [20]. There are numerous reports of the effects of paramylon derived from *E. gracilis* EOD-1 in diet-induced obese mice, such as the ability to prevent an increase in blood glucose levels during glucose load and reduce blood cholesterol and intraperitoneal fat weight [21]. In addition, paramylon has been shown to reduce cellular oxidative injury, improve liver disorder due to oxidative induction in rats, and increase reductase activity in the liver [22].

The abovementioned findings suggest that paramylon derived from *E. gracilis* EOD-1 reduces fatigue and fatigue sensation by maintaining immune function and reducing cellular oxidative injury. We have previously reported that food containing paramylon derived from *E. gracilis* EOD-1 (EOD-1PM) reduces physical fatigue sensation in daily life and increases pedometer steps and outing duration [23]. However, it has not been confirmed whether EOD-1PM actually reduces fatigue or not. Moreover, the mechanism of action has not been verified.

Against this backdrop, we conducted a randomized, double-blind, placebo-controlled, parallel-group trial in healthy individuals between the ages of 20 and 64 years to examine whether EOD-1PM reduces fatigue and fatigue sensation and to clarify the mechanism of action.

2. Materials and Methods

2.1. Subjects

Subjects were healthy men and women aged 20 to 64 years at the time of consent. The eligibility criteria were as follows: (1) those who had fatigue sensation in daily life, (2) those who agreed with the purpose of this study after receiving an explanation of the trial prior to participating, and (3) those who provided written informed consent. The exclusion criteria were the following 14 items: (1) those with severe cardiovascular disorders, liver dysfunction, renal dysfunction, respiratory disturbance, endocrine disorders, metabolism disorders, or those with a history of these conditions; (2) those with chronic fatigue syndrome or those who were deemed by the principal investigator to have severe fatigue, such as idiopathic chronic fatigue; (3) outpatients with chronic diseases; (4) those with mental illnesses such as depression/schizophrenia; (5) those with sleep disorders requiring treatment; (6) those who may have had an allergic reaction to the test food; (7) those who regularly used medical products and quasi-drugs with functional claims related to fatigue sensation that promote recovery from fatigue or nutritional supplements during physical fatigue; (8) those who regularly used Foods with Function Claims for reducing fatigue sensation; (9) heavy drinkers (those with pure alcohol consumption of

60 g/day or more); (10) those with a Body Mass Index (BMI) less than 17 or 30 or more; (11) those who provided more than 200 mL of blood within 1 month or 400 mL of blood within 3 months (donation, etc.) before the start of this trial; (12) those who participated in other clinical trials in the past 3 months or those who were currently participating in another clinical trial; (13) those who were pregnant or breastfeeding, or those who were planning to become pregnant; and (14) those who were deemed ineligible by the principal investigator. Sixty-six subjects who satisfied the abovementioned conditions were randomly assigned to two groups using random numbers by a staff member in charge of subject allocation (Statcom Co., Ltd., Tokyo, Japan) who did not participate in the trials directly. At the time of subject assignment, the staff in charge of subject allocation ensured that there were no significant differences in gender, age during the informed consent process, the score of overall fatigue sensation on the Visual Analogue Scale (VAS) [24], Pittsburgh Sleep Quality Index score [25–27], total power and low-frequency/high-frequency (LF/HF) of evaluation of autonomic nerve function [28], and the number of trials of the advanced trail making test (ATMT) [11]. The staff in charge of subject allocation sealed the allocation table immediately after the completion of subject allocation and safely stored it in a locked cabinet. The number of subjects was determined based on the results of similar trials that evaluated the anti-fatigue effects of food ingredients and compositions such as sesamin and astaxanthin [29], aged garlic extract [30], and lemon citric acid [31].

2.2. Test Food

Regarding the test food and placebo, EOD-1PM was used as the test food to be compared with the placebo. The dose of the *E. gracilis* EOD-1 (trade name: *Euglena gracilis* EOD-1, Kobelco Eco-Solutions Co., Ltd., Kobe, Japan) powder per capsule was adjusted so that the test food would contain 175 mg of paramylon derived from *E. gracilis* EOD-1, whereas the placebo contained dextrin. The test food and placebo were approved by the Institutional Review Board after passing an indistinguishability test.

2.3. Study Design and Intake Method

This study was a randomized, double-blind, placebo-controlled, parallel-group trial. The study design is shown in Figure 1. The trial period consisted of the examination day at 0 weeks (ED0W), 4 weeks of test food intake, and the examination day at 4 weeks after the start of the intake (ED4W). A fatigue load was applied on the examination day, which was followed by a recovery period. An evaluation was performed in each period. Subjects were instructed to keep a diary from the instruction day to the ED4W. During the test food period, the subjects consumed two test foods or placebos once a day at dinner. During the trial period, subjects were not allowed to take drugs, quasi-drugs, and Foods with Function Claims related to fatigue sensation that promote recovery from fatigue or nutritional supplements during physical fatigue. In addition, they were instructed to maintain their normal lifestyle (e.g., alcohol, diet, exercise and smoking). On the day before the examination, the subjects had a designated dinner and were prohibited from consuming alcohol or caffeine-rich foods and beverages. On the day of the test, the subject had a designated breakfast and lunch. They were instructed to take only their designated food and water until the end of the test. This study was performed in accordance with the Declaration of Helsinki after the approval of the Institutional Review Board of Fukuda Clinic (approval number: IRB-20190216-5). The trial protocol was implemented after being registered in the University Hospital Medical Information Network Clinical Trials Registry (UMIN-CTR) (UMIN trials ID: UMIN000036314). This trial was carried out in Osaka, Japan.

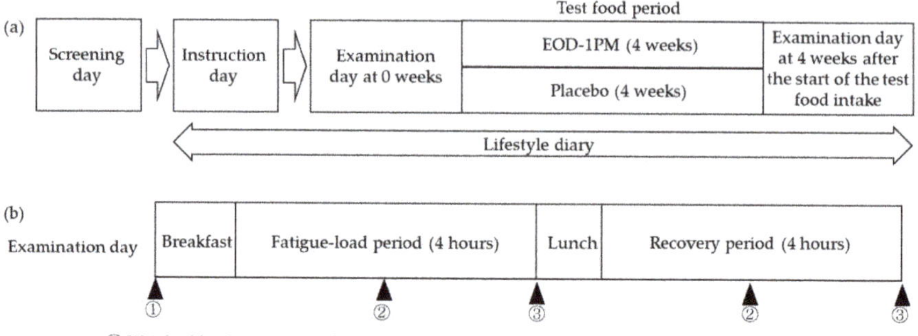

Figure 1. Study design and endpoints: (**a**) overall trial schedule, (**b**) examination schedule. EOD-1PM, food containing paramylon derived from *Euglena gracilis* EOD-1; VAS, Visual Analogue Scale.

2.4. Method of Fatigue Loading

Fatigue loading consisted of 4 sets (4 h) of 30-min 2-back tasks (simple working memory tasks) [4,32] and 30-min of the ATMT (selective attention and spatial working memory tasks) [4,11] (1 h in total). In the 2-back tasks, in which one letter of the alphabet was displayed every 3 s on the laptop screen, the subjects were asked to right-click the mouse if the letter matched the second-to-last letter, Otherwise, they were asked to left-click the mouse. In the ATMT (with three types of tasks, A, B, and C, in which 25 numbers from 1 to 25 were randomly presented on the laptop screen), subjects were instructed to left-click in order from 1 to 25. Tasks A, B, and C were repeated in order. In task A, the positions of 1 to 25 were fixed. Task B was the same as task A, except that when the subjects clicked the number, the number disappeared and a new number appeared at any position between 26 and 49. In task C, the arrangement of all numbers changed each time the target number was clicked. Regarding the fatigue loading, an instruction day was added to reduce the effect of the timing of fatigue loading.

2.5. Evaluation Method

The primary endpoint was the VAS [24] (overall fatigue sensation, physical fatigue sensation, mental fatigue sensation). Secondary endpoints were the evaluation of autonomic nerve function, work efficiency, blood oxidative stress, antioxidant markers, and safety.

Overall fatigue sensation, physical fatigue sensation, and mental fatigue sensation were evaluated by the VAS upon arrival to the facility, 2 and 4 h after fatigue loading, and 2 and 4 h after recovery on the ED0W and ED4W. The scores were evaluated according to the Guideline of Clinical Evaluation of Anti-fatigue by the Japanese Society of Fatigue Science [33]. Autonomic nerve function was evaluated by frequency analysis of the a-a wave intervals (maximum entropy method) by measuring accelerated plethysmography using ARTETT CDN (U-MEDICA, Inc. Co., Osaka, Japan) [28]. Measurements were taken upon arrival to the facility, 2 and 4 h after fatigue loading, and 2 and 4 h after recovery on the ED0W and ED4W. The endpoints were low-frequency component (LF, 0.04–0.15 Hz), high-frequency component (HF, 0.15–0.40 Hz), total power, and LF/HF.

Work efficiency was evaluated by the ATMT during fatigue loading on the ED0W and ED4W [4,11]. The endpoints [i.e., mean reaction time, standard deviation of reaction time, coefficient of variation (CV) of reaction time, and the number of errors in each task] were calculated as the total for the 1st to 4th sets.

Blood oxidative stress and antioxidant markers were evaluated during blood collection upon arrival to the facility, 4 h after fatigue loading, and 4 h after recovery on the ED0W and ED4W. The FREE CARRIO DUO system (Diacron International, Grosseto, Italy) was used to measure serum derivatives of reactive oxygen metabolites (d-ROMs) as an oxidative stress marker and biological antioxidant potential (BAP) as an antioxidant marker [34–38].

To evaluate safety, a blood test was performed upon arrival to the facility on the ED0W and ED4W. The endpoints were white blood cell count, differential leukocyte count (neutrophils, eosinophils, basophils, monocytes, lymphocytes), total protein, albumin, albumin/globulin ratio, aspartate aminotransferase, alanine aminotransferase, γ-glutamyl transpeptidase, creatinine, glucose, total cholesterol, high-density lipoprotein cholesterol, low-density lipoprotein cholesterol, triglyceride, creatine phosphokinase, uric acid, urea nitrogen, alkaline phosphatase, lactate dehydrogenase, Na, K, Cl, Ca, Mg, P, total bilirubin, and high-sensitivity C-reactive protein. The evaluation of the above endpoints was performed by BML, Inc. (Kawagoe, Japan), using a conventional method. An additional safety evaluation was conducted, and blood pressure, pulse rate, and body weight were measured, and medical interviews were performed upon arrival to the facility on the ED0W and ED4W. During the trial period, subjects were instructed to keep a diary on their lifestyle.

2.6. Statistical Analysis

For each endpoint, the values for each time point, changes before and after the intake of the test food, and changes from the time of facility arrival were evaluated using the following method. Regarding the intergroup comparisons, the Student's t-test was used to analyze data with a normal distribution, whereas the Mann–Whitney U test was used to analyze data with a non-normal distribution. Regarding the intragroup changes in the values before and after the intake, a paired t-test was used since the data followed a normal distribution. For all analyses, the statistical software package IBM SPSS ver. 22.0 (IBM Japan Ltd., Tokyo, Japan) was used. $p < 0.05$ (two-tailed) was considered statistically significant. Values are presented as the mean ± standard deviation.

3. Results

3.1. Subjects

Figure 2 shows a flowchart depicting the process from the subject screening to the analysis. A total of 205 subject candidates provided informed consent and underwent screening. The screening for recruitment of healthy volunteers was done with the data of routine blood test and blood biochemistry test including the number of red blood cells and hemoglobin concentration in plasma. None of the volunteers entered in this study showed any abnormal values of them and other routine biochemistry data. They have no signs of anemia and other metabolic diseases. Sixty-six subjects, who were included in this study trial, were randomly assigned to two groups by a staff member in charge of subject allocation. The following three subjects were excluded from the analysis: (1) one subject who voluntarily withdrew from the trial for personal reasons, (2) one subject whose participation was discontinued at the discretion of the principal investigator after developing a chronic migraine during fatigue loading on the ED0W, and (3) one subject whose participation was discontinued at the discretion of the principal investigator after vomiting during fatigue loading on the ED0W (The subject did not receive the test food at all). A total of 63 subjects, after excluding the abovementioned three subjects, were included in the full analysis set for safety evaluation. Moreover, 52 subjects, after further excluding one subject whose participation was discontinued at the discretion of the principal investigator due to symptoms such as vertigo and nausea upon arrival to the facility on the ED4W and 10 subjects who were deemed by the principal investigator to be ineligible for analysis, were included in the per protocol set (PPS) for efficacy analysis. All the endpoint data after the start of fatigue loading for the following subjects were partially excluded from the analysis: three subjects who vomited during fatigue loading on the ED0W and two subjects who vomited during fatigue loading on the ED4W. In 62 patients, after excluding four patients who withdrew from this trial, the intake rate was as follows: 100% (n = 26) and 96.4% (n = 4) for the EOD-1PM group and 100% (n = 30) and 96.4% (n = 2) for the placebo group. Regarding the following subject backgrounds, no significant differences were found between the two groups for age, gender, BMI, VAS (overall fatigue sensation) at the time of the screening test, the Pittsburgh Sleep Quality Index, autonomic nerve function (LF/HF, total tower), and the number of ATMT trials at the time of instruction (Tables 1 and 2).

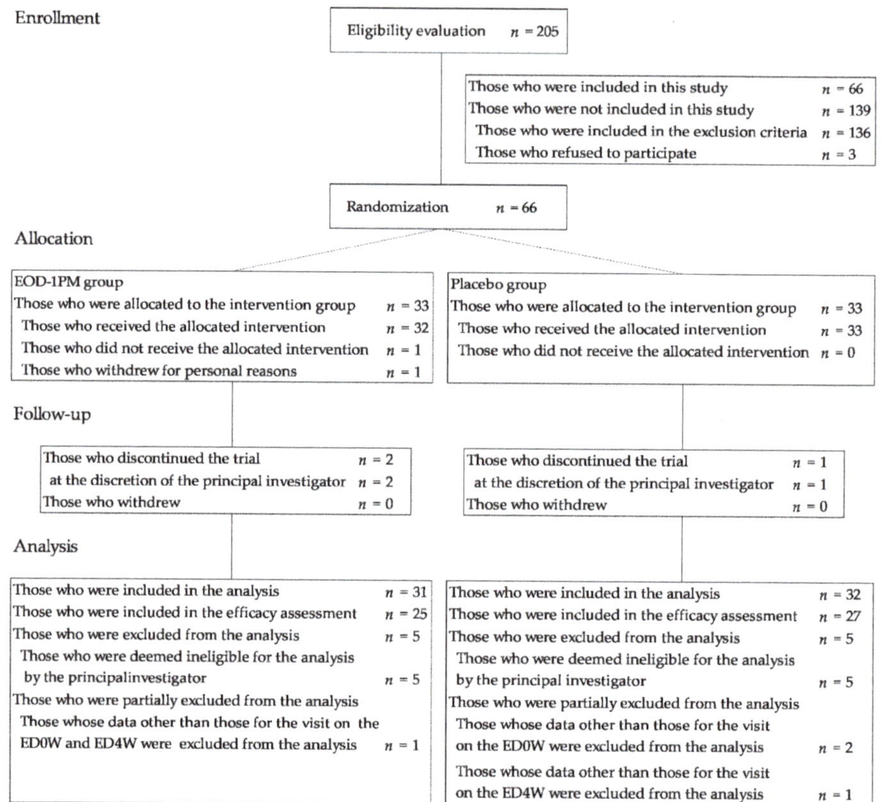

Figure 2. Flow diagram of participants. ED0W, examination day at 0 weeks; ED4W, examination day at 4 weeks after the start of the intake.

Table 1. Subject backgrounds.

	n	Men	Women	Age (Years)	BMI (kg/m^2)
EOD-1PM	25	13	12	49.2 ± 6.5	23.4 ± 2.4
Placebo	27	12	15	48.9 ± 10.3	22.7 ± 2.6

The data are presented as the mean ± standard deviation. BMI, Body Mass Index; EOD-1PM, food containing paramylon derived from Euglena gracilis EOD-1.

Table 2. Subject backgrounds (test score and evaluated autonomic nerve function).

	Values at the Screening Test				
	VAS (Overall Fatigue Sensation) Score	Pittsburgh Sleep Quality Index Score	Evaluated Autonomic Nerve Function (Total Power)	Evaluated Autonomic Nerve Function (LF/HF)	Number of ATMT Trials at the Time of Instruction
EOD-1PM	59.4 ± 8.0	5.2 ± 1.9	1350.9 ± 961.9	1.202 ± 1.232	797 ± 128
Placebo	59.9 ± 7.2	5.1 ± 1.9	1447.2 ± 732.9	1.200 ± 1.253	803 ± 129

The data are presented as the mean ± standard deviation; VAS, Visual Analogue Scale; ATMT, advanced trail making test; LF/HF, low-frequency/high-frequency.

3.2. Subjective Evaluation by the VAS

Figure 3 shows the results of the evaluation of fatigue sensation by using VAS. The values of overall fatigue sensation and physical fatigue sensation upon arrival to the facility on the ED4W were significantly lower than on the ED0W, in the EOD-1PM group ($p = 0.001$ for both). On the other hand, no significant difference was observed in the placebo group (Figure 3a,b). Regarding changes before and after intake, physical fatigue sensation and mental fatigue sensation upon arrival to the facility were significantly lower in the EOD-1PM group than in the placebo group ($p = 0.037$ for both, Figure 3e,f). Overall fatigue sensation in the EOD-1PM group tended to be lower, but not statistically significantly more so than in the placebo group ($p = 0.099$) (Figure 3d).

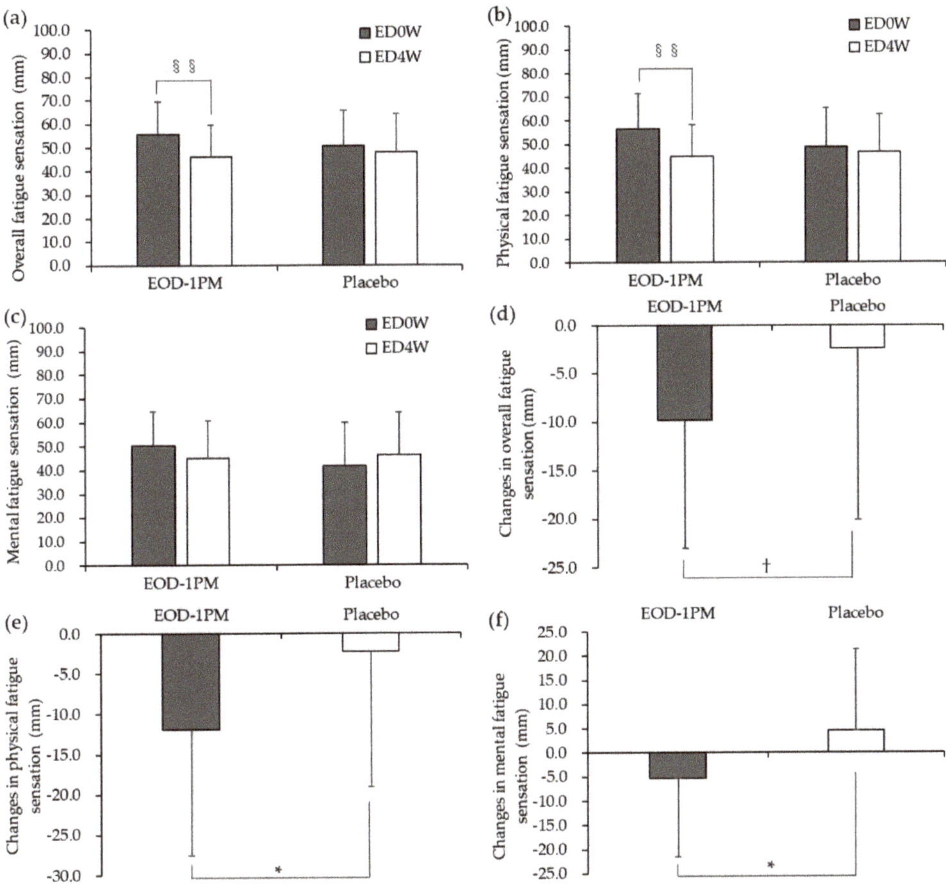

Figure 3. Evaluation of fatigue sensation by using VAS upon arrival to the facility on the examination day. (**a**) Comparison of overall fatigue sensation on the examination day at 0 weeks (ED0W) and at 4 weeks after the start of the intake (ED4W); (**b**) comparison of physical fatigue sensation on the ED0W and ED4W; (**c**) comparison of mental fatigue sensation on the ED0W and ED4W; (**d**) comparison of changes in overall fatigue sensation before and after the intake; (**e**) comparison of changes in physical fatigue sensation before and after the intake; and (**f**) comparison of changes in mental fatigue sensation before and after the intake. EOD-1PM group: $n = 25$, placebo group: $n = 27$. §§ $p < 0.01$. Comparison with that at facility arrival on the ED0W. * $p < 0.05$, † $p < 0.1$. Comparison between the placebo and EOD-1PM groups. The data are presented as the mean ± standard deviation.

3.3. Blood Cells and Blood Biochemistry Data

Statistically significant changes could not be observed before and after 4-week ingestion of EOD-1PM on blood cell counts especially the counts of leukocyte subpopulations and other routine blood biochemistry data indicating liver enzymes, kidney functions, and metabolic markers.

3.4. Oxidative Stress and Antioxidant Markers

In the evaluation of the serum d-ROMs values, BAP levels, and the BAP/d-ROMs ratio, BAP levels and the BAP/d-ROMs ratio upon arrival to the facility on the ED4W were significantly higher than those on the ED0W ($p < 0.001$, $p = 0.014$, respectively). On the other hand, no significant difference was observed in any of the items in the placebo group (Table 3). Furthermore, BAP levels and the BAP/d-ROMs ratio upon arrival to the facility on the ED4W were significantly higher in the EOD-1PM group than in the placebo group ($p = 0.035$, 0.01, respectively) (Table 3), and changes in BAP levels upon arrival to the facility before and after the intake period were significantly higher in the EOD-1PM group than in the placebo group ($p = 0.043$) (Table 3).

Table 3. The concentration of oxidative stress and antioxidant markers at facility arrival on the examination day.

		ED0W EOD-1PM (*n* = 25) Placebo (*n* = 27)	ED4W EOD-1PM (*n* = 25) Placebo (*n* = 27)			Changes before and after Intake EOD-1PM (*n* = 25) Placebo (*n* = 27)	
d-ROMs (U.CARR)	EOD-1PM	307 ± 58	309 ± 54			3 ± 21	
	Placebo	326 ± 41	329 ± 39			2 ± 31	
BAP (mmol/L)	EOD-1PM	2085 ± 131	2238 ± 130	*	§§	153 ± 171	*
	Placebo	2097 ± 190	2147 ± 168			50 ± 187	
BAP/d-ROMs	EOD-1PM	7.0 ± 1.4	7.4 ± 1.4	**	§	0.4 ± 0.8	
	Placebo	6.5 ± 0.9	6.6 ± 0.8			0.1 ± 0.9	

** $p < 0.01$, * $p < 0.05$. Comparison between the placebo and EOD-1PM groups. §§ $p < 0.01$, § $p < 0.05$. Comparison with that upon arrival to the facility on the examination day at the start of the intake. The data are presented as the mean ± standard deviation. d-ROMS, derivatives of reactive oxygen metabolites; BAP, biological antioxidant potential; ED0W, examination day at 0 weeks; ED4W, examination day at 4 weeks after the start of the intake.

3.5. Evaluation of Fatigue Load and Work Efficiency

On the ED0W and ED4W, beyond the examinations of "daily-life fatigue" before fatigue-load tasks, we added the mental tasks with 2-back task and ATMT tasks (as shown in Figure 1). The overall, mental, and physical fatigue VAS scores were increased significantly after 4-h load ($p < 0.01$ for all scores before and after fatigue load), but the difference of the overall, mental, and physical fatigue VAS scores with fatigue load tolerance was not statistically significant. Such subjective scores were not altered statistically, but the objective markers of fatigue were changed as follows. In addition, we made the subtraction between the measurement values of each item at ED0W and ED4W and the subtracted values (Changes before and after intake of EOD-1PM or placebo) are also incorporated in the evaluation process as we had informed in the IRB protocol.

Table 4 shows the mean reaction time, the standard deviation of reaction time, the CV of reaction time, the number of errors of the total for all sets in the ATMT trials, and the changes before and after the intake for each ATMT task.

Table 4. Evaluation of work efficiency by the ATMT.

		ED0W Total for All Sets EOD-1PM (*n* = 24) Placebo (*n* = 25)		**ED4W Total for All Sets EOD-1PM (*n* = 24) Placebo (*n* = 26)**		**Changes before and after Intake Total for All Sets EOD-1PM (*n* = 24) Placebo (*n* = 24)**	
Task A							
Mean reaction time (sec)	EOD-1PM Placebo	1.673 ± 0.532 1.639 ± 0.337		1.836 ± 0.684 1.993 ± 0.752		0.164 ± 0.241 0.164 ± 0.208	
Standard deviation of reaction time (sec)	EOD-1PM Placebo	1.213 ± 0.457 1.345 ± 0.492		1.445 ± 0.693 2.053 ± 1.096	*	0.233 ± 0.382 0.476 ± 0.535	
CV of reaction time	EOD-1PM Placebo	72.8 ± 19.8 80.8 ± 19.3		78.6 ± 24.5 100.2 ± 33.3	*	5.8 ± 11.3 17.9 ± 20.9	*
Number of errors	EOD-1PM Placebo	25.0 ± 27.1 35.0 ± 37.4		29.1 ± 32.7 38.2 ± 38.5		4.0 ± 10.7 4.5 ± 23.6	
Task B							
Mean reaction time (sec)	EOD-1PM Placebo	2.230 ± 0.616 2.220 ± 0.465		2.392 ± 0.714 2.693 ± 0.984		0.162 ± 0.242 0.274 ± 0.446	
Standard deviation of reaction time (sec)	EOD-1PM Placebo	1.893 ± 0.554 2.196 ± 0.751		2.183 ± 0.842 3.034 ± 1.600	*	0.289 ± 0.530 0.638 ± 1.261	
CV of reaction time	EOD-1PM Placebo	85.8 ± 16.8 97.7 ± 20.0	*	91.2 ± 21.7 109.6 ± 30.4	*	5.4 ± 12.7 11.9 ± 23.2	
Number of errors	EOD-1PM Placebo	23.6 ± 27.2 38.7 ± 36.0		31.0 ± 35.4 41.4 ± 40.0		7.4 ± 13.4 5.0 ± 17.9	
Task C							
Mean reaction time (sec)	EOD-1PM Placebo	3.303 ± 0.587 3.341 ± 0.571		3.534 ± 0.735 3.963 ± 1.047		0.231 ± 0.402 0.390 ± 0.433	
Standard deviation of reaction time (sec)	EOD-1PM Placebo	2.099 ± 0.598 2.359 ± 0.773		2.434 ± 0.904 3.502 ± 1.651	**	0.335 ± 0.673 0.832 ± 0.840	*
CV of reaction time	EOD-1PM Placebo	63.0 ± 11.3 69.5 ± 14.7		67.8 ± 14.8 85.3 ± 25.6	**	4.8 ± 9.3 13.7 ± 15.2	*
Number of errors	EOD-1PM Placebo	31.0 ± 33.5 35.8 ± 36.5		32.4 ± 38.4 44.8 ± 40.3		1.5 ± 13.0 10.7 ± 17.6	*

** $p < 0.01$, * $p < 0.05$. Comparison between the placebo and EOD-1PM groups. The data are presented as the mean ± standard deviation. ATMT, advanced trail making test; CV, coefficient of variation.

The comparison of values for the ED4W between the EOD-1PM and placebo groups showed the following results. In task A, the standard deviation and CV of reaction time were significantly lower in the EOD-1PM group than in the placebo group ($p = 0.023$ and 0.013, respectively). In task B, the standard deviation of reaction time was significantly lower in the EOD-1PM group than in the placebo group ($p = 0.022$). The CV of reaction time were also significantly lower in the EOD-1PM group than in the placebo group ($p = 0.018$). However, the CV of the reaction time on the ED0W were also significantly lower in the EOD-1PM group than in the placebo group ($p = 0.029$). In task C, the standard deviation and CV of reaction time were significantly lower in the EOD-1PM group than in the placebo group ($p = 0.0066$ and 0.0047, respectively).

Moreover, comparisons of the changes before and after the intake showed the following results: significantly lower CV of reaction time in task A, standard deviation and CV of reaction time, and number of errors in task C in the EOD-1PM group than in the placebo group ($p = 0.018$, 0.029, 0.020, and 0.044, respectively).

Table 5 shows the rate of correct responses in the 2-back tasks. No significant difference was observed for the ED0W and ED4W.

Table 5. The rate of correct responses in the 2-back tasks.

		ED0W Total for All Sets EOD-1PM (n = 24) Placebo (n = 25)	**ED4W Total for All Sets EOD-1PM (n = 24) Placebo (n = 26)**
Rate of correct responses	EOD-1PM	0.863 ± 0.102	0.853 ± 0.127
	Placebo	0.850 ± 0.093	0.841 ± 0.102

The data are presented as the mean ± standard deviation.

3.6. *Evaluation of Autonomic Nerve Function*

To evaluate LF, HF, total power, and LF/HF, frequency analysis of the a-a wave intervals of accelerated plethysmography (as an indicator of autonomic nerve function) was performed. In intergroup comparisons of values, a significant difference and similar tendency were observed for total power for both ED0W and ED4W. Furthermore, intergroup comparisons of the values showed no significant differences in LF, HF, and LF/HF.

Regarding the changes from the time of arrival to the facility (difference in pre- and post-intake), LF/HF at 2 h after recovery was significantly lower in the EOD-1PM group than in the placebo group (p = 0.031) (Table 6).

Table 6. Evaluation of autonomic nerve function: changes from the time of facility arrival (difference in pre- and post-intake).

		2 h after Load EOD-1PM (n = 24) Placebo (n = 24)	**4 h after Load EOD-1PM (n = 24) Placebo (n = 24)**	**2 h after Recovery EOD-1PM (n = 24) Placebo (n = 24)**		**4 h after Recovery EOD-1PM (n = 24) Placebo (n = 24)**
LF (msec^2)	EOD-1PM	9.8 ± 373.5	41.9 ± 571.1	−34.8 ± 401.4		−88.1 ± 370.0
	Placebo	−112.7 ± 509.0	−89.1 ± 889.0	9.7 ± 512.9		−21.5 ± 675.2
HF (msec^2)	EOD-1PM	98.0 ± 270.3	72.0 ± 426.6	183.7 ± 344.7		−50.0 ± 366.5
	Placebo	30.7 ± 405.0	50.1 ± 535.9	38.1 ± 365.4		−8.9 ± 583.1
Total Power (msec^2)	EOD-1PM	469.9 ± 1780.1	460.9 ± 1548.5	421.2 ± 1281.5		−163.0 ± 1248.0
	Placebo	−60.0 ± 1435.1	154.4 ± 2270.9	121.4 ± 1967.4		437.0 ± 2888.1
LF/HF	EOD-1PM	−0.623 ± 1.903	−0.719 ± 1.555	−0.389 ± 1.024	*	−0.364 ± 1.187
	Placebo	−0.209 ± 1.913	−0.302 ± 2.379	0.463 ± 1.568		0.025 ± 1.400

* $p < 0.05$. Comparison between the placebo and EOD-1PM groups. The data are presented as the mean ± standard deviation. LF, low frequency component; HF, high frequency component.

3.7. *Safety Evaluation*

There were no statistically significant differences in the results of physical examination between the EOD-1PM and placebo groups. Although the blood tests showed a slight statistically significant difference, since the values were within the standard values (the range of daily changes), the principal investigator determined that there were no clinically relevant changes. Furthermore, in this trial, no adverse events were attributed to the test food.

4. Discussion

This trial examined the effect of a 4-week intake of EOD-1PM on the reduction of fatigue and fatigue sensation in healthy adult men and women between the ages of 20 and 64 years.

The results showed a significant reduction in physical and mental fatigue sensation upon arrival to the facility before and after the 4-week intake in the EOD-1PM group compared to the placebo group. These changes may indicate that there was an effect on fatigue during the 4-week intake, suggesting that EOD-1PM reduced physical and mental fatigue sensation in daily life. In addition, the intra-group comparison showed a significant reduction in the values of overall fatigue sensation before and after the intake in the EOD-1PM group, whereas the placebo group showed no significant difference. Regarding changes before and after intake, overall fatigue sensation upon arrival to the facility tended to be lower in the EOD-1PM group than in the placebo group. The above results suggest that EOD-1PM also reduced overall fatigue sensation in daily life. In the previous study, we reported that four-week administration of EOD-1PM improves physical fatigue sensation in daily life [23]. The present study demonstrated the effect of EOD-1PM on the reduction of mental fatigue sensation and suggested that EOD-1PM was able to reduce overall fatigue sensation.

The relationship between fatigue and oxidative stress is well known. It has been reported that an oxidative stress marker, d-ROMs, and an antioxidant marker, BAP, are useful as biomarkers of fatigue [12]. The evaluation of the serum markers showed a significant increase in BAP and the BAP/d-ROMs ratio upon arrival to the facility on the ED4W of EOD-1PM and significant changes in BAP before and after the intake of EOD-1PM. On the other hand, no significant change was observed in d-ROMs. BAP, which measures the ability to reduce Fe^{3+} to Fe^{2+}, is an index of antioxidant potential [35]. d-ROMs, which imply hydroperoxide levels, have been reported to be a useful marker of oxidative stress [34]. BAP, by its nature, may increase as oxidative stress increases. This study demonstrated an increase in antioxidant potential despite no increase in oxidative stress in the EOD-1PM group during the four-week study period. Therefore, EOD-1PM may reduce fatigue by permanently improving antioxidant potential without increasing oxidative stress. In addition, paramylon has been reported to increase reductase activities such as superoxide dismutase, catalase, and glutathione peroxidase in the rat liver [22]. EOD-1PM may also enhance these reductase activities and improve antioxidant potential. In most cases, studies in healthy subjects using food ingredients that have antioxidant effects show significant differences in the BAP/d-ROMs ratio [39,40]. Ingredients that change BAP alone are extremely rare. This is one of the characteristic properties of EOD-1PM. Therefore, EOD-1PM may have a considerably strong antioxidant effect.

It has been speculated that EOD-1PM is less susceptible to degradation by intestinal bacteria and is less likely to be absorbed through the intestines in mice [19,21]. Therefore, EOD-1PM may not be directly involved in improving serum antioxidant potential. Interestingly, EOD-1PM has been reported to induce the expression of genes, such as peroxisome proliferator-activated receptor-α, in the liver. EOD-1PM may be indirectly involved in gene expression in tissues by stimulating intestinal cells [21]. Regarding the findings of the effect of improving antioxidant potential in this trial, EOD-1PM may improve reductase activity through the same indirect action from the intestinal tract.

The evaluation of the results of the ATMT showed significant suppression of an increase in the standard deviation and CV of reaction time in tasks A and C and the standard deviation of reaction time in task B on the ED4W in the EOD-1PM compared to the placebo group. Regarding the changes before and after intake, the results also showed more significant suppression of an increase in the standard deviation of reaction time and CV of reaction time in tasks A and C and the number of errors in task C in the EOD-1PM group than in the placebo group. Therefore, it was found that EOD-1PM maintains work efficiency for a longer period of time by reducing the variation and changes in reaction time and the number of errors in the tasks.

Fatigue is defined as a decline in the ability and efficiency of mental and/or physical activities [2,3]. In other words, maintenance of work efficiency indicates that fatigue is being suppressed [4,41–43]. There was no significant difference in the rate of correct responses in the 2-back tasks between the two groups, suggesting that the fatigue loading due to work was evenly distributed between the two groups and that EOD-1PM intake reduced fatigue due to workload. In addition, work efficiency in the ATMT confirmed in this trial was especially maintained in task C. Tasks A and B, in which the position of numbers does not change, assess both spatial working memory and selective attention as cognitive functions. On the other hand, task C, in which the position of numbers changes, assesses selective attention only (not spatial working memory) [4,11]. Because a marked effect that assessed selective attention was confirmed in task C, EOD-1PM may prevent the deterioration of selective attention at work, leading to the maintenance of work efficiency and the reduction of fatigue.

In the evaluation of autonomic nerve function (changes from the time of arrival to the facility [difference in pre- and post-intake]), LF/HF at 2 h after recovery was significantly lower in the EOD-1PM group than in the placebo group. It is known that LF mainly reflects sympathetic nerve activity, and HF reflects parasympathetic nerve activity [44], and that LF/HF, a relative indicator of sympathetic nerve activity, increases during fatigue [45]. EOD-1PM intake may expedite recovery from fatigue by promoting parasympathetic nerve activity early in the post-fatigue recovery period.

Moreover, it has been reported that the phenomenon of fatigue is closely related to immune dysfunction and that immunity decreases with stress and fatigue. A study of salivary secretory immunoglobulin A (sIgA), which is considered as one of the biomarkers of fatigue, showed a negative correlation between middle-age stress experiences and the secretion rate of salivary sIgA [46]. Furthermore, studies have shown that intense sports training and competition reduce the concentration and secretion rate of salivary sIgA [47,48]. In a previous study, we reported that daily intake (4 weeks) of *E. gracilis* EOD-1 increased both the salivary sIgA concentration and sIgA secretion rate [20]. *E. gracilis* EOD-1 may reduce fatigue by suppressing the reduction of immunity in fatigue.

This trial demonstrated that continuous intake of EOD-1PM reduces fatigue and fatigue sensation. The phenomenon of fatigue is said to be due to reduced homeostasis (i.e., poor neural, endocrine, and immune interactions) [1,3,6–10]. EOD-1PM has been found to recover parasympathetic nerve activity from fatigue and may also promote the maintenance of immune function. In other words, EOD-1PM may maintain homeostasis and reduce fatigue by affecting the autonomic nervous system and immune system. Fatigue is thought to be caused by oxidative injury due to reactive oxygen species produced by cell overactivity [9,17]. This trial suggested that EOD-1PM permanently increases antioxidant potential and reduces fatigue by reducing cellular oxidative injury. These results suggest that regular intake of EOD-1PM may lead to a fatigue-resistant body. Furthermore, the daily accumulation of fatigue is thought to lead to chronic fatigue [49]. EOD-1PM, which can reduce accumulation of fatigue, may have a preventive effect on chronic fatigue.

Regarding the safety, no adverse events were reported in this study from the volunteers who made the daily reports of ingestion, timing, and adverse events if any by the intake of EOD-1PM. Other trials of longer-term intake of EOD-1PM have also showed its safety [23,50]. In addition, the volunteers reported no stimulating and awakening effects of EOD-1PM unlike the intake of caffeine which causes an awakening effect sometimes confused with anti-fatigue effects.

This trial had several limitations. First, the sample size was limited. Second, the analysis was performed on the PPS. Third, this trial was considered to be within the range of daily fatigue. However, because the degree of fatigue could not be clearly shown, it was not possible to show how effective the EOD-1PM was in terms of fatigue levels. Fourth, this trial confirmed the effect of EOD-1PM by using long-term work on a personal computer as a fatigue load. However, the effect of EOD-1PM on fatigue due to high-intensity physical load, such as exercise, was unknown. Therefore, future studies should further explore these issues.

5. Conclusions

In this randomized, double-blind, placebo-controlled, parallel-group trial, a 4-week intake of EOD-1PM in healthy subjects was found to reduce physical and mental fatigue sensation in daily life. The results showed that EOD-1PM reduces fatigue, suggesting that EOD-1PM reduces fatigue sensation by preventing the daily accumulation of fatigue. The mechanism by which EOD-1PM reduces fatigue and fatigue sensation is believed to be by improving antioxidant potential, and sustaining selective attention.

Author Contributions: Conceptualization, J.N., M.N., S.K., T.S., Y.W.; methodology, J.N., M.N., S.K., T.S., Y.W.; formal analysis, T.K., J.N., M.N., S.K.; investigation, T.K., J.N., M.N., S.K., T.S., Y.W.; resources, N.N., M.T.; writing-original draft preparation, T.K.; writing-review and editing, J.N., M.N., S.K., T.S., Y.W.; supervision, Y.W. All authors have read and agreed to the published version of the manuscript.

Funding: This study was funded by Kobelco Eco-Solutions Co., Ltd.

Acknowledgments: We would like to express our sincere gratitude to Statcom. Co., Ltd. for their contribution to the randomization and allocation concealment in this trial.

Conflicts of Interest: Kobelco Eco-Solutions Co., Ltd. is the sole funder of this study and the merchandiser of the Euglena powder. The food substance used in this study was provided by Kobelco Eco-solutions Co., Ltd. This trial was outsourced to Soiken Inc. by Kobelco Eco-Solutions Co., Ltd., and Soiken Inc. received contract research expenses from Kobelco Eco-Solutions Co., Ltd. Five of the authors (T.K., J.N., M.N., N.N., and M.T.) are employees of and receive their salaries from Kobelco Eco-Solutions Co., Ltd. S.K. is an employee of and receives salary from Soiken Inc. T.S. is the Representative Director of Soiken Inc. Y.W. is both the advisor and the academic collaborator for all steps of this study to decide the protocols, to advise for data analyses, and to write the manuscript of this study, and received honorarium from Kobelco Eco-Solutions Co., Ltd.

References

1. Chaudhuri, A.; Behan, P.O. Fatigue in neurological disorders. *Lancet* **2004**, *363*, 978–988. [CrossRef]
2. Kitani, T. Term Committee of Japanese Society of Fatigue Science. *Nihon Hirougakkaishi* **2011**, *6*, 1.
3. Ishii, A.; Tanaka, M.; Watanabe, Y. Neural mechanisms of mental fatigue. *Rev. Neurosci.* **2014**, *25*, 469–479. [CrossRef]
4. Tanaka, M.; Mizuno, K.; Tajima, S.; Sasabe, T.; Watanabe, Y. Central nervous system fatigue alters autonomic nerve activity. *Life Sci.* **2009**, *84*, 235–239. [CrossRef]
5. Chaudhuri, A.; Behan, P.O. Fatigue and basal ganglia. *J. Neurol. Sci.* **2000**, *179*, 34–42. [CrossRef]
6. Nozaki, S.; Tanaka, M.; Mizuno, K.; Ataka, S.; Mizuma, H.; Tahara, T.; Sugino, T.; Shirai, T.; Eguchi, A.; Okuyama, K.; et al. Mental and physical fatigue-related biochemical alterations. *Nutrition* **2009**, *25*, 51–57. [CrossRef]
7. Watanabe, Y. What is fatigue?-Statistics of fatigue in Japan and what are the problems to be solved by fatigue science? *J. Clin. Exp. Med.* **2009**, *228*, 593–597.
8. Watanabe, Y. Preface and Mini-review: Fatigue Science for Human Health. In *Fatigue Science for Human Health*; Springer: New York, NY, USA, 2008; pp. 5–11.
9. Watanabe, Y.; Kuratsune, H.; Kajimoto, O. Biochemical Indices of Fatigue for Anti-fatigue Strategies and Products. In *The Handbook of Operator Fatigue*; CRC Press: Boca Raton, FL, USA, 2018; pp. 209–224.
10. Silverman, M.N.; Heim, C.M.; Nater, U.M.; Marques, A.H.; Sternberg, E.M. Neuroendocrine and immune contributors to fatigue. *PM R* **2010**, *2*, 338–346. [CrossRef]
11. Mizuno, K.; Watanabe, Y. Utility of an Advanced Trail Making Test as a Neuropsychological Tool for an Objective Evaluation of Work Efficiency during Mental Fatigue. In *Fatigue Science for Human Health*; Springer: New York, NY, USA, 2008; pp. 47–54.
12. Fukuda, S.; Nojima, J.; Motoki, Y.; Yamaguti, K.; Nakatomi, Y.; Okawa, N.; Fujiwara, K.; Watanabe, Y.; Kuratsune, H. A potential biomarker for fatigue: Oxidative stress and anti-oxidative activity. *Biol. Psychol.* **2016**, *118*, 88–93. [CrossRef]
13. Kume, S.; Yamato, M.; Tamura, Y.; Jin, G.; Nakano, M.; Miyashige, Y.; Eguchi, A.; Ogata, Y.; Goda, N.; Iwai, K.; et al. Potential biomarkers of fatigue identified by plasma metabolome analysis in rats. *PLoS ONE* **2015**, *10*, e0120106. [CrossRef]
14. Yamano, E.; Sugimoto, M.; Hirayama, A.; Kume, S.; Yamato, M.; Jin, G.; Tajima, S.; Goda, N.; Iwai, K.; Fukuda, S.; et al. Index markers of chronic fatigue syndrome with dysfunction of TCA and urea cycles. *Sci. Rep.* **2016**, *6*, 34990. [CrossRef] [PubMed]
15. Barclay, J.K.; Hansel, M. Free radicals may contribute to oxidative skeletal muscle fatigue. *Can. J. Physiol. Pharmacol.* **1991**, *69*, 279–284. [CrossRef] [PubMed]
16. Sivonova, M.; Zitnanova, I.; Hlincikova, L.; Skodacek, I.; Trebaticka, J.; Durackova, Z. Oxidative stress in university students during examinations. *Stress* **2004**, *7*, 183–188. [CrossRef] [PubMed]
17. Watanabe, Y. Molecular and neural mechanisms of fatigue-previous hypotheses and present concepts. *J. Clin. Exp. Med.* **2009**, *228*, 598–604.
18. Takahashi, M.; Kawashima, J.; Nishida, N.; Onaka, N. Beta-Glucan Derived from Euglena (Paramylon). In *Trends in Basic Research and Applied Sciences of Beta-Glucan*; CMC Publishing Co., Ltd.: Tokyo, Japan, 2018; pp. 174–182.
19. Anraku, M.; Iohara, D.; Takada, H.; Awane, T.; Kawashima, J.; Takahashi, M.; Hirayama, F. Morphometric analysis of paramylon particles produced by Euglena gracilis EOD-1 using FIB/SEM tomography. *Chem. Pharm. Bull.* **2020**, *68*, 100–102. [CrossRef] [PubMed]

20. Ishibashi, K.I.; Nishioka, M.; Onaka, N.; Takahashi, M.; Yamanaka, D.; Adachi, Y.; Ohno, N. Effects of Euglena gracilis EOD-1 ingestion on salivary IgA reactivity and health-related quality of life in humans. *Nutrients* **2019**, *11*, 1144. [CrossRef]
21. Aoe, S.; Yamanaka, C.; Koketsu, K.; Nishioka, M.; Onaka, N.; Nishida, N.; Takahashi, M. Effects of paramylon extracted from Euglena gracilis EOD-1 on parameters related to metabolic syndrome in diet-induced obese mice. *Nutrients* **2019**, *11*, 1674. [CrossRef]
22. Sugiyama, A.; Suzuki, K.; Mitra, S.; Arashida, R.; Yoshida, E.; Nakano, R.; Yabuta, Y.; Takeuchi, T. Hepatoprotective effects of paramylon, a beta-1, 3-D-glucan isolated from Euglena gracilis Z, on acute liver injury induced by carbon tetrachloride in rats. *J. Vet. Med. Sci.* **2009**, *71*, 885–890. [CrossRef]
23. Kawano, T.; Naito, J.; Nishioka, M.; Nishida, N.; Takahashi, M.; Ando, T.; Ebihara, S.; Watanabe, Y. Effects of consumption of food containing paramylon derived from Euglena gracilis EOD-1 on attenuating feelings of fatigue in daily life—A randomized, double-blind, placebo-controlled, parallel-group comparison study. *Jpn. Pharmacol. Ther.* **2019**, *47*, 1851–1859.
24. Leung, A.W.; Chan, C.C.; Lee, A.H.; Lam, K.W. Visual analogue scale correlates of musculoskeletal fatigue. *Percept. Mot. Skills* **2004**, *99*, 235–246. [CrossRef]
25. Buysse, D.J.; Reynolds, C.F., 3rd; Monk, T.H.; Berman, S.R.; Kupfer, D.J. The Pittsburgh Sleep Quality Index: A new instrument for psychiatric practice and research. *Psychiatry Res.* **1989**, *28*, 193–213. [CrossRef]
26. Doi, Y.; Minowa, M.; Uchiyama, M.; Okawa, M.; Kim, K.; Shibui, K.; Kamei, Y. Psychometric assessment of subjective sleep quality using the Japanese version of the Pittsburgh Sleep Quality Index (PSQI-J) in psychiatric disordered and control subjects. *Psychiatry Res.* **2000**, *97*, 165–172. [CrossRef]
27. Doi, Y.; Minowa, M.; Okawa, M.; Uchiyama, M. Development of the Japanese version of the Pittsburgh Sleep Quality Index. *Jpn. J. Psychiatr. Treat.* **1998**, *13*, 755–763.
28. Takada, M.; Ebara, T.; Sakai, Y. The Acceleration Plethysmography System as a new physiological technology for evaluating autonomic modulations. *Health Eval. Promot.* **2008**, *35*, 373–377. [CrossRef]
29. Imai, A.; Oda, Y.; Ito, N.; Seki, S.; Nakagawa, K.; Miyazawa, T.; Ueda, F. Effects of dietary supplementation of astaxanthin and sesamin on daily fatigue: A randomized, double-blind, placebo-controlled, two-way crossover study. *Nutrients* **2018**, *10*, 281. [CrossRef] [PubMed]
30. Nagata, Y.; Nakamura, A.; Hashiguchi, K.; Yamasaki, K.; Takashige, H.; Araki, E.; Saito, M. A randomized, double-blind, placebo-controlled parallel study of effects of food containing aged garlic extract on fatigue. *Jpn. Pharmacol. Ther.* **2017**, *45*, 405–421.
31. Kajimoto, O.; Mieda, H.; Hiramitsu, M.; Sakaida, K.; Sugino, T.; Kajimoto, Y. Effect of a drink containing lemon citric acid on people frequently feeling fatigue. *Jpn. Pharmacol. Ther.* **2007**, *35*, 809–817.
32. Kirchner, W.K. Age differences in short-term retention of rapidly changing information. *J. Exp. Psychol.* **1958**, *55*, 352–358. [CrossRef]
33. Hirayama, Y. Development of guideline of clinical evaluation of anti-fatigue. *J. Clin. Exp. Med.* **2009**, *228*, 733–736.
34. Alberti, A.; Bolognini, L.; Macciantelli, D.; Caratelli, M. The radical cation of N,N-diethyl-para-phenylendiamine: A possible indicator of oxidative stress in biological samples. *Res. Chem. Intermed.* **2020**, *26*, 253–267. [CrossRef]
35. Benzie, I.F.; Strain, J.J. The ferric reducing ability of plasma (FRAP) as a measure of "antioxidant power": The FRAP assay. *Anal. Biochem.* **1996**, *239*, 70–76. [CrossRef] [PubMed]
36. Sanchez-Rodriguez, M.A.; Mendosa-Nunez, V.M. Oxidative Stress Indexes for Diagnosis of Health or Disease in Humans. *Oxid. Med. Cell. Longev.* **2019**, *2019*, 4128152. [CrossRef] [PubMed]
37. Lubrano, V.; Pingitore, A.; Traghella, I.; Storti, S.; Parri, S.; Berti, S.; Ndreu, R.; Andrenelli, A.; Palmieri, C.; Iervasi, G.; et al. Emerging Biomarkers of Oxidative Stress in Acute and Stable Coronary Artery Disease: Levels and Determinants. *Antioxidants* **2019**, *8*, 115. [CrossRef] [PubMed]
38. Alvarez-Satta, M.; Berna-Erro, A.; Carrasco-Garcia, E.; Alberro, A.; Saenz-Antonanzas, A.; Vergara, I.; Otaegui, D.; Matheu, A. Relevance of oxidative stress and inflammation in frailty based on human studies and mouse models. *Aging* **2020**, *12*, 9982–9999. [CrossRef] [PubMed]
39. Katada, S.; Watanabe, T.; Mizuno, T.; Kobayashi, S.; Takeshita, M.; Osaki, N.; Katsuragi, Y. Effects of chlorogenic acid-enriched and hydroxyhydroquinone-reduced coffee on postprandial fat oxidation and antioxidative capacity in healthy men: A randomized, double-blind, placebo-controlled, crossover trial. *Nutrients* **2018**, *10*, 525. [CrossRef]

40. Omi, N.; Shiba, H.; Nishimura, E.; Tsukamoto, S.; Maruki-Uchida, H.; Oda, M.; Morita, M. Effects of enzymatically modified isoquercitrin in supplementary protein powder on athlete body composition: A randomized, placebo-controlled, double-blind trial. *J. Int. Soc. Sports Nutr.* **2019**, *16*, 39. [CrossRef] [PubMed]
41. Kuratsune, D.; Tajima, S.; Koizumi, J.; Yamaguti, K.; Sasabe, T.; Mizuno, K.; Tanaka, M.; Okawa, N.; Mito, H.; Tsubone, H.; et al. Changes in reaction time, coefficient of variance of reaction time, and autonomic nerve function in the mental fatigue state caused by long-term computerized Kraepelin test workload in healthy volunteers. *World J. Neurosci.* **2012**, *2*, 113–118. [CrossRef]
42. Kajimoto, O.; Shiraichi, Y.; Kuriyama, A.; Ohtsuka, M.; Kadowaki, T.; Sugino, T. Effects of newly developed LED lighting on living comfort in an indoor environment (Part 3)—Evaluation of indoor work performance in the room under new LED lighting. *Jpn. J. Complement. Altern. Med.* **2013**, *10*, 25–37.
43. Basner, M.; Rubinstein, J. Fitness for duty: A 3-minute version of the Psychomotor Vigilance Test predicts fatigue-related declines in luggage-screening performance. *J. Occup. Environ. Med.* **2011**, *53*, 1146–1154. [CrossRef]
44. Malliani, A.; Pagani, M.; Lombardi, F.; Cerutti, S. Cardiovascular neural regulation explored in the frequency domain. *Circulation* **1991**, *84*, 482–492. [CrossRef]
45. Tanaka, M.; Mizuno, K.; Yamaguti, K.; Kuratsune, H.; Fujii, A.; Baba, H.; Matsuda, K.; Nishimae, A.; Takesaka, T.; Watanabe, Y. Autonomic nervous alterations associated with daily level of fatigue. *Behav. Brain Funct.* **2011**, *7*, 46. [CrossRef] [PubMed]
46. Phillips, A.C.; Carroll, D.; Evans, P.; Bosch, J.A.; Clow, A.; Hucklebridge, F.; Der, G. Stressful life events are associated with low secretion rates of immunoglobulin A in saliva in the middle aged and elderly. *Brain Behav. Immun.* **2006**, *20*, 191–197. [CrossRef] [PubMed]
47. Fahlman, M.M.; Engels, H.J. Mucosal IgA and URTI in American college football players: A year longitudinal study. *Med. Sci. Sports Exerc.* **2005**, *37*, 374–380. [CrossRef] [PubMed]
48. Penailillo, L.; Maya, L.; Nino, G.; Torres, H.; Zbinden-Foncea, H. Salivary hormones and IgA in relation to physical performance in football. *J. Sports Sci.* **2015**, *33*, 2080–2087. [CrossRef] [PubMed]
49. Mizuno, K.; Tanaka, M.; Tajima, K.; Okada, N.; Rokushima, K.; Watanabe, Y. Effects of mild-stream bathing on recovery from mental fatigue. *Med. Sci. Monit.* **2010**, *16*, CR8–14.
50. Onaka, N.; Nishioka, M.; Naito, J.; Nishida, N.; Takahashi, M.; Namba, T.; Ebihara, S.; Ishibashi, K.; Ohno, N. Safety evaluation of long-term intake of *Euglena gracilis* EOD-1—A randomized, double-blind, placebo-controlled, parallel-group comparison study. *Jpn. Pharmacol. Ther.* **2020**, *48*, 665–673.

Article

Circulatory and Urinary B-Vitamin Responses to Multivitamin Supplement Ingestion Differ between Older and Younger Adults

Pankaja Sharma [1,2], Soo Min Han [1], Nicola Gillies [1,2], Eric B. Thorstensen [1], Michael Goy [1], Matthew P. G. Barnett [2,3], Nicole C. Roy [2,3,4,5], David Cameron-Smith [1,2,6] and Amber M. Milan [1,3,4,*]

1 The Liggins Institute, University of Auckland, Auckland 1023, New Zealand; p.sharma@auckland.ac.nz (P.S.); clara.han@auckland.ac.nz (S.M.H.); n.gillies@auckland.ac.nz (N.G.); e.thorstensen@auckland.ac.nz (E.B.T.); michael_goy@waters.com (M.G.); dcameron_smith@sics.a-star.edu.sg (D.C.-S.)
2 Riddet Institute, Palmerston North 4474, New Zealand; matthew.barnett@agresearch.co.nz (M.P.G.B.); nicole.roy@otago.ac.nz (N.C.R.)
3 Food & Bio-based Products Group, AgResearch, Palmerston North 4442, New Zealand
4 High-Value Nutrition National Science Challenge, Auckland 1023, New Zealand
5 Department of Human Nutrition, University of Otago, Dunedin 9016, New Zealand
6 Singapore Institute for Clinical Sciences, Agency for Science, Technology, and Research, Singapore 117609, Singapore
* Correspondence: a.milan@auckland.ac.nz; Tel.: +64-(0)9-923-4785

Received: 23 October 2020; Accepted: 13 November 2020; Published: 17 November 2020

Abstract: Multivitamin and mineral (MVM) supplements are frequently used amongst older populations to improve adequacy of micronutrients, including B-vitamins, but evidence for improved health outcomes are limited and deficiencies remain prevalent. Although this may indicate poor efficacy of supplements, this could also suggest the possibility for altered B-vitamin bioavailability and metabolism in older people. This open-label, single-arm acute parallel study, conducted at the Liggins Institute Clinical Research Unit in Auckland, compared circulatory and urinary B-vitamer responses to MVM supplementation in older (70.1 ± 2.7 y, n = 10 male, n = 10 female) compared to younger (24.2 ± 2.8 y, n = 10 male, n = 10 female) participants for 4 h after the ingestion of a single dose of a commercial MVM supplement and standardized breakfast. Older adults had a lower area under the curve (AUC) of postprandial plasma pyridoxine ($p = 0.02$) and pyridoxal-5′phosphate ($p = 0.03$) forms of vitamin B_6 but greater 4-pyridoxic acid AUC ($p = 0.009$). Urinary pyridoxine and pyridoxal excretion were higher in younger females than in older females (time × age × sex interaction, $p < 0.05$). Older adults had a greater AUC increase in plasma thiamine ($p = 0.01$), riboflavin ($p = 0.009$), and pantothenic acid ($p = 0.027$). In older adults, there was decreased plasma responsiveness of the ingested (pyridoxine) and active (pyridoxal-5′phosphate) forms of vitamin B_6, which indicated a previously undescribed alteration in either absorption or subsequent metabolic interconversion. While these findings cannot determine whether acute B_6 responsiveness is adequate, this difference may have potential implications for B_6 function in older adults. Although this may imply higher B vitamin substrate requirements for older people, further work is required to understand the implications of postprandial differences in availability.

Keywords: B-vitamins supplement; vitamin B_6; ageing; B-vitamin bioavailability; ultra-high performance liquid chromatography coupled with mass spectrometry; excretion

1. Introduction

B-vitamins are indispensable for human health given their roles in cell metabolism. These water-soluble vitamins exist in various coenzyme forms (vitamers) which are essential for normal function of enzymes linked to energy metabolism in the tricarboxylic acid (TCA) cycle [1], and in the one carbon metabolism pathway for DNA synthesis and repair, epigenetic regulation, and homocysteine regulation [2]. While the central roles of folate and vitamin B_{12} in regulating one-carbon metabolism are well described [3], the B_6-vitamer pyridoxal 5′-phosphate (PLP) is also required for the irreversible transsulfuration of homocysteine to cysteine, a precursor of the antioxidant glutathione [4], as well as being important in cellular protein metabolism [5]. Thiamine, riboflavin, niacin, pantothenic acid, and biotin are precursors of coenzymes in mitochondrial energy synthesis [1]. Hence, it is well recognized that an adequate dietary intake of B-vitamers is fundamentally important for cellular functioning and health maintenance, including cognitive function [6], mediation of systemic acute phase inflammation [7,8], and cardiovascular health [9].

Nutritional sufficiency is an increasing challenge for the growing global aging population [10,11]. The aging process can either adversely impact nutrient intake and/or impair digestive function, which collectively increases the risks for micronutrient inadequacy [12]. In addition, some water-soluble B-group vitamins inherently have limited absorption [13], are transient in the body [14,15], and are excreted in the urine [16]. Decreased dietary intake in aging is therefore widely assumed to be the major the risk of deficiency [17,18]. However, there is also evidence that increased malabsorption or impaired metabolism, well documented for vitamin B_6 and B_{12}, may elevate the risks for deficiency [13,19,20].

Although consuming a healthy varied and balanced diet is encouraged by organizations including the U.S. Department of Agriculture (USDA) [21] and the New Zealand Ministry of Health [22], many seniors regularly consume multivitamin and mineral supplements (MVM) [23] or choose vitamin fortified foods with the aim of improving health status [24]. Yet, the efficacy of MVMs in older populations remains controversial [10]. Higher doses of B-vitamins may be necessary for older adults to normalize status [25,26], which may similarly point towards an altered bioavailability of the ingested B-vitamins with advancing age. However, it remains unclear whether the postprandial availability of B-vitamins is altered with increasing age. Post-ingestion vitamin B_6 bioavailability has previously been shown not to differ with age [13]. Most acute investigations have been undertaken in younger individuals [27,28], and provide limited information regarding acute postprandial responses of the range of B-vitamin and vitamers in healthy older people. In addition to implications for MVM efficacy, any age-related postprandial alterations may have relevant implications to postprandial metabolic flux such as glucose homeostasis, one carbon metabolism [29], or cognitive function [14].

The present study examined the postprandial responses of a wide range of B vitamins and vitamers to a single MVM supplement ingested with a test meal in older compared to younger adults. We hypothesized that postprandial vitamin B_{12} and B_6 availability is less responsive in healthy older adults, Plasma and urine samples were assessed, and men and women included to identify any sex-specific effects. A profile of 14 vitamins and vitamers was achieved using a recently validated ultra-high performance liquid chromatography coupled with mass spectrometry (UHPLC-MS/MS) technique [30] complemented by immunoenzymatic methods.

2. Materials and Methods

2.1. Study Design

The study was an open-label, single-arm acute parallel trial. Participants presented to the clinic after an overnight fast to ingest a single tablet of MVM (Centrum Advance General Multi, Pfizer, New York, NY, USA) along with a standardized breakfast. Blood samples were collected at fasting and hourly for 4 h following the meal and supplement. Urine was collected at fasting and all subsequent urine produced during the 4 h.

2.2. Participants

The participants, 20 younger adults and 20 older adults, were recruited from the community through advertisements internal to the University of Auckland, in local newspapers in Auckland, New Zealand, and via social media. An equal number of males and females were recruited. Eligible participants were required to be between age ranges 19–30 years or 65–76 years, healthy, with a BMI ranging from 18–30 kg/m^2, with no major medical conditions and non-smokers. Exclusion criteria included: current consumption of multivitamin and mineral supplements (within 3 weeks); any present or recent history of gastrointestinal diseases including celiac, Crohn's, colitis; medical history of myocardial infarction, angina, stroke, or cancer; pre-existing metabolic disease or diabetes; self-reported alcohol intake > 28 units per week. Similarly, individuals with food allergies or intolerances to the intervention foods or who were taking any medications likely to impact on digestive or metabolic function, including proton pump inhibitors and thyroid medications, were excluded.

2.3. Ethics Approval

Ethics approval was obtained from The University of Auckland Human Participants Ethics Committee (UAHPEC; Reference No. 019392). The trial was registered prospectively at Australia New Zealand Clinical Trials Registry (ANZCTR) ID: ACTRN12617000969369. Written informed consent was obtained from all participants. The clinical trial was conducted between July-September 2017 at the Clinical Research Unit (CRU) based at the Liggins Institute, the University of Auckland.

2.4. Intervention: Standardized Breakfast Meal

The breakfast items were purchased from a local supermarket (Countdown, Progressive Enterprises, Auckland, New Zealand) and prepared onsite prior to the trial day. The breakfast meal consisted of two slices of white bread (74 g prior to light toasting), butter (10 g), honey (20 g), apple sauce (100 g), and orange juice (250 mL), based on a previously described meal [31]. Participants were instructed to consume all of the items provided. Unconsumed items were weighed and recorded. The amounts of each food item consumed by the participants are presented in Table 1 and the estimated intakes of B-vitamins from the supplement and the test meal are presented in Table 2. On average, the younger adults left more unconsumed portions of honey and applesauce compared to the older adults ($p = 0.02$, for both).

Table 1. Breakfast items consumed by the study participants.

Breakfast Items	Older Adults		Younger Adults	
	M ($n = 10$)	F ($n = 10$)	M ($n = 10$)	F ($n = 10$)
Toast (g)	76 ± 1.3	73.6 ± 2.4	75.9 ± 1.5	72.5 ± 0.7
Butter (g)	9.3 ± 0.4	9.5 ± 0.3	9.1 ± 0.6	8.4 ± 1.0
Honey (g) *	18.9 ± 0.6	19.2 ± 0.5	13.8 ± 2.1	17.3 ± 1.5
Applesauce (g)*	100 ± 0	100 ± 0	77.3 ± 12.1	82 ± 12.1
Orange Juice (mL)	242.1 ± 7.9	250 ± 0	250 ± 0	235 ± 15

Values are means ± standard error of the mean (SEM)s. All the food items were weighed before and after (unconsumed portions) eaten by the participants. *, significant difference in consumed portions between older and younger adults ($p < 0.05$ for the items indicated). M, males; F, females.

2.5. Outcomes

The study was powered to investigate the maximum concentration (C_{max}) of B_{12} between younger and older subjects, while also investigating the postprandial responses of vitamin B_1 (thiamine), B_2 (riboflavin), B_3 (niacin), B_5 (pantothenic acid), B_6 (pyridoxine), B_7 (biotin), folic acid and their derivatives in plasma, serum, and urine. Other secondary outcomes, not reported here, were postprandial circulating levels of plasma minerals, metabolites, and urinary minerals and metabolites.

2.5.1. Anthropometric Measurements

Anthropometric measurements, including body weight, height, waist circumference, and resting blood pressure, were recorded as an average of two measurements. Body weight was measured to the nearest 0.1 kg with light clothing and without shoes using a digital weighing scale (Wedderburn WM206, Taiwan). Height was measured to the nearest 0.1 cm without shoes using a stadiometer (Holtain Ltd., Crymych, Dyfed, UK), waist circumference was measured using a non-extensible measuring tape following an international standardized protocol [32] and resting blood pressure was measured using an automatic blood pressure monitor (Heart Sure BP100, Omron, Kyoto, Japan).

Table 2. Estimated B-vitamin intake from a single multivitamin tablet and the standard breakfast test meal by the study participants.

Vitamins	Supplement	Test Meal	Total Intake
Vitamin B_{12} (μg)	22.00	0.00	22.00
Total Folate (μg)	0.00	79.01	79.01
Folic Acid (μg)	400.00	0.00 [1]	400.00
Vitamin B_6 (mg)	6.00	0.38	6.38
Thiamine (mg)	2.18	0.43	2.61
Pantothenic acid (mg)	10.80	0.00 [1]	10.80
Riboflavin (mg)	3.20	0.16	3.36
Niacin (mg)	15.00	0.78	15.78
Biotin (μg)	45.00	0.00 [1]	45.00

B-vitamin content of multivitamin supplement tablet was obtained from nutrition information available from Centrum (Pfizer, New York, NY, USA). Foodworks software was used to extract the nutrient composition of test meals from the New Zealand food composition database. [1] Values not present in the database [33]; folic acid was not present as fortified foods were not present in the test meals.

2.5.2. Dietary and Meal Intake Analysis

Dietary B-vitamin intake was determined from self-reported 3-day food dairy records. Food records and test meals were analyzed by a New Zealand Registered Dietitian using Foodworks 8 Professional (Xyris Software PTY Ltd., Brisbane, Australia) using the New Zealand Food Composition Database (FOODfiles™ 2016 Version 1).

2.5.3. Biochemical Measures

Serum glucose and triglyceride, fasting cholesterol, low density lipoprotein cholesterol (LDL-C), high density lipoprotein cholesterol (HDL-C), and urinary creatinine measurements were performed using a Roche Cobas c311 clinical chemistry autoanalyzer (Roche Diagnostics GmbH, D-68298 Mannheim, Germany) based on enzymatic colorimetric assays. Serum insulin was measured using a Roche Cobas e411 autoanalyzer (Roche Diagnostics, Mannheim, Germany) utilizing an electrochemiluminescence immunoassay.

2.5.4. Analysis of B-Vitamins

All blood samples were collected in light-protected ethylenediaminetetraacetic acid (EDTA) vacutainers for plasma and additive-free vacutainers for serum. Serum tubes were left to clot at room temperature for 15 min and both plasma and serum tubes were centrifuged at 1500× *g* for 15 min at 4 °C. Aliquots were stored in amber vials at −80 °C until analysis.

Vitamin B_{12} and folate were measured in serum samples using the Roche Cobas e411 autoanalyzer. These assay methods were not optimized for analysis in urine samples, nor are these vitamin forms standardly measured in urine [34], so they were only applied to serum samples.

All other B-vitamins and vitamers were measured using UHPLC-MS/MS with slight modifications to the previously published method [30]. This included thiamine, riboflavin and its vitamer flavin mononucleotide (FMN), nicotinic acid (niacin) and its vitamers (nicotinamide and nicotinuric acid), pantothenic acid, vitamin B_6 vitamers (pyridoxine, pyridoxal, pyridoxamine, PLP, and 4-pyridoxic acid (4-PA) in plasma), and urine. Biotin and folic acid concentrations were also measured in urine samples. The technical details of plasma samples preparation and UHPLC-MS/MS analysis have been reported elsewhere [35]. In brief, an automated liquid handling robotic system Eppendorf epMotion® 5075 automated liquid handling system (Eppendorf, AG, Hamburg, Germany) was used for plasma sample preparation. Plasma protein was precipitated using 400 μL of methanol containing 0.3% acetic acid and 2.5% H_2O, pipetted into a 2 mL square 96-well Impact™ Protein Precipitation filter plate (Phenomenex, Torrance, CA, USA). As certified reference materials were not available for this assay, matrix-matched quality controls of pooled human plasma spiked with known concentrations of the vitamins and vitamers of interest were used as quality controls (QCs) [30]. All samples, QCs, and standards were spiked with 10 μL of internal standard mix, mixed thoroughly, and filtered into a 96-well square collection plate by applying vacuum pressure using the robot. This was followed by solvent evaporation and reconstitution of the concentrated samples with 200 μl reconstitution solvent made up of water containing 5% acetic acid, 0.2% heptafluorobutyric acid, and 1% ascorbic acid, then placed into the autosampler of the UHPLC-MS/MS for injection. Urine samples were diluted 10-fold with the reconstitution solvent prior to centrifugation and injection to capture the measures within the calibration range. The urinary concentrations of B-vitamins were reported per urinary creatinine concentration to account for differences in urine output.

2.6. Sample Size, Data Interpretation, and Statistical Methods

There are limited data reporting acute vitamin bioavailability across a range of B-vitamins and vitamers, as previous studies have primarily focused on vitamin B_{12}, vitamin B_6, and folate [14], yet expected ranges for vitamers are scarce. Further, few comparisons between younger and older subjects exist. Sample size calculations were therefore based on published data of supplemental B_{12} bioavailability; an expected C_{max} after ingestion is 549 ± 128 pg/mL [27]. A sample size of 20 subjects per group was estimated for a between group difference of 20% to identify significant differences ($\alpha = 0.05$ and $\beta = 0.8$) in the proposed primary outcome measure (vitamin B_{12}).

Data are presented as means ± SEM. The incremental area under the curve (AUC) was obtained after correction for baseline concentrations. Homeostatic model assessment of insulin resistance (HOMA-IR) was calculated from fasting glucose and insulin concentrations using the equation from Matthews et al. [36] Statistical significance was tested using SPSS (IBM SPSS Statistics 25). To determine the effects of time, a general linear mixed model was used with time as repeated measures and group (age) and sex as fixed factors. All other age and sex comparisons were done using the general linear model. In case of an interaction, post hoc tests were performed using Sidak–Holm adjustments for pairwise comparisons. The alpha was set at 0.05.

3. Results

3.1. Baseline Characteristics

A total of 40 participants completed the study; $n = 20$ older, $n = 20$ younger (Figure 1 and Table 3). The average age (mean ± SEM) of younger and older adults was 24.2 ± 0.6 and 70.1 ± 2.5 years, respectively. Older subjects had higher systolic and diastolic blood pressure, fasting glucose, LDL-C, total cholesterol, and triglycerides ($p < 0.01$), compared to the younger group. However, HDL-C, insulin, and HOMA-IR did not differ by age.

3.2. Estimated B-Vitamin Intake from 3-Day Food Record

Generally, the mean vitamin intake of participants met the recommended dietary intake (RDI); the exceptions were dietary folate equivalents (DFEs) and thiamin in younger females. Older individuals had lower habitual vitamin B_{12} intake than younger adults (main age effect, $p = 0.003$; Table 4). Differences in intake for many B-vitamins between the older and younger adults were sex-dependent.

Total folate, vitamin B_6, riboflavin, and niacin intakes were higher in younger males (age × sex interaction, $p < 0.05$) compared to all other groups. Regardless of age, males had greater thiamine intake than females (main sex effect, $p = 0.009$).

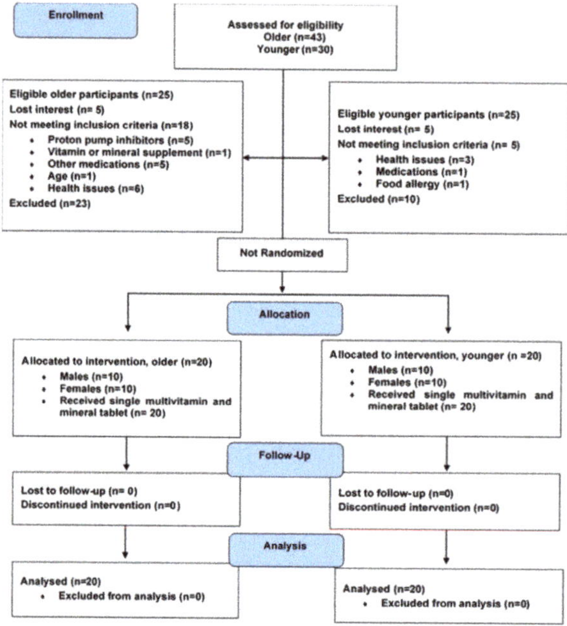

Figure 1. Consort flow diagram of study participant recruitment, intervention, follow-up, and analysis.

Table 3. Baseline characteristics and biochemistry of the study participants at fasting.

Variable	Older Adults (n = 20)		Younger Adults (n = 20)		Effect		
	M (n = 10)	F (n = 10)	M (n = 10)	F (n = 10)	Age	Sex	Age × Sex
Age (years)	71.1 ± 0.9	69.1 ± 0.7	23.3 ± 1.1	25.0 ± 0.5	<0.001 *	0.857	0.031 *
Weight (kg)	80.4 ± 4.1	69.4 ± 3.3	77.9 ± 3.6	62.7 ± 2.4	0.185	<0.001 *	0.539
Height (cm)	176.0 ± 2.3	165.4 ± 1.6	174.4 ± 1.6	165.3 ± 2.0	0.653	<0.001 *	0.696
BMI (kg/m^2)	25.9 ± 1.0	25.3 ± 1.0	25.6 ± 1.2	23.0 ± 0.9	0.226	0.135	0.321
Waist circumference (cm)	90.5 ± 3.5	86.3 ± 3.1	83.1 ± 3.0	75.3 ± 2.0	0.004 *	0.051	0.545
Systolic BP (mmHg)	148 ± 4	128 ± 4	121 ± 3	110 ± 2	<0.001 *	<0.001 *	0.189
Diastolic BP (mmHg)	79 ± 3	73 ± 3	67 ± 1	71 ± 2	0.004 *	0.723	0.059
HOMA-IR	1.31 ± 0.32	1.47 ± 0.22	1.97 ± 0.41	1.38 ± 0.28	0.358	0.490	0.234
Fasting Serum Measures							
Glucose (mmol/L)	5.27 ± 0.18	4.74 ± 0.09	4.78 ± 0.15	4.49 ± 0.13	0.014 *	0.007 *	0.405
Insulin (μU/mL)	8.24 ± 1.65	6.51 ± 1.38	6.24 ± 1.46	7.39 ± 1.05	0.690	0.838	0.310
Cholesterol (mmol/L)	5.75 ± 0.39	6.61 ± 0.42	4.70 ± 0.33	4.57 ±0.27	<0.001 *	0.318	0.172
HDL-C (mmol/L)	1.40 ± 0.18	1.85 ± 0.09	1.47 ± 0.15	1.77 ± 0.14	0.960	0.006 *	0.559
LDL-C (mmol/L)	3.85 ± 0.39	4.40 ± 0.39	2.84 ± 0.28	2.63 ± 0.26	<0.001 *	0.624	0.263
Triglyceride (mmol/L)	1.51± 0.19	1.31 ± 0.21	1.06 ± 0.15	0.79 ± 0.07	0.005 *	0.156	0.803

Values are means ± SEMs. Data was compared using general linear multivariate analysis of variance with age and sex as fixed factors. * Significant main effects or interactions, $p < 0.05$. BMI, body mass index; BP, blood pressure; HOMA-IR, homeostatic model assessment of insulin resistance; HDL-C, high density lipoprotein cholesterol; LDL-C, low density lipoprotein cholesterol; M, males; F, females.

Table 4. Estimated habitual B-vitamin intake of the study participants from 3-day dietary intake records.

B-Vitamins	RDI^{o}	Older Adults		RDI^{y}	Younger Adults		Effect		
	M, F	M (n = 10)	F (n = 10)	M, F	M (n = 9) [1]	F (n = 9) [1]	Age	Sex	Age × Sex
Vitamin B_{12} (μg)	2.4	2.95 ± 0.4	3.12 ± 0.2	2.4	6.34 ± 1.1	4.31 ± 1.0	0.003 *	0.209	0.140
Total Folate (μg)		290.57 ± 62.5	377.45 ± 51.0		443.96 ± 105.8	168.40 ± 36.8	0.685	0.174	0.012 *
Folate, total DFE (μg)	400	437.81 ± 91.8	448.93 ± 83.3	400	465.11 ± 116.2	178.04 ± 37.8	0.172	0.123	0.097
Vitamin B_6 (mg)	1.7, 1.5	2.33 ± 0.3	2.69 ± 0.3	1.3	3.83 ± 0.6	1.38 ± 0.2	0.782	0.005 *	<0.001 *
Thiamin (mg)	1.2, 1.1	1.60 ± 0.2	1.37 ± 0.1	1.2, 1.1	1.74 ± 0.4	0.77 ± 0.1	0.3	0.009 *	0.098
Riboflavin (mg)	1.3, 1.1 [2]	1.88 ± 0.2	1.99 ± 0.2	1.3, 1.1	2.48 ± 0.4	1.20 ± 0.2	0.74	0.036 *	0.013 *
Niacin (mg)		19.35 ± 2.4	18.30 ± 1.5		32.66 ± 4.3	12.48 ± 2.2	0.18	<0.001 *	0.001 *
Niacin E (mg)	16, 14	36.88 ± 3.1	33.59 ± 2.2	16, 14	63.59 ± 7.3	24.84 ± 3.4	0.044 *	<0.001 *	<0.001 *

Foodworks software was used to extract the nutrient composition from the New Zealand food composition database. Data was compared using general linear multivariate analysis of variance with age and sex as fixed factors. Sidak post hoc test was applied if significant age × sex interaction was present. * Significant main effects or interactions, $p < 0.05$ DFE, dietary folate equivalent; Niacin E, niacin equivalent; RDI, recommended dietary intake; o, older; y, younger; M, males; F, females. [1] Dietary intake records for one younger male and one female could not be obtained. [2] RDI presented for 51–70 years. RDI for >70 years is 1.6 and 1.3 mg for males and females, respectively.

3.3. Urine Output and Creatinine Concentration

Urine volume, both at fasting and postprandially (accumulated over 1–4 h) during the trial period, was not different between any age or sex groups ($p > 0.05$, Supplemental Table S1). However, fasted urinary creatinine concentrations were greater in younger than older adults (main age effect, $p = 0.005$), and the postprandial concentration was greater in males compared to females (main sex effect, $p = 0.012$).

3.4. Fasting Circulating Vitamin Status and Urinary Concentrations

Fasting serum B_{12} concentrations were different between younger and older adults, but this difference was sex-specific (Table S2), with lower concentrations in older males (age × sex interaction, $p = 0.005$) than younger males ($p = 0.002$) or older females ($p = 0.009$). Older compared to younger individuals had higher serum folate (16.2 ± 1.7 vs. 9.5 ± 0.8 mg/mL) and plasma thiamine (1.7 ± 0.3 vs. 0.6 ± 0.1 nmol/L) concentrations at baseline (main age effect, $p = 0.001$ and $p < 0.001$, respectively). Fasting plasma pantothenic acid concentration was higher in older females (age × sex interaction, $p = 0.014$) than younger females ($p = 0.004$) and older males ($p = 0.009$). Fasting plasma concentration of nicotinuric acid, the excretory metabolite of vitamin B_3, was lower in older females (age × sex interaction, $p = 0.016$) compared to younger females ($p = 0.014$) but was not different significantly between old and young males or between sexes. Fasting urinary riboflavin concentration was higher in older compared to younger adults (337.8 ± 41.5 vs. 188.7 ± 39.8 nmol/mmol creatinine; main age effect, $p = 0.015$) (Table S2), whereas the riboflavin-vitamer FMN was not different. Similarly, older adults had greater urinary nicotinic acid compared to younger adults (82.7 ± 10.1 vs. 47.1 ± 6.1 nmol/mmol creatinine; main age effect, $p = 0.003$), whereas nicotinamide and nicotinuric acid concentrations differed between age groups depending on participant sex (age × sex interactions $p < 0.05$, Table S2). Younger females excreted more nicotinuric acid than older females ($p = 0.007$) and more nicotinuric acid and nicotinamide than younger males ($p = 0.004$ and $p = 0.013$, respectively). No other measured vitamins or vitamers differed between older and younger adults, but several sex differences were observed at fasting (Table S2).

3.5. Acute Postprandial Response of B-Vitamins and Vitamers Following MVM Supplement Ingestion

Serum vitamin B_{12} (Figure 2A) was responsive to single supplement ingestion, as were almost all B-vitamins and vitamers, demonstrating increased postprandial concentrations. These included serum folate (Figure 2B); plasma and urinary B_6-vitamers pyridoxine, pyridoxal, pyridoxamine, PLP and 4-PA (Figures 3 and 4); plasma thiamine, riboflavin, FMN and pantothenic acid (Figure 5), urinary biotin and folic acid (Figure 6), plasma and urinary nicotinamide nicotinuric acid and urinary nicotinic acid (Figures 7 and 8). Only postprandial plasma nicotinic acid concentration was unresponsive to supplement ingestion (Figure 7A).

3.6. Impact of Age on Acute Circulating Vitamin B_{12} and Folate Response to MVM Supplement Ingestion

Serum vitamin B_{12} concentrations were increased at 1 h following MVM ingestion (main time effect, $p < 0.001$). The older adults tended to have lower serum vitamin B_{12} (main age effect, $p = 0.053$) with a C_{max} of 493.88 ± 176.07 ng/mL compared to 601.26 ± 164.43 ng/mL in the younger adults. However, the postprandial serum vitamin B_{12} concentrations and overall appearance (AUC) were not different between older and younger subjects (Figure 2A). Sex specific difference between age groups were present (described below with sex-specific results).

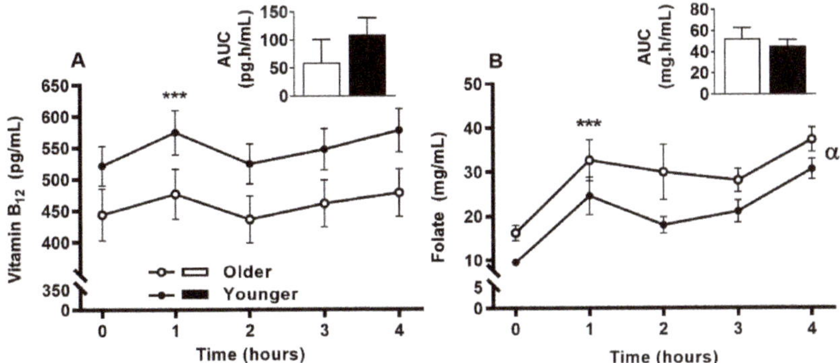

Figure 2. Serum (**A**) vitamin B_{12} and (**B**) folate at fasting (time, 0), hourly until 4 h (time, 1, 2, 3, 4) and the four hour incremental area under the curve (AUC, represented by bar-graphs) following a single multivitamin and mineral supplement ingestion in older and younger adults. Data for multiple time points was compared using general linear mixed model with time as repeated and age as fixed factor. AUC data was compared with general linear univariate analysis of variance with age as fixed factor. There was a main time effect for serum vitamin B_{12} concentration ($p < 0.001$) and main time ($p < 0.001$) and main age effect ($p = 0.022$) for serum folate. Both vitamin B_{12} and folate AUC were not different between age groups. ***, $p < 0.001$, significant increase following supplement intake relative to baseline in both age groups; α, $p < 0.05$, main age effect.

3.7. Impact of Age on Acute Plasma and Urinary B_6-Vitamer Responses to MVM Supplement Ingestion

All the B_6-vitamers, pyridoxine, pyridoxal, pyridoxamine, PLP and 4-PA concentrations measured in both plasma (Figure 3A–E) and urine (Figure 4A–E) samples were responsive to MVM supplement ingestion. The older adults had a smaller postprandial AUC for both plasma pyridoxine and PLP concentrations than the younger group ($p < 0.05$; bar graphs, Figure 3A,D), and a larger AUC for the circulating excretory metabolite 4-PA (bar graph, Figure 3E). While postprandial plasma pyridoxine concentrations were not significantly different between older and younger individuals at any specific time point (Figure 3A, time x age interaction $p > 0.05$), younger subjects had higher postprandial pyridoxine concentration (age effect $p = 0.035$). Unlike younger adults, postprandial plasma PLP concentrations did not increase in the older group (Figure 3D, time x age interaction; $p = 0.026$; $p > 0.05$, older group baseline to postprandial time points), and remained lower than the younger group throughout the postprandial period ($p < 0.05$). In contrast plasma 4-PA concentration was higher in older than younger subjects from 3 h (Figure 3E, age x time, $p = 0.002$; $p < 0.01$). Only postprandial urinary concentrations of pyridoxine (time x age interaction $p = 0.037$; Figure 4) was higher in younger compared to older adults post-supplement ingestion. Although urinary concentrations of the other B_6-vitamers (Figure 4) were not different between overall age groups, there were sex specific differences between age groups in pyridoxine and pyridoxal concentrations (detailed below in sex specific section).

3.8. Impact of Age on Acute Postprandial Plasma and Urinary Thiamine, Pantothenic Acid, Riboflavin, FMN, and Urinary Biotin Responses to Single MVM Supplement Ingestion

Postprandial plasma thiamine, pantothenic acid, and riboflavin concentrations were highly responsive to supplementation, while the riboflavin vitamer FMN showed a slower response (Figure 5). Interestingly, older adults had greater postprandial thiamine, pantothenic acid and riboflavin plasma concentrations than younger adults indicated by AUC (main age effect, $p < 0.05$), whereas FMN AUC did not differ between age groups (bar graphs, Figure 5A–D). Although thiamine (Figure 5A), pantothenic acid (Figure 5B), and riboflavin (Figure 5C) plasma concentrations were not significantly different between age groups at any specific time points (age x time interaction, $p > 0.05$), these vitamins were higher in the older group (age effect, $p < 0.05$). In contrast to riboflavin, plasma FMN (Figure 5D) showed a slow baseline to postprandial increase at 4 h (time effect, $p < 0.001$) and remained lower in older than younger adults (age effect, $p = 0.001$). However, postprandial urinary concentrations of these vitamins (Figure 6A–D) including folic acid (Figure 6E) and biotin (Figure 6F) were not different between any age groups.

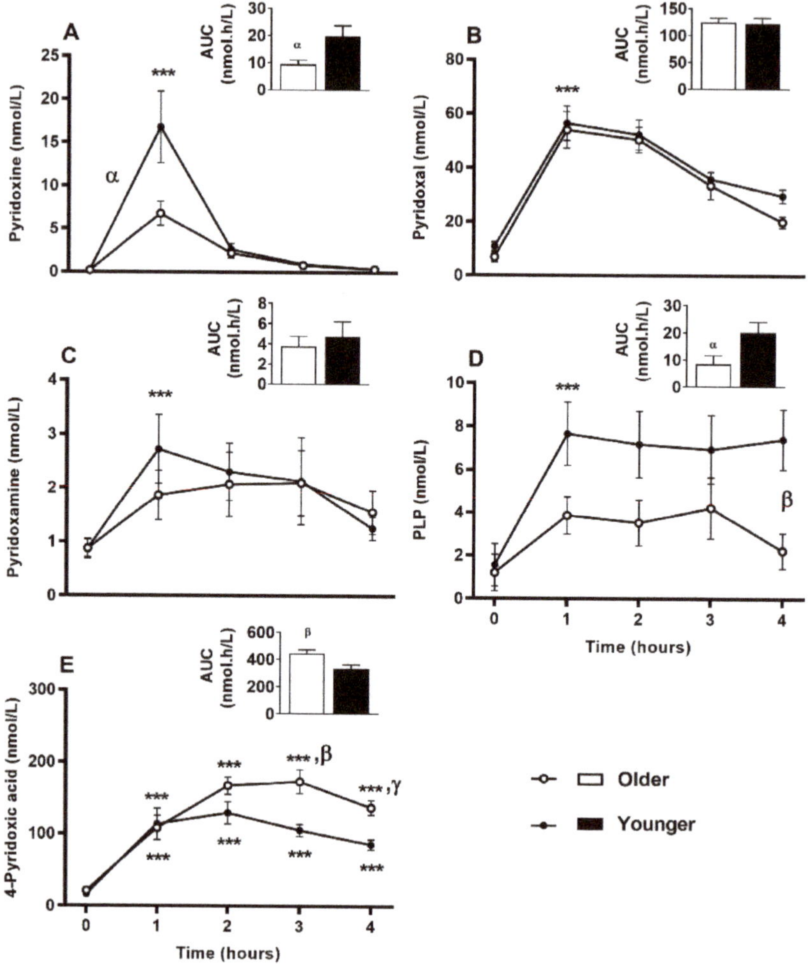

Figure 3. Plasma response of B_6 vitamers at fasting (time, 0), hourly until 4 h (time, 1, 2, 3, 4) and the four hour incremental area under the curve (AUC, represented by bar-graphs) following a single multivitamin and mineral supplement ingestion in older and younger adults. (**A**), Pyridoxine (**B**), pyridoxal (**C**), pyridoxamine (**D**), pyridoxal 5′-phosphate (PLP), and (**E**) 4-pyridoxic acid. Data for multiple time points was compared using general linear mixed model with time as repeated and age as fixed factor. AUC data was compared with general linear univariate analysis of variance with age as fixed factor. Sidak post hoc test was applied for pairwise comparisons if significant interaction was present. ***, $p < 0.001$, significant increase following supplement intake relative to baseline in both age groups. The older adults had decreased postprandial increase in pyridoxine (age effect $p < 0.035$, time effect $p < 0,001$) and PLP (age effect $p = 0.009$, time effect $p < 0.001$) but increased 4-pyridoxic acid (age × time, $p = 0.002$) than the younger adults (α, $p < 0.05$; β, $p < 0.01$; γ, $p < 0.001$, main age difference or difference with age × time interaction at particular time points).

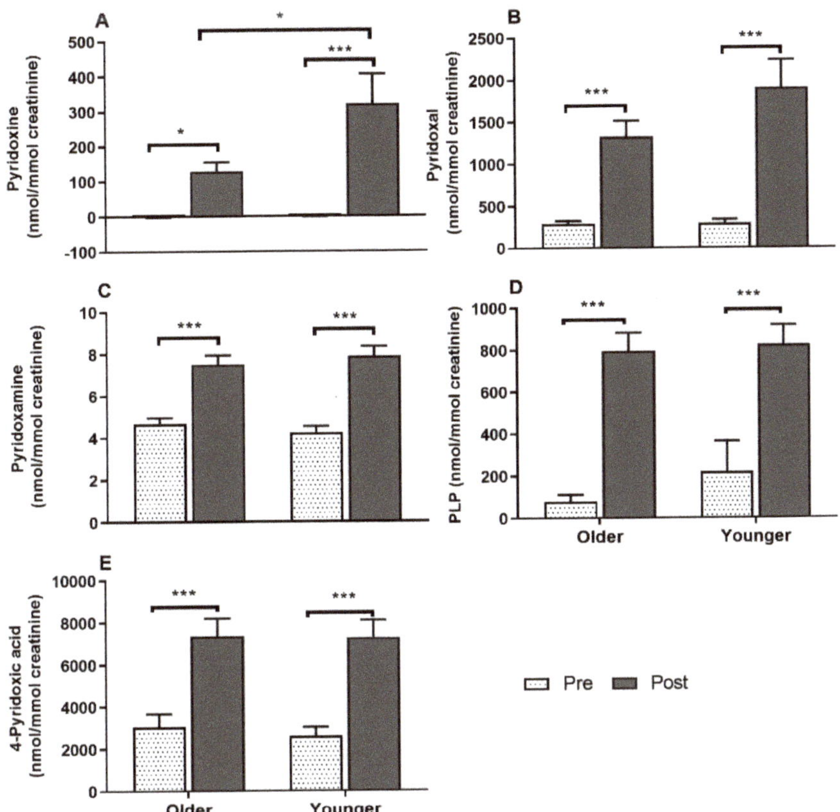

Figure 4. Urinary concentrations of the vitamers of vitamin B_6, (**A**), pyridoxine (**B**), pyridoxal (**C**), pyridoxamine (**D**) pyridoxal 5′-phosphate (PLP), and (**E**) 4-pyridoxic acid at baseline (pre) and after (post) supplement ingestion in older and younger adults. Bar plots and error bars represent means and ± SEM, respectively, $n = 20$ in each age groups and concentrations are normalized for urinary creatinine (nmol/mmol creatinine). Data for pre and post supplement ingestion was compared using generalized linear repeated measures analysis of variance with time as within subject and age as between subject factor. Sidak post hoc test was applied for pairwise comparisons in case of an interaction. All the urinary vitamers increased following supplement ingestion compared to baseline (main time effect, *, $p < 0.05$; *** $p < 0.001$). A time x age interaction ($p = 0.037$) showed a greater urinary concentration of pyridoxine in younger compared to older adults post-supplement ingestion. There were no significant differences in urinary excretion of other vitamers between older and younger adults.

3.9. Impact of Age on Acute Plasma and Urinary B_3-Vitamer Responses to Single MVM Supplement Ingestion

There was no significant difference between the older and younger subjects in plasma AUC for any of the B_3-vitamers (nicotinic acid, nicotinamide, and nicotinuric acid, bar graphs, Figure 7A–C). Although none of the B_3-vitamers differed between age groups at any specific time points, older adults had lower plasma nicotinamide concentrations (age effect, $p = 0.038$, 7B), but no difference in nicotinic acid (age effect, $p = 0.052$, 7A) or nicotinuric acid concentrations (age effect, $p = 0.326$, 7C). The postprandial urinary concentration of B_3-vitamers were not different between overall age groups (Figure 8A–C); however, nicotinuric acid differed between older and younger females (detailed below in sex specific section).

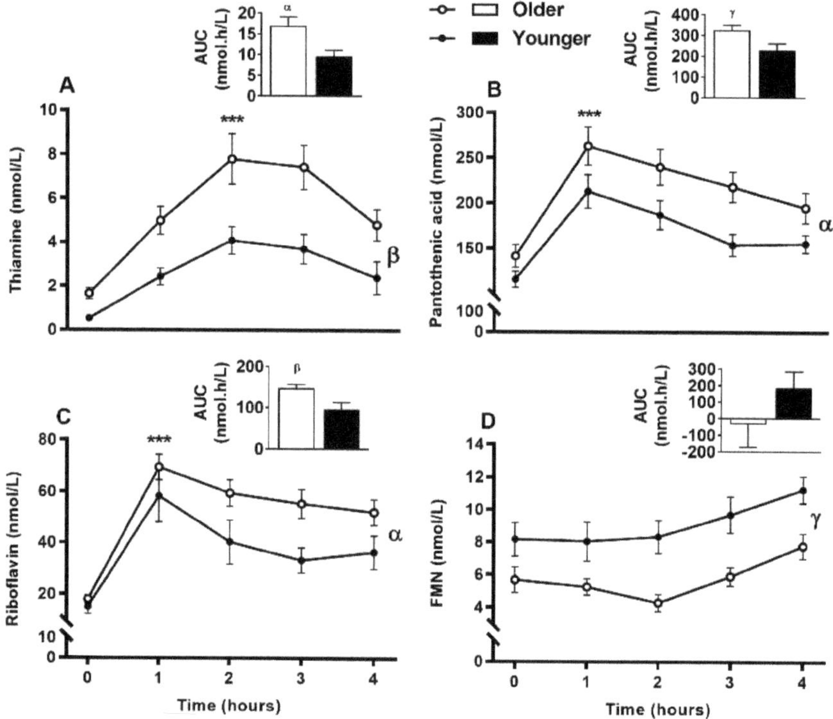

Figure 5. Plasma response of (**A**), thiamine (**B**), pantothenic acid (**C**), riboflavin, and its vitamer (**D**), flavin mononucleotide (FMN) at fasting (time, 0), hourly until four hours (time, 1, 2, 3, 4) and the four hour incremental area under the curve (AUC, represented by bar-graphs) following a single multivitamin and mineral supplement ingestion in older and younger adults. Data for multiple time points was compared using general linear mixed model with time as repeated and age as fixed factors. AUC data was compared with general linear univariate analysis of variance with age as fixed factor. ***, $p < 0.001$, significant increase following supplement intake relative to baseline in both age groups. The older adults had increased postprandial thiamine (age effect $p = 0.001$, time effect $p < 0{,}001$), pantothenic acid (age effect $p = 0.008$, time effect $p < 0.001$) and increased riboflavin (age effect $p = 0.026$, time effect $p < 0.001$) than the younger adults (α, $p < 0.05$; β, $p < 0.01$; γ, $p < 0.001$, main age difference).

3.10. Sex-Specific Effects on Acute Circulating and Urinary B-Vitamin and Vitamer Responses to Supplement Ingestion

Several postprandial circulating B-vitamin responses demonstrated sex-specific differences within or between age groups, including vitamin B_{12}, pyridoxal, riboflavin, thiamine, and nicotinuric acid (Figure S1). Postprandial serum vitamin B_{12} concentrations differed between age groups among the males and between sexes among the older group, with lower increases in older males (age × sex interaction; $p = 0.007$) than younger males ($p = 0.002$) and older females ($p = 0.019$).

Similarly, postprandial plasma B_6-vitamers pyridoxal (Figure S1) and pyridoxamine, although not different between overall age groups, showed sex-specific age effects. Older females had lower postprandial plasma pyridoxal concentrations than younger females ($p = 0.032$; age × sex interaction, $p = 0.043$) although AUC was not different. Whereas no sex-specific differences were found in plasma pyridoxine, PLP and 4-PA concentrations were found across the age groups. Postprandial urinary excretion of pyridoxal and pyridoxine in younger females was also higher compared to older females and younger males (Figure S2; age x time x sex interactions $p = 0.013$ and $p = 0.005$, respectively; age comparison $p = 0.001$ and $p = 0.002$ and sex comparisons $p < 0.001$ each, respectively). Females, regardless of age, excreted more pyridoxamine, PLP and 4-PA compared to males (time x sex interaction $p < 0.05$ each, respectively; sex comparison $p < 0.01$ each, respectively).

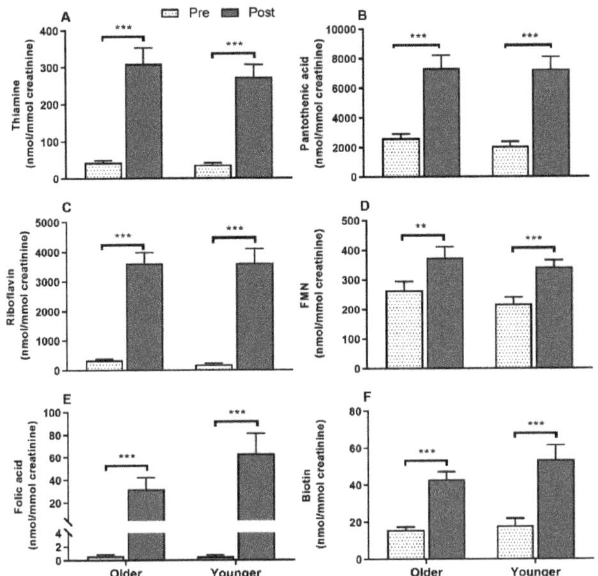

Figure 6. Urinary concentrations of (**A**) thiamine, (**B**), pantothenic acid (**C**), riboflavin, the vitamer (**D**), flavin mononucleotide (FMN), (**E**) folic acid, and (**F**), biotin following supplement ingestion in older and younger adults. Bar plots and error bars represent means and ± SEM respectively, $n = 20$ in each age groups and concentrations were normalized for urinary creatinine (nmol/mmol creatinine). Data for pre and post supplement ingestion was compared using generalized linear repeated measures analysis of variance with time as within subject and age as between subject factor. Sidak post hoc test was applied for pairwise comparisons in case of an interaction. There were no interactions between any independent factors (age and time). There were main time effects (**, $p < 0.01$; ***, $p < 0.001$) for all the vitamins.

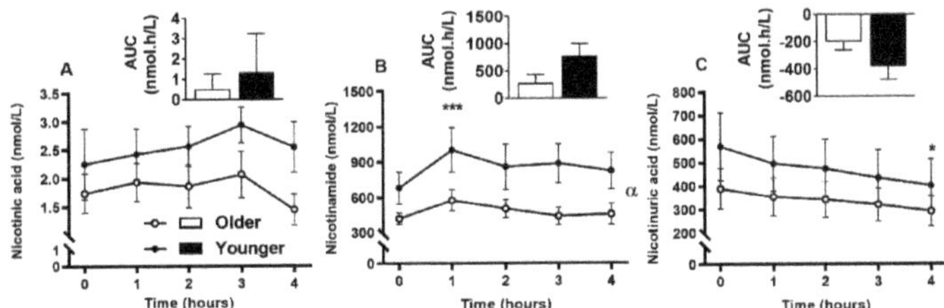

Figure 7. Plasma response of B_3 vitamers, (**A**), nicotinic acid (**B**), nicotinamide, and (**C**), nicotinuric acid at baseline (fasting, time, 0), hourly until four hours (time, 1, 2, 3, 4) and the four hour incremental area under the curve (AUC, represented by bar-graphs) following a single multivitamin and mineral supplement ingestion in older and younger adults. Data for multiple time points was compared using general linear mixed model with time as repeated and age as fixed factors. AUC data was compared with general linear univariate analysis of variance with age as fixed factor. *, $p < 0.05$; ***, $p < 0.001$, significant change following supplement intake relative to baseline in both age groups. No significant difference in postprandial change between the age groups. Nicotinic acid concentration did not increase compared to baseline (time effect, $p = 0.370$), whereas nicotinamide increased (time effect, $p < 0.001$) with lower concentration in older than the younger adults (age effect α, $p = 0.038$) and nicotinuric acid decreased continuously throughout the 4-h postprandial period (time effect, $p = 0.002$).

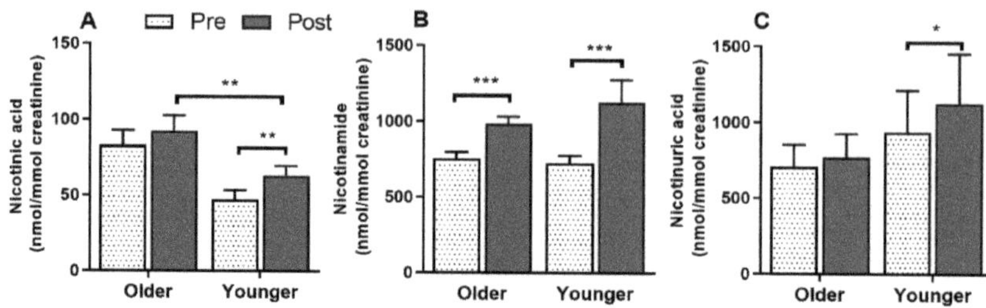

Figure 8. Urinary concentrations of the vitamers of vitamin B_3, (**A**), nicotinic acid, (**B**), nicotinamide, and (**C**) nicotinuric acid at baseline (pre) and after (post) supplement ingestion in older and younger adults. Bar plots and error bars represent means and ± SEM respectively, $n = 20$ in each age groups and concentrations are normalized for urinary creatinine (nmol/mmol creatinine). Data for pre and post supplement ingestion was compared using generalized linear repeated measures analysis of variance with time as within subject and age as between subject factor. Sidak post hoc test was applied for pairwise comparisons in case of an interaction. There were no interactions between any independent factors (age and time). There were main time effects (*, $p < 0.05$; **, $p < 0.01$; ***, $p < 0.001$) for all the vitamers and a main age effect for nicotinic acid (**, $p = 0.007$).

Sex-specific differences were also seen for postprandial plasma (Figure S1) and urinary (Figure S2) riboflavin and thiamine responses and urinary pantothenic acid, folic acid and biotin (the latter two were measurable in urine but poorly detected in plasma). Younger males had lower plasma riboflavin responses (time × age × sex interaction, $p = 0.033$) than younger females while the plasma 4-h riboflavin AUC was, regardless of age, lower in males than females (sex effect, $p = 0.033$). Similarly, younger males had a lower plasma thiamine response (time × age × sex interaction $p = 0.009$) than older males ($p = 0.012$) at 1 and 2 h following the supplement; at 3 h, a similar age difference was found in females ($p = 0.006$). Interestingly, the younger males showed no significant increase in postprandial plasma thiamine concentrations at any timepoints relative to baseline ($p > 0.05$), whereas in all the other age and sex groups significant increases occurred at all 4 timepoints ($p < 0.05$). However, there was no impact of sex on the overall plasma thiamine AUC. Postprandial urinary concentrations of riboflavin, thiamine, pantothenic acid (Figure S2), folic acid, and biotin were greater in females (time x sex interaction $p < 0.05$) compared to males ($p < 0.05$). In urine, the riboflavin-vitamer FMN concentration was greater in females compared to males (sex effect $p = 0.007$).

Among the plasma B_3-vitamers, a sex-specific effect was present only for postprandial nicotinuric acid, with lower concentrations in older females (age × sex interaction $p = 0.015$) than younger females ($p = 0.017$). However, the 4-h nicotinuric acid AUC was still lower in younger females (age × sex interaction, $p = 0.041$) compared to both younger males ($p = 0.025$) and older females ($p = 0.011$). The postprandial urinary excretion of both nicotinamide and nicotinuric acid were sex-dependent. Younger females had greater urinary nicotinamide excretion compared to younger males ($p = 0.009$) but not significantly different to older females ($p = 0.060$), whereas nicotinuric acid was greater in younger females compared to both younger males ($p = 0.003$) and older females ($p = 0.005$).

4. Discussion

Despite the greater risk for B-vitamin deficiency in older populations, the impact of aging on the postprandial response and dynamic regulation of individual vitamers is not well understood. The current study therefore investigated the postprandial response of a range of B-vitamins and their vitamers in older and younger subjects following a single MVM supplement ingestion in order to better understand the changes in absorption and metabolism of B-vitamins associated with aging. This study demonstrated that older men had a transiently different vitamin B_{12} response, with suppressed supplement-mediated increases compared to older females and younger males and females, but this did not impact the AUC over the 4 h studied. Older adults had lower postprandial plasma concentrations of the B_6-vitamers pyridoxine and PLP, whereas they had higher 4-pyridoxic acid, thiamine, pantothenic acid, and riboflavin concentrations. While lower urinary pyridoxine excretion in the older adults matches circulating response, none of the other urinary metabolites corresponded with the plasma response. The decreased availability of ingested and active forms of B_6-vitamers in older subjects suggests possible alterations in the absorption or subsequent metabolic interconversion of vitamin B_6 in older people.

Serum vitamin B_{12} concentrations were lower in older males compared to both younger and older females both at fasting and postprandially. This is consistent with our hypothesis that acute postprandial vitamin B_{12} bioavailability is altered with increasing age. Age and sex-dependent variations in fasting vitamin B_{12} status have been previously reported, with lower vitamin B_{12} status reported in older adults [31,37], and in males [38], which is consistent with the current findings. Lower B_{12} has been attributed to genetic variations, as no significant associations with dietary habits or hormones could be established [38]. However, capturing the genetic variation was beyond the scope of the present study and not assessed in our population. Malabsorption of vitamin B_{12}, due to impaired gastrointestinal acid secretion is frequently reported in older populations [20,39] even when nutritional status is adequate [37,40]. However, although our fasting results suggest greater risk of B_{12} deficiency in older males, the total appearance (AUC) of postprandial vitamin B_{12} was not different between age groups. Hence, despite differences in fasting B_{12} status in older men, the current findings do not indicate that total acute appearance of B_{12} following a single MVM in these subjects is a contributing factor to low B_{12} status.

Of the B-vitamins and vitamers measured, the B_6 vitamers responded most differently in older adults, suggesting alterations in postprandial B_6 metabolism. Older adults showed a blunted increase in postprandial plasma pyridoxine and PLP; PLP is integral to the regulation of one-carbon metabolism and other diverse enzymatic reactions [2]. In contrast, the concentration of the metabolized end-product, 4-PA, which does not contribute to the co-enzyme functions of B_6, was higher in plasma, but not urine in these subjects. The lower pyridoxine response in older adults could implicate either altered intestinal or hepatic metabolic conversion to the active coenzyme form PLP [41]; however, the absence of an age difference in postprandial plasma pyridoxal concentrations does not support the idea of inadequate enzymatic pyridoxine conversion. As urinary excretion of pyridoxine was lower in older adults, malabsorption, rather than greater excretion of absorbed pyridoxine, may be one possible explanation. Previous studies have found that factors such as inflammation and high protein intake alter circulating vitamin B_6 distribution [42,43], such that PLP concentrations decrease during both acute and chronic adjuvant-arthritic inflammation [42] while increasing during high protein intake [43]. However, the enzymatic conversion of pyridoxine has not been previously reported with respect to the aging process outside of age-related altered protein metabolism [44]. Although low PLP has been implicated in systemic inflammatory responses including cardiovascular [45] and chronic kidney diseases [46], available evidence describes long term rather than postprandial implications. While in practice, these findings may suggest that older people have higher B_6 substrate requirements, this would require further confirmation to understand the relationship between postprandial differences in availability and long-term requirements. Indeed, while acute responsiveness of the B vitamins studied here could be interpreted positively in the context of absorption or metabolic flexibility, it is similarly important to note that without an understanding of appropriate postprandial targets for these compounds, it is difficult to interpret whether age-related differences observed have any relevant physiological consequences.

Age-related enzymatic conversion and excretion differences may explain both a blunted pyridoxine and PLP response, and greater circulating 4-PA concentrations in the older subjects. Aldehyde oxidase 1 (AOX 1) facilitates this conversion [47]; as older adults had higher 4-PA in plasma post-ingestion, but not at baseline or in urinary excretion, it is possible that activity of AOX1 differed between older and younger subjects. While there are limited studies reporting age differences in B_6 enzyme activity, including AOX1 [48–50], these findings align with several reports of decreased plasma PLP and increased 4-PA concentrations with advancing age [48,50], yet contradict another postprandial supplement study in men which found no age difference indices of vitamin B_6 metabolism [19]. Equally conflicting, evidence from animal models has suggested vitamin B_6 metabolizing enzymes are unaltered by age, despite decreased PLP concentrations [49], but others have demonstrated an age-related increase in pyridoxal kinase (PDXK) activity in humans [48,50], which would decrease pyridoxine but increase, rather than decrease, PLP concentrations. These authors concluded that the decline in PLP concentrations may be due to the changes in intermediary PLP-binding skeletal muscle protein metabolism during the metabolic switch from anabolism to catabolism during aging [48,51]. However, without further understanding alterations in intermediary enzymes, it is unclear whether the current observed differences in B_6 vitamers are due to altered absorption or conversion. Regardless, given the importance of PLP as a coenzyme in diverse cellular mechanisms, including one carbon metabolism [5], the low postprandial response in circulating plasma in older adults may have functional impacts on successful aging.

Unlike the response of B_6, the older age group demonstrated higher postprandial concentrations of some B-vitamins including thiamine, riboflavin, and pantothenic acid. The differences in thiamine and riboflavin responses between age groups were sex-specific to males, with younger males having lower concentrations than older males, while pantothenic acid differences were seen in both sexes. These contrasting findings contradict a simple alteration in absorption [52] or even cellular uptake [53]. A similar lack of age-related difference in circulating pantothenic acid concentrations was reported in a longer term intervention in an animal model [54], yet evidence for age-related differences in these vitamins is limited. The comparatively lower plasma response of both riboflavin and thiamine in younger males than the rest of the subjects was unexpected, and a non-significant but higher FMN in younger adults, regardless of gender, may suggest age-related differences in riboflavin conversion. However, this remains speculative as this study did not analyze FMN beyond 4 h post-ingestion. Nevertheless, a similar age-related difference in postprandial FMN response to a single riboflavin rich meal has been previously reported by our laboratory [35], suggesting altered riboflavin conversion to FMN. Other factors not assessed in the current study, such as physical activity, are known to increase requirements of thiamine, riboflavin and vitamin B_6, that may accelerate the cellular uptake of these vitamins, impacting circulating concentrations [53].

Urinary excretion of all the B_6-vitamers and riboflavin, thiamine, and pantothenic acid was greater in females than males, with nicotinamide and nicotinuric acid concentrations also being greater in females, but only in the younger group, which contrasted with the absence of sex-specific differences in circulating concentrations of these vitamins. Others have similarly reported greater urinary excretion of thiamine, pantothenic acid, folate, and vitamin B_{12}, but not of riboflavin and B_3-vitamers in females following a 7-day controlled diet [55]. Healthy females may have lower requirements due to lower body weight and energy requirements than males [56], which may contribute to greater urinary excretion. Moreover, postprandial urinary creatinine concentration was lower in females, which would have amplified this difference as urinary B-vitamin measurements were corrected for urine output. However, more studies are required to understand the basis of these sex-specific differences in excretion, including the use of 24-h urinary collection to extend the data from the time-limited analysis in the current study. In particular, the lack of correlation between plasma and urinary concentrations highlights that measures of status and acute circulating response may differ from measures of elimination acutely, and may be useful in offering insight into differences in saturation of B vitamin uptake. However, like absorption, excretion of several vitamers (including thiamine and nicotinamide metabolites) has also been shown to be influenced by habitual intake [57], which was not controlled for in the current study.

The fasting concentrations of circulating thiamine, pantothenic acid, folate, and nicotinuric acid were different between older and younger subjects, which may have contributed to the age-related postprandial differences. Except for fasting thiamine and serum folate concentrations, which were higher in older adults, this relationship was sex-specific. Older compared to younger females had lower nicotinuric acid concentrations whereas males had lower fasting pantothenic acid than females of the older group. Although there have been reports of inadequate intake of B vitamins in older New Zealanders in community dwelling [58,59] and indigenous New Zealand Māori populations [60], and indications of inadequate status in residential care settings [61], these trends do not appear to be mirrored in this study population. Contrary to concerns regarding greater risk of thiamine deficiency with aging [62], and available assessments of circulating status [63], the present study found higher thiamine status in older subjects, corresponding with the trend in greater habitual thiamine intake. However, other age and sex specific differences, including fasting thiamine, riboflavin, and niacin concentrations were not in line with the habitual dietary assessments collected in this study. This could be related to complex interactions across dietary habits [64], food choices [65], or metabolism [66], as well as lifestyle factors such as physical activity [53], smoking, and alcohol consumption [67], which were not thoroughly explored in the current study. Longer term dietary assessment or other biomarker measures for specific B-vitamins status could elucidate this further, but were not a primary focus of this study.

The present study represented a simple but necessary approach to understand the key fundamental changes in the acute postprandial response to supplementation in aging. Although this study was not without limitations, these responses may be integral to the prevalence of low B-vitamin status found in older populations [2]. This was accomplished by a comprehensive analysis of B-vitamins including many vitamers, although further work should include plasma biotin and folate vitamers (5-methyltetrahydrofolate, tetrahydrofolate, and 5–10, methylenetetrahydrofolate), and should also consider the use of certified reference materials rather than plasma matrix QCs. Some commonly reported status measures of urinary excretion of vitamin B_{12} and folate were not reported (e.g., methylmalonic acid (MMA) and holotranscobalamin) [68], yet these are less appropriate for acute bioavailability as MMA is a functional biomarker and holotranscobalamin represents long term intake [69]. Changes in many tissue-specific vitamers, such as phosphorylated thiamine vitamers [70], may occur during a later phase of postprandial metabolism [71]; although these may not have been captured, additional work is required to determine their contribution to circulating pools. A further limitation of evaluating acute vitamin responses is the impact of habitual intestinal exposure in regulation of vitamin absorption; indeed, carrier-mediated mechanisms of B vitamins (including thiamine, riboflavin, and folate, but not vitamin B_6) have been shown to be upregulated during periods of deficiency, and downregulated with excess supplementation [72,73]. It is unclear whether recent intake [74], which was uncontrolled prior to the study, or differences in habitual vitamin intake observed in the current study may have impacted absorption within the intestine differently across groups. There is also uncertainty regarding the actual vitamin content ingested, as this was based on the manufacturer's information for the MVM tablet and food table estimates; both are prone to error, and may not reflect the ingested amounts [75]. Although inclusion of healthy participants may limit the generalizability of the results to typical aging populations with comorbidities including reduced gastrointestinal function [76], alcoholism [67], and other chronic diseases [77,78], this approach eliminated potential confounding factors that would have impacted the results. Impact of aging being the primary comparison, the study may not be sufficiently powered to confirm the sex-specific findings with low sample size, however, it provides preliminary data on the effect of sex on acute B-vitamin responses.

5. Conclusions

The present study demonstrated that although vitamin B_{12} response was not different between age groups, older males had lower B_{12} appearance in circulation from acute ingestion of a MVM supplement compared to younger males and older females. The significance for long-term B_{12} status in older males is unclear. There was a suppressed postprandial rise in the abundance of the active B_6-vitamers, with higher excretory B_6-vitamers. These data suggest a previously undescribed change in vitamin B_6 absorption, metabolism, or interconversion in healthy older adults. Further studies investigating other related metabolites and enzymes involved in the B-vitamin metabolic pathway are required to understand the possible mechanisms. As the study may face limitations due to the small sample size, acute assessment period, lack of tissue measures in comparison to circulating concentrations, and uncertainty of the implications for adequate intake and status, the results should therefore be interpreted with caution. However, these current findings are suggestive of differences in the postprandial handling of differing members of the B-vitamin family and therefore contribute to the complexity of current understanding of the possible differing B-vitamin requirements for aging individuals.

Supplementary Materials: The following are available online at http://www.mdpi.com/2072-6643/12/11/3529/s1, Table S1: Urine output and creatinine concentration of the study participants at fasting and 1–4 h following multivitamin and mineral supplement ingestion; Figure S1: Postprandial circulating B-vitamin concentrations with sex-specific effects following supplement ingestion; Figure S2: Sex-specific effects on the urinary concentrations of B-vitamins and vitamers following supplement ingestion.

Author Contributions: A.M.M. and D.C.-S. designed and supervised the research and reviewed the paper, P.S. and S.M.H. enrolled participants and conducted the research, E.B.T. and M.G. provided technical support for mass spectrometry analysis, P.S. analyzed data and conducted the statistical analysis, N.G. analyzed dietary intake data and reviewed the paper, P.S. wrote the paper. M.P.G.B., N.C.R., and D.C.-S. designed the overall research strategy that supported this PhD project and obtained funding. A.M.M. obtained funding and had primary responsibility for final content. All authors have read and agreed to the published version of the manuscript.

Funding: This research was funded by AgResearch Ltd. through the Strategic Science Investment Fund (SSIF) Project "Nutritional Strategies for an Ageing Population" (Contract A21246), Maurice and Phyllis Paykel Trust (3716230), and a Faculty Research and Development Fund from the University of Auckland (3716936).

Acknowledgments: We thank Sarah Mitchell, Aahana Shrestha, Cameron J Mitchell, Julia Ängemalm, Sofia Egelrud and Franny Markstedt for their clinical assistance. Thank you to Mark H. Vickers for proof-reading and PhD supervision.

Conflicts of Interest: The authors declare no conflict of interest. The funders had no role in the design of the study; in the collection, analyses, or interpretation of data; in the writing of the manuscript, or in the decision to publish the results.

References

1. Janssen, J.J.E.; Grefte, S.; Keijer, J.; De Boer, V.C.J. Mito-nuclear communication by mitochondrial metabolites and its regulation by B-vitamins. *Front. Physiol.* **2019**, *10*, 1–23. [CrossRef]
2. Porter, K.; Hoey, L.; Hughes, C.; Ward, M.; McNulty, H. Causes, consequences and public health implications of low B-vitamin status in ageing. *Nutrients* **2016**, *8*, 725. [CrossRef]
3. Froese, D.S.; Fowler, B.; Baumgartner, M.R. Vitamin B12, folate, and the methionine remethylation cycle—biochemistry, pathways, and regulation. *J. Inherit. Metab. Dis.* **2019**, *42*, 673–685. [CrossRef] [PubMed]
4. Ye, X.; Maras, J.E.; Bakun, P.J.; Tucker, K.L. Dietary intake of vitamin B6, plasma pyridoxal 5′-phosphate and homocysteine in Puerto Rican adults. *J. Am. Diet. Assoc.* **2010**, *110*, 1660–1668. [CrossRef]
5. Parra, M.; Stahl, S.; Hellmann, H. Vitamin B6 and its role in cell metabolism and physiology. *Cells* **2018**, *84*, 28. [CrossRef]
6. An, Y.; Feng, L.; Zhang, X.; Wang, Y.Y.; Wang, Y.Y.; Tao, L.; Qin, Z.; Xiao, R. Dietary intakes and biomarker patterns of folate, vitamin B6, and vitamin B12 can be associated with cognitive impairment by hypermethylation of redox-related genes NUDT15 and TXNRD1. *Clin. Epigenetics* **2019**, *11*, 139. [CrossRef] [PubMed]
7. Morris, M.S.; Sakakeeny, L.; Jacques, P.F.; Picciano, M.F.; Selhub, J. Vitamin B-6 intake is inversely related to, and the requirement is affected by, inflammation status. *J. Nutr.* **2010**, *140*, 103–110. [CrossRef] [PubMed]
8. Abbenhardt, C.; Miller, J.W.; Song, X.; Brown, E.C.; David Cheng, T.-Y.; Wener, M.H.; Zheng, Y.; Toriola, A.T.; Neuhouser, M.L.; Beresford, S.A.; et al. Nutritional epidemiology biomarkers of one-carbon metabolism are associated with biomarkers of inflammation in women. *J. Nutr.* **2014**, *144*, 714–721. [CrossRef] [PubMed]
9. Zhong, J.; Trevisi, L.; Urch, B.; Lin, X.; Speck, M.; Coull, B.A.; Liss, G.; Thompson, A.; Wu, S.; Wilson, A.; et al. B-vitamin supplementation mitigates effects of fine particles on cardiac autonomic dysfunction and inflammation: A pilot human intervention trial. *Sci. Rep.* **2017**, *7*, 45322. [CrossRef]

10. Blumberg, J.; Bailey, R.; Sesso, H.; Ulrich, C. The Evolving Role of Multivitamin/Multimineral Supplement Use among Adults in the Age of Personalized Nutrition. *Nutrients* **2018**, *10*, 248. [CrossRef]
11. Power, S.; Jeffery, I.; Ross, R.; Stanton, C.; O'Toole, P.; O'Connor, E.; Fitzgerald, G. Food and nutrient intake of Irish community-dwelling elderly subjects: Who is at nutritional risk? *J. Nutr. Health Aging* **2014**, *18*, 561–572. [CrossRef] [PubMed]
12. Ter Borg, S.; Verlaan, S.; Hemsworth, J.; Mijnarends, D.M.; Schols, J.M.G.A.; Luiking, Y.C.; De Groot, L.C.P.G.M. Micronutrient intakes and potential inadequacies of community-dwelling older adults: A systematic review. *Br. J. Nutr.* **2015**, *113*, 1195–1206. [CrossRef] [PubMed]
13. Ferroli, C.E.; Trumbo, P.R. Bioavailability of vitamin B-6 in young and older men. *Am. J. Clin. Nutr.* **1994**, *60*, 68–71. [CrossRef] [PubMed]
14. Kennedy, D.O. B vitamins and the brain: Mechanisms, dose and efficacy—A review. *Nutrients* **2016**, *8*, 68. [CrossRef]
15. Shibata, K.; Sugita, C.; Sano, M.; Fukuwatari, T. Urinary excretion of B-group vitamins reflects the nutritional status of B-group vitamins in rats. *J. Nutr. Sci.* **2013**, *2*, 1–7. [CrossRef]
16. Said, H.M.M. Intestinal absorption of water-soluble vitamins in health and disease. *Biochem. J.* **2011**, *437*, 357–372. [CrossRef]
17. Morley, J.E. Decreased Food Intake With Aging. *J. Gerontol. Ser. A* **2001**, *56*, 81–88. [CrossRef]
18. Conzade, R.; Koenig, W.; Heier, M.; Schneider, A.; Grill, E.; Peters, A.; Thorand, B. Prevalence and predictors of subclinical micronutrient deficiency in German older adults: Results from the population-based KORA-Age Study. *Nutrients* **2017**, *9*, 1276. [CrossRef]
19. Kant, A.K.; Moser-Veillon, P.B.; Reynolds, R.D. Effect of age on changes in plasma, erythrocyte, and urinary B-6 vitamers after an oral vitamin B-6 load. *Am. J. Clin. Nutr.* **1988**, *48*, 1284–1290. [CrossRef]
20. Kozyraki, R.; Cases, O. Vitamin B12 absorption: Mammalian physiology and acquired and inherited disorders. *Biochimie* **2013**, *95*, 1002–1007. [CrossRef]
21. U.S. Department of Health and Human Services and U.S. Department of Agriculture. *2015–2020 Dietary Guidelines for Americans*, 8th ed.; U.S. Department of Health and Human Services and U.S. Department of Agriculture: Washington, DC, USA. Available online: http://health.gov/dietaryguidelines/2015/guidelines/ (accessed on 31 December 2015).
22. Ministry of Health. *Food and Nutrition Guidelines for Healthy Older People: A Background Paper*; Ministry of Health: Wellington, New Zealand, 2013; ISBN 9780478393996.
23. Sebastian, R.S.; Cleveland, L.E.; Goldman, J.D.; Moshfegh, A.J. Older adults who use vitamin/mineral supplements differ from nonusers in nutrient intake adequacy and dietary attitudes. *J. Am. Diet. Assoc.* **2007**, *107*, 1322–1332. [CrossRef] [PubMed]
24. Douglas, J.W.; Lawrence, J.C.; Knowlden, A.P. The use of fortified foods to treat malnutrition among older adults: A systematic review, Quality in Ageing and Older Adults. *Ageing Older Adults* **2017**, *18*, 104–119. [CrossRef]
25. Eussen, S.J.P.M.; de Groot, L.C.P.G.M.; Clarke, R.; Schneede, J.; Ueland, P.M.; Hoefnagels, W.H.L.; van Staveren, W.A. Oral Cyanocobalamin Supplementation in Older People With Vitamin B12 Deficiency. *Arch. Intern. Med.* **2005**, *165*, 1167. [CrossRef] [PubMed]
26. Bradbury, K.E.; Williams, S.M.; Green, T.J.; McMahon, J.A.; Mann, J.I.; Knight, R.G.; Skeaff, C.M. Differences in Erythrocyte Folate Concentrations in Older Adults Reached Steady-State within One Year in a Two-Year, Controlled, 1 mg/d Folate Supplementation Trial. *J. Nutr.* **2012**, *142*, 1633–1637. [CrossRef]
27. Kalman, D.S.; Lou, L.; Schwartz, H.I.; Feldman, S.; Krieger, D.R. A pilot trial comparing the availability of vitamins C, B6, and B12 from a vitamin-fortified water and food source in humans. *Int. J. Food Sci. Nutr.* **2009**, *60*, 114–124. [CrossRef]
28. Navarro, M.; Wood, R.J. Plasma changes in micronutrients following a multivitamin and mineral supplement in healthy adults. *J. Am. Coll. Nutr.* **2003**, *22*, 124–132. [CrossRef]
29. Mayengbam, S.; Virtanen, H.; Hittel, D.S.; Elliott, C.; Reimer, R.A.; Vogel, H.J.; Shearer, J. Metabolic consequences of discretionary fortified beverage consumption containing excessive vitamin B levels in adolescents. *PLoS ONE* **2019**, *14*, e0209913. [CrossRef]
30. Meisser Redeuil, K.; Longet, K.; Bénet, S.; Munari, C.; Campos-Giménez, E. Simultaneous quantification of 21 water soluble vitamin circulating forms in human plasma by liquid chromatography-mass spectrometry. *J. Chromatogr. A* **2015**, *1422*, 89–98. [CrossRef]

31. Kwok, T.; Lee, J.; Ma, R.C.; Wong, S.Y.; Kung, K.; Lam, A.; Ho, C.S.; Lee, V.; Harrison, J.; Lam, L. A randomized placebo controlled trial of vitamin B12 supplementation to prevent cognitive decline in older diabetic people with borderline low serum vitamin B12. *Clin. Nutr.* **2017**, *36*, 1509–1515. [CrossRef]
32. WHO Expert Consultation. *Waist Circumference and Waist–Hip Ratio: Report of a WHO Expert Consultation, Geneva, 8–11 December, 2008*; WHO: Geneva, Switzerland, 2011.
33. FZANZ. *AUSNUT 2011-13 Food Nutrient Database*; FSANZ, Food Standards Australia New Zealand: Canberra, ACT, Australia, 2014.
34. Gregory, J.F.; Swendseid, M.E.; Jacob, R.A. Urinary excretion of folate catabolites responds to changes in folate intake more slowly than plasma folate and homocysteine concentrations and lymphocyte DNA methylation in postmenopausal women. *J. Nutr.* **2000**, *130*, 2949–2952. [CrossRef]
35. Sharma, P.; Gillies, N.; Pundir, S.; Pileggi, C.A.; Markworth, J.F.; Thorstensen, E.B.; Cameron-Smith, D.; Milan, A.M. Comparison of the acute postprandial circulating B-vitamin and vitamer responses to single breakfast meals in young and older individuals: Preliminary secondary outcomes of a randomized controlled trial. *Nutrients* **2019**, *11*, 2893. [CrossRef] [PubMed]
36. Matthews, D.R.; Hosker, J.R.; Rudenski, A.S.; Naylor, B.A.; Treacher, D.F.; Turner, R.C. Homeostasis model assessment: Insulin resistance and β-cell function from fasting plasma glucose and insulin concentrations in man. *Diabetologia* **1985**, *28*, 412–419. [CrossRef] [PubMed]
37. Pannérec, A.; Migliavacca, E.; De Castro, A.; Michaud, J.; Karaz, S.; Goulet, L.; Rezzi, S.; Ng, T.P.; Bosco, N.; Larbi, A.; et al. Vitamin B12 deficiency and impaired expression of amnionless during aging. *J. Cachexia. Sarcopenia Muscle* **2018**, *9*, 41–52. [CrossRef] [PubMed]
38. Margalit, I.; Cohen, E.; Goldberg, E.; Krause, I. Vitamin B12 deficiency and the role of gender: A cross-sectional study of a large cohort. *Ann. Nutr. Metab.* **2018**, *72*, 265–271. [CrossRef]
39. Stover, P.J. Vitamin B12 and older adults. *Curr. Opin. Clin. Nutr. Metab. Care* **2010**, *13*, 24–27. [CrossRef]
40. Araújo, D.; Noronha, M.; Cunha, N.; Abrunhosa, S.; Rocha, A.; Amaral, T. Low serum levels of vitamin B12 in older adults with normal nutritional status by mini nutritional assessment. *Eur. J. Clin. Nutr.* **2016**, *70*, 859–862. [CrossRef]
41. Albersen, M.; Bosma, M.; Knoers, N.V.V.A.M.; De Ruiter, B.H.B.; Ne, E.; Diekman, F.; De Ruijter, J.; Visser, W.F.; De Koning, T.J.; Verhoeven-Duif, N.M. The intestine plays a substantial role in human vitamin B6 metabolism: A Caco-2 cell model. *PLoS ONE* **2013**, *8*, e54113. [CrossRef]
42. Chiang, E.-P.; Smith, D.E.; Selhub, J.; Dallal, G.; Wang, Y.-C.; Roubenoff, R. Inflammation causes tissue-specific depletion of vitamin B6. *Arthritis Res. Ther.* **2005**, *7*, R1254–R1262. [CrossRef]
43. Pannemans, D.L.E.; van den Berg, H.; Westerterp, K.R. The influence of protein intake on vitamin B-6 metabolism differs in young and elderly humans. *J. Nutr.* **1994**, *124*, 1207–1214. [CrossRef]
44. Bode, W.; Mocking, J.A.J.J.; Van den Berg, H. Influence of age and sex on vitamin B-6 vitamer distribution and on vitam B-6 metabolizing enzymes in Wistar rats. *J. Nutr.* **1991**, *121*, 318–329. [CrossRef]
45. Midttun, Ø.; Ulvik, A.; Ringdal Pedersen, E.; Ebbing, M.; Bleie, Ø.; Schartum-Hansen, H.; Nilsen, R.M.; Nygård, O.; Ueland, P.M. Low plasma vitamin B-6 status affects metabolism through the kynurenine pathway in cardiovascular patients with systemic inflammation. *J. Nutr.* **2011**, *141*, 611–617. [CrossRef] [PubMed]
46. Chen, C.-H.; Yeh, E.-L.; Chen, C.-C.; Huang, S.-C.; Huang, Y.-C. Vitamin B-6, independent of homocysteine, is a significant factor in relation to inflammatory responses for chronic kidney disease and hemodialysis patients. *BioMed Res. Int.* **2017**, *2017*, 1–8. [CrossRef] [PubMed]
47. Schmidt, A.; Schreiner, M.G.; Mayer, H.K. Rapid determination of the various native forms of vitamin B6 and B2 in cow's milk using ultra-high performance liquid chromatography. *J. Chromatogr. A* **2017**, *1500*, 89–95. [CrossRef] [PubMed]
48. van den Berg, H.; Bode, W.; Mocking, J.; Löwik, M. Effect of aging on vitamin B6 status and metabolism. *Ann. N. Y. Acad. Sci.* **1990**, *585*, 96–105. [CrossRef]
49. Bode, W.; Van Den Berg, H. Pyridoxal-5′-phosphate and pyridoxal biokinetics in aging Wistar rats. *Exp. Gerontol.* **1991**, *26*, 589–599. [CrossRef]
50. van den Berg, H. Vitamin B 6 status and requirements in older adults. *Br. J. Nutr.* **1999**, *81*, 175–176. [CrossRef]
51. Roubenoff, R. Catabolism of aging: Is it an inflammatory process? *Curr. Opin. Clin. Nutr. Metab. Care* **2003**, *6*, 295–299. [CrossRef]

52. Russell, R.M. Factors in aging that effect the bioavailability of nutrients. *Bioavailab. Nutr. Other Bioact. Components Diet. Suppl.* **2001**, *131*, 1359S–1361S. [CrossRef]
53. Manore, M.M. Effect of physical activity on thiamine, riboflavin, and vitamin B-6 requirements. *Am. J. Clin. Nutr.* **2000**, *72*, 598S–606S. [CrossRef]
54. Ukuwatari, T.F.; Ada, H.W.; Hibata, K.S. Age-Related Alterations of B-Group Vitamin Contents in Urine, Blood and Liver from Rats. *J. Nutr. Sci. Vitaminol.* **2008**, *54*, 357–362. [CrossRef]
55. Shibatai, K.; Fukuwatari, T.; Ohta, M.; Okamoto, H.; Watanabe, T.; Fukui, T.; Nishimuta, M.; Totani, M.; Kimura, M.; Ohishi, N.; et al. Values of water-soluble vitamins in blood and urine of Japanese young men and women consuming a semi-purified diet based on the Japanese dietary reference intakes. *J. Nutr. Sci. Vitaminol.* **2005**, *51*, 319–328. [CrossRef] [PubMed]
56. Institute of Medicine (US) Subcommittee on Interpretation and Uses of Dietary Reference Intakes; Institute of Medicine (US) Standing Committee on the Scientific Evaluation of Dietary Reference Intakes. Using Dietary Reference Intakes in Planning Diets for Individuals. In *Dietary Reference Intakes: Applications in Dietary Planning*; National Academies Press (US): Washington, DC, USA, 2003.
57. Ukuwatari, T.F.; Oshida, E.Y.; Akahashi, K.T.; Hibata, K.S. Effect of Fasting on the Urinary Excretion of Water-Soluble Vitamins in Humans and Rats. *J. Nutr. Sci. Vitaminol.* **2010**, *56*, 19–26. [CrossRef]
58. Fernyhough, L.K.; Horwath, C.C.; Campbell, A.J.; Robertson, M.C.; Busby, W.J. Changes in dietary intake during a 6-year follow-up of an older population. *Eur. J. Clin. Nutr.* **1999**, *53*, 216–225. [CrossRef] [PubMed]
59. Gillies, N.; Cameron-Smith, D.; Pundir, S.; Wall, C.; Milan, A. Exploring trajectories in dietary adequacy of the B vitamins folate, riboflavin, vitamins B 6 and B 12, with advancing older age: A systematic review. *Br. J. Nutr.* **2020**, 1–32, (online ahead of print). [CrossRef] [PubMed]
60. Wham, C.; Teh, R.; Moyes, S.A.; Rolleston, A.; Muru-Lanning, M.; Hayman, K.; Kerse, N.; Adamson, A. Micronutrient intake in advanced age: Te Puāwaitanga o Ngā Tapuwae Kia ora Tonu, Life and Living in Advanced Age: A Cohort Study in New Zealand (LiLACS NZ). *Br. J. Nutr.* **2016**, *116*, 1754–1769. [CrossRef] [PubMed]
61. Ongena, K.P.S. B-Vitamin Status and Homocysteine in Older Adults in Residential Aged Care Facilities in New Zealand: A Cross-Sectional Study. Ph.D. Dissertation, University of Otago, Otago, New Zealand, 6 April 2020.
62. O'Keeffe, S.T. Thiamine deficiency in elderly people. *Age Ageing* **2000**, *29*, 99–101. [CrossRef] [PubMed]
63. Wilkinson, T.J.; Hanger, H.C.; George, P.M.; Sainsbury, R. Is thiamine deficiency in elderly people related to age or co-morbidity? *Age Ageing* **2000**, *29*, 111–116. [CrossRef]
64. Leblanc, V.; Bégin, C.; Corneau, L.; Dodin, S.; Lemieux, S. Gender differences in dietary intakes: WHAT is the contribution of motivational variables? *J. Hum. Nutr. Diet.* **2015**, *28*, 37–46. [CrossRef]
65. Manippa, V.; Padulo, C.; Van Der Laan, L.N.; Brancucci, A.; Herrmann, M.J. Gender differences in food choice: Effects of superior temporal sulcus stimulation. *Front. Hum. Neurosci.* **2017**, *11*, 597. [CrossRef]
66. Tarnopolsky, M.A. Gender differences in metabolism; nutrition and supplements. *J. Sci. Med. Sport* **2000**, *3*, 287–298. [CrossRef]
67. Veitch, R.L.; Lumeng, L.; Li, T.K. Vitamin B6 metabolism in chronic alcohol abuse. The effect of ethanol oxidation on hepatic pyridoxal 5′ phosphate metabolism. *J. Clin. Investig.* **1975**, *55*, 1026–1032. [CrossRef] [PubMed]
68. Iqbal, N.; Azar, D.; Yun, Y.M.; Ghausi, O.; Ix, J.; Fitzgerald, R.L. Serum methylmalonic acid and holotranscobalamin-II as markers for vitamin B12 deficiency in end-stage renal disease patients. *Ann. Clin. Lab. Sci.* **2013**, *43*, 243–249. [PubMed]
69. Herrmann, W.; Obeid, R.; Schorr, H.; Geisel, J. The Usefulness of Holotranscobalamin in Predicting Vitamin B12 Status in Different Clinical Settings. *Curr. Drug Metab.* **2005**, *6*, 47–53. [CrossRef] [PubMed]
70. Ihara, H.; Matsumoto, T.; Shino, Y.; Hashizume, N. Assay values for thiamine or thiamine phosphate esters in whole blood do not depend on the anticoagulant used. *J. Clin. Lab. Anal.* **2005**, *19*, 205–208. [CrossRef] [PubMed]
71. Portari, G.V.; Vannucchi, H.; Jordao, A.A. Liver, plasma and erythrocyte levels of thiamine and its phosphate esters in rats with acute ethanol intoxication: A comparison of thiamine and benfotiamine administration. *Eur. J. Pharm. Sci.* **2013**, *48*, 799–802. [CrossRef]
72. Said, H.M.; Mohammed, Z.M.; Hamid, S.M.; Mohammed, Z.M.; Said, H.M.; Mohammed, Z.M. Intestinal absorption of water-soluble vitamins: An update. *Curr. Opin. Gastroenterol.* **2006**, *22*, 140–146. [CrossRef]

73. Said, H.M.; Rose, R.; Seetharam, B. Intestinal absorption of water-soluble vitamins: Cellular and molecular aspects. In *Gastrointestinal Transport: Molecular Physiology; Current Topics in Membranes*; Barnett, K.E., Donwitz, M., Eds.; Academic Press: San Diego, CA, USA, 2000; Volume 50, pp. 35–75. ISBN 0121533506.
74. Laird, E.; Casey, M.C.; Ward, M.; Hoey, L.; Hughes, C.F.; McCarroll, K.; Cunningham, C.; Strain, J.J.; McNulty, H.; Molloy, A.M. Dairy intakes in older Irish adults and effects on vitamin micronutrient status: Data from the TUDA study. *J. Nutr. Health Aging* **2017**, *21*, 954–961. [CrossRef]
75. Andrews, K.W.; Roseland, J.M.; Gusev, P.A.; Palachuvattil, J.; Dang, P.T.; Savarala, S.; Han, F.; Pehrsson, P.R.; Douglass, L.W.; Dwyer, J.T.; et al. Analytical ingredient content and variability of adult multivitamin/mineral products: National estimates for the Dietary Supplement Ingredient Database. *Am. J. Clin. Nutr.* **2017**, *105*, 526–539. [CrossRef]
76. Lin, S.K.; Lambert, J.R.; Wahlqvist, M.L. Nutrition and gastrointestinal disorders. *Asia Pac. J. Clin. Nutr.* **1992**, *1*, 37–42.
77. Jankowska, M.; Rutkowski, B.; Dębska-Ślizień, A. Vitamins and microelement bioavailability in different stages of chronic kidney disease. *Nutrients* **2017**, *9*, 282. [CrossRef]
78. Iwakawa, H.; Nakamura, Y.; Fukui, T.; Fukuwatari, T.; Ugi, S.; Maegawa, H.; Doi, Y.; Shibata, K. Concentrations of water-soluble vitamins in blood and urinary excretion in patients with diabetes mellitus. *Nutr. Metab. Insights* **2016**, *9*, 85–92. [CrossRef] [PubMed]

Article

Characteristics of Websites Presenting Parenteral Supplementation Services in Five European Countries: A Cross-Sectional Study

Mikołaj Kamiński [1,*], Matylda Kręgielska-Narożna [1,†], Monika Soczewka [2,†], Agnieszka Wesołek [2,†], Paulina Rosiejka [2], Sara Szuman [2] and Paweł Bogdański [1]

1 Department of the Treatment of Obesity and Metabolic Disorders, and of Clinical Dietetics, Poznań University of Medical Sciences, Szamarzewskiego 82/84, 60-569 Poznań, Poland; matylda-kregielska@wp.pl (M.K.-N.); pawelbogdanski73@gmail.com (P.B.)

2 Student Scientific Club of Clinical Dietetics, Department of the Treatment of Obesity and Metabolic Disorders, and of Clinical Dietetics, Poznań University of Medical Sciences, Szamarzewskiego 82/84, 60-569 Poznań, Poland; monika.soczewka2@gmail.com (M.S.); aa.wesolek@gmail.com (A.W.); paulinarosiejka@gmail.com (P.R.); saraszuman@onet.eu (S.S.)

* Correspondence: mikolaj.w.kaminski@gmail.com; Tel.: +48-516-268-563

† These authors contributed equally to this work.

Received: 22 September 2020; Accepted: 23 November 2020; Published: 25 November 2020

Abstract: We aimed to characterize the parenteral supplementation services in Czechia, Ireland, Italy, Poland, and the United Kingdom based on their websites. We generated a list of websites by searching Google using the term "vitamin infusion" and selected cities with 250,000 citizens from each analyzed country. All search inputs were performed using the native language. Data on the features of services, indications, contraindications, offered parenteral supplements, and social media activity were obtained. We analyzed 317 websites representing 371 active facilities. Only 6 (1.9%) facilities cited the scientific sources on parenteral supplementation, but these reference were highly biased; 17.4% did not provide information regarding their personnel, while 11.9% indicated the different contraindications. The most common indications were fatigue (62.5%), immunity enhancement (58.0%), anti-aging, and physical activity (51.5%). Approximately, 11.6% of facilities claimed that some parenteral supplements can help manage certain malignancies, while 2.2% claimed that they can help manage fertility problems. The most offered intravenous supplements were vitamins C (57.4%), B12 (47.7%), and B6 (42.3%). The parenteral supplementation market offers numerous ingredients as treatment for general health problems and serious health conditions. Many analyzed websites lacked essential information, which creates concerns for regarding the quality and reliability of the services.

Keywords: alternative medicine; dietary supplements; public health; Internet; vitamin

1. Introduction

"Vitamin drips" and "vitamin injections" are complementary medicine services consisting of intravenous or subcutaneous administration of vitamins, minerals, or other dietary supplements [1,2]. In this paper, we used the term "parenteral supplementation" [2]. Firstly, most of the ingredients used in the "cocktails" are typical constituents of popular dietary supplements, e.g., minerals, vitamins. Secondly, the mixtures can be provided intravenously as well as subcutaneously. Finally, the companies claim that the intravenous/subcutaneous form of supplementation offers unique advantages over the oral intake [1,3]. The effects of such interventions were poorly investigated; to date, no study has provided high-quality evidence to support the efficacy of parental supplementation [4,5]. At the end of 2019, the NHS medical director warned the public of the potential dangers of these infusions [6].

However, it did not stop companies or celebrities from promoting vitamin intravenous infusions as a panacea [1,7]. Despite the intensive promotion of parenteral supplementation, the phenomenon remains under-researched [2]. The medical, psychological, and legal aspects of commercial infusions are poorly investigated. Most of the reports, also describing fatalities after the administration of supplement drips, were anecdotal and based on journal investigations [8,9]. The scientific reports and recommendations mainly concern the potential efficacy of the intervention using certain substances [10,11] or safety of the drips, in general [12]. Recently, we published the results of a cross-sectional survey study on the experience of Polish individuals using parenteral supplementation services [2]. Most of the respondents used parenteral supplementation to boost their immunity and endurance or deal with fatigue. In most cases, the overall experience with the infusions was described as "good" or "very good". However, the study group was limited to only 17 individuals.

The analysis of web-derived data provides insights into the under-researched fields of medicine [13]. The most popular source of e-data on health issues are search engine statistics, social media [14], e-forums [15], and websites [16]. The interest of Google users in various dietary supplements has increased in the last years [17]. Several researchers previously investigated the branches of complementary medicine by identifying websites that provide information on alternative treatments [18–21]. To date, no study focused on companies offering parenteral supplementation services. There is no systematic report on a variety of indications of parenteral supplementation as well as ingredients used in cocktails. We hypothesized that careful analysis of the content of the websites representing facilities offering parenteral supplementation might help understand this branch of complementary medicine. The acquired knowledge may pave the way for further research on this new public health phenomenon.

We aimed to characterize parenteral supplementation services in five European countries based on their websites.

2. Materials and Methods

2.1. Data Collection

The study was approved by the Bioethical Committee of Poznan University of Medical Sciences (nr of consent 227/20). The study was performed from January to July 2020.

We only included facilities in the cities with at least 250,000 citizens in the Opendatasoft database [22]. The cities with at least 250,000 citizens represent urban centers sizes from L to Global city according to the Organisation for Economic Cooperation and Development and European Commission definitions [23]. Finally, we included 3 cities from Czechia (Brno, Ostrava, and Prague), 1 from Ireland (Dublin), 12 from Italy (Bari, Bologna, Catania, Florence, Bologna, Milan, Naples, Palermo, Rome, Turin, Venice, and Verona), 11 from Poland (Białystok, Bydgoszcz, Gdańsk, Katowice, Kraków, Lublin, Łódź, Poznań, Szczecin, Warsaw, and Wrocław), and 29 from the United Kingdom (Belfast, Birkenhead, Birmingham, Bradford, Bristol, Cardiff, Coventry, Derby, Edinburgh, Glasgow, Islington, Kingston upon Hull, Leeds, Leicester, Liverpool, London, Luton, Manchester, Newport, Nottingham, Plymouth, Preston, Reading, Sheffield, Southend-on-Sea, Stoke-on-Trent, Sunderland, Swansea, and Wolverhampton). These countries were chosen arbitrarily: We wanted to include European countries only, but the research team had limited language skills. To our best knowledge, no study focused on companies offering parenteral services in these countries.

Furthermore, we collected the links to the websites of facilities offering parental supplementation using the Google search engine. Initially, we used Google Ads to compare the search volumes of terms "vitamin infusion," "vitamin drip," and "intravenous supplements" in the previous months (July–December 2019). According to Google Ads Keyword Planner, the "vitamin infusion" and its translations in Czech (translated by SS), Italian (translated by SS and PR), and Polish (translated by MK and MS) were the most popular keyword used for searching. In the Google search engine, we typed the term "vitamin infusion" and the city name, in native language, for all analyzed countries.

We excluded the links from inactive websites or websites that did not represent facilities offering parenteral supplementation, such as personal websites or blogs. We screened all links generated by Google. The workflow of the links collection is presented in Figure S1.

We prepared a researcher's form that contained a collection of information from chosen websites. Initially, we used the R-programming language to randomly choose five websites from each country. After analyzing the content of five randomly chosen websites from Poland, three researchers (MK, MS, and MKN) jointly prepared form describing the websites. Furthermore, the form's initial version was used to collect data from the other five randomly chosen websites representing facilities in Czechia, Ireland, Italy, and the United Kingdom. The other researchers (AW, PR, and SS) independently suggested corrections or new items to the form. The final version of the form is presented in Table S1. The final version of the form was used to subsequently collect the data from all links generated by Google (including websites used previously for the preparation of the initial and final versions of the form). The final form included data on the general characteristics of the facilities (country, link to the website, name of the company, affiliation to the commercial network, address, phone number, and email), social media activity (information on ratings on Google maps, Facebook, Instagram, or Twitter profiles), a feature of the services (experience, information on initial visits, laboratory investigations, location where the service can be performed, and citations of scientific researches), staff, indications and contraindications for parenteral supplementation, offered ingredients, and other services. In this study, the term "indications" referred to the recommendations provided in the websites for parenteral supplementations (both single and multi-ingredient) for different conditions as supportive and/or curative treatment. In the study, we presented other services offered in at least 5% of facilities in at least one country. One website provided information on a chain of facilities. For this reason, we multiplied the record by the number of facilities represented by a website and changed the details of the facilities affiliated to the chains that were different such as address or services offered.

The collection of data from different websites was carried out from January to May 2020. The websites were analyzed by the researchers with excellent/native knowledge of the following languages: Czech (SS), English (AW), Italian (PR), and Polish (MS). All doubts on the characteristics of websites were consulted with MK or MKN.

2.2. Data Analysis

All statistical analyses were performed using custom R 3.6.3 code (R Foundation, Vienna, Austria). The data analysis was performed in June–July 2020. The categorical data were presented as number (percentage), while numerical data were expressed as median (interquartile range). We visualized the total number and number of facilities per 100,000 citizens in each included city. Moreover, we computed the total number of facilities per 100,000 citizens in all analyzed cities for each country. Only indications, contraindications, and ingredients that occurred in at least 5% of websites in one country were included in the main results, while the rest is presented in the Supplementary Materials.

3. Results

3.1. Number of Facilities

We included a total of 317 websites, representing 371 facilities. For Italy, we matched 76 webpages of physicians, which mentioned parenteral supplementation as one of the procedures performed by physicians, but these websites were excluded from the study. Most of the included facilities were located in the United Kingdom (n = 154), followed by Poland (n = 121), Italy (n = 58), Czechia (n = 24), and Ireland (n = 14) (Figure 1). The highest number of facilities per 100,000 citizens were observed in Katowice (Poland) (3.06 facilities/100,000 citizens), Gdańsk (Poland) (2.57 facilities/100,000 citizens), and Leeds (the United Kingdom) (2.42 facilities/100,000 citizens). Considering the total population of the included towns, the highest number of facilities was reported in Czechia (1.62 facilities/100,000 citizens), followed by the United Kingdom (1.46 facilities/100,000 citizens), Ireland (1.37 facilities/100,000 citizens), Poland (1.35 facilities/100,000 citizens), and Italy (0.72 facilities/100,000 citizens).

Figure 1. Map of analyzed countries with the number of facilities per 100,000 citizens in each analyzed city.

3.2. Social Media

The general characteristics and social media activity of the analyzed facilities are presented in Table 1. In Czechia, Ireland, and Italy, more than 20% of facilities were affiliated to a commercial chain; in Poland and the United Kingdom, more than 40% were affiliated to a commercial chain. Overall, parenteral supplementation companies obtained a good consumer rating on Google maps with a median between 4 and 5. The analyzed facilities were active on different social media platforms, mostly in Facebook, followed by Instagram and Twitter.

3.3. Features, Staff, and Scientific Information

The features of services and their staff are presented in Table 2. Overall, 6 (1.9%) of all facilities support the use of parenteral supplementation by citing scientific evidence on their websites. In one Polish website, in the section describing offered ingredients, several books related to the use of glutathione [24–26] and review on reactive oxygen species were cited [27]. In the second Polish website, in the "Scientific research" section (pl. "Badania naukowe"), the following references were cited: "Publications available on the website: https://www.ncbi.nlm.nih.gov/pubmed and on the website https://www.pum.edu.pl/biblioteka after entering the author's names" (pl. "Publikacje dostępne na stronie: https://www.ncbi.nlm.nih.gov/pubmed oraz na stronie https://www.pum.edu.pl/biblioteka po wpisaniu nazwisk autora"). However, no article related to parenteral supplementation can be found after typing the names of physicians working in the facility. One facility from Dublin provided links to 22 abstracts from Medline on vitamin C. Among them, there were clinical studies investigating the effects of intravenous vitamin C on the quality of life of individuals with malignancy [28,29], cancer markers [30,31], sepsis [32,33], and C-reactive protein level in hemodialysis patients [34]. Most of the other articles were reviews or basic science research. Other websites mostly cited reviews on offered micronutrients and meta-analyses on the effects of zinc lozenges on common cold [35]. In two websites from the United Kingdom, we found a reference regarding "Myers' cocktail" [36]. In one case, the scientific section was inaccessible during the preparation of our manuscript (July 2020), and no citation was found. The substantial number (17.4%) of companies did not provide information regarding their personnel.

3.4. Indications and Contraindications

The most popular indications were fatigue (62.5%), immunity enhancement (58.0%), anti-aging, and physical activity (51.5%) (Table 3). Less than one of the eight companies provided information on contraindications to the procedure; among them, the most commonly described were hypersensitivity, pregnancy, and renal insufficiency (Table 3).

Table 1. General information of facilities offering parenteral supplementation, and their social media activity. Data presented as number (%)/median (interquartile range).

Variables	All n = 317	Czechia n = 24	Ireland n = 14	Italy n = 58	Poland n = 121	United Kingdom n = 154
Facilities in commercial chain (n)	144 (38.8%)	7 (29.2%)	3 (21.4%)	13 (22.4%)	54 (44.6%)	67 (43.5%)
Information on address (n)	345 (93.0%)	24.00 (100.0%)	14 (100.0%)	58 (100.0%)	103 (85.1%)	146 (94.8%)
Information on e-mail address (n)	335 (90.3%)	21.00 (87.5%)	13 (92.9%)	54 (93.1%)	121 (100.0%)	141 (91.6%)
Information on phone number (n)	369 (99.5%)	24.00 (100.0%)	13 (92.9%)	58 (100.0%)	106 (87.6%)	153 (99.4%)
Social Media						
Review in Google Maps (n)	283 (76.3%)	16 (66.7%)	10 (71.4%)	51 (87.9%)	83 (68.6%)	123 (79.9%)
Rating in Google Maps (n)	5.0 (5.0–5.0)	4.0 (3.5–4.4)	5.0 (4.7–5.0)	4.1 (4.5–5.0)	4.6 (4.3–5.0)	4.7 (4.5–5.0)
Number of Google ratings per facility (n)	20 (8–48)	19 (7–27)	8 (24–119)	12 (6–32)	19 (8–46)	28 (10–100)
Number of Facebook pages (n)	313 (84.4%)	12 (50.0%)	12 (85.7%)	55 (94.8%)	97 (80.1%)	137 (89.0%)
Facebook likes (n)	1149 (4295–2846)	1607 (762–31,500)	2071 (309–5488)	1527 (763–3776)	1000 (332–1963)	1070 (477–3322)
Facebook followers (n)	1249 (438–2906)	1634 (783–31,170)	2081 (315–5470)	1539 (782–3804)	1042 (337–1995)	1101 (514–3363)
Instagram profiles (n)	229 (61.7%)	7 (29.2%)	10 (71.4%)	30 (51.7%)	61 (50.4%)	121 (78.6%)
Instagram followers (n)	1289 (402–5460)	12,325 (1293–20,000)	781 (245–9570)	783 (402–5081)	536 (97–1048)	2542 (754–9715)
Instagram posts (n)	261 (70–542)	561 (78–864)	253 (49–498)	179 (63–461)	74 (23–180)	436 (213–629)
Twitter pages (n)	139 (37.5%)	2 (8.3%)	8 (57.1%)	9 (15.5%)	4 (3.3%)	113 (73.4%)
Twitter followers (n)	430 (80–1663)	87 (47–127)	177 (76–762)	14 (9–69)	19 (0–1974)	81 (146–1386)

Table 2. Features of the services offering parenteral supplementation and their staff. Data presented as number (%)/median (interquartile range).

Variables	All n = 317	Czechia n = 24	Ireland n = 14	Italy n = 58	Poland n = 121	United Kingdom n = 154
			Feature of the Services			
Information on years of experience (n)	142 (38.3%)	9 (37.5%)	8 (57.1%)	13 (22.4%)	59 (48.8%)	53 (34.4%)
Experience (years)	9.0 (3.0–15.0)	9.0 (9.0–9.0)	11.5 (9.0–16.3)	18.0 (10.0–20.0)	4.0 (2.0–13.5)	10.0 (5.0–14.0)
Initial visit (n)	165 (44.5%)	11 (45.8%)	6 (42.9%)	9 (15.5%)	54 (44.6%)	85 (55.2%)
Lab investigations (n)	82 (22.1%)	6 (25.0%)	4 (28.6%)	9 (15.5%)	33 (27.3%)	30 (19.5%)
Service in own premise (n)	341 (91.9%)	24 (100.0%)	14 (100.0%)	58 (100.0%)	96 (79.3%)	149 (96.8%)
Services at the place indicated by the customer (n)	53 (14.3%)	1 (4.2%)	0 (0.0%)	3 (5.2%)	35 (28.9%)	14 (9.1%)
Citation of scientific evidence on websites (n)	6 (1.9%)	2 (8.3%)	1 (7.1%)	0 (0.0%)	2 (1.7%)	3 (2.0%)
			Staff			
Information on personnel (n)	310 (83.6%)	24 (100.0%)	11 (78.6%)	58 (100.0%)	81 (66.9%)	136 (88.3%)
Physician (n)	248 (66.9%)	24 (100.0%)	8 (57.14%)	54 (93.1%)	72 (59.5%)	90 (58.4%)
Nurse and/or midwife (n)	74 (19.95%)	4 (16.7%)	1 (7.1%)	7 (12.1%)	25 (20.7%)	37 (24.0%)
Dietician (n)	45 (12.13%)	0 (0.0%)	0 (0.0%)	17 (29.3%)	18 (14.9%)	10 (6.5%)
Cosmetologist (n)	32 (8.63%)	0 (0.0%)	2 (14.3%)	5 (8.6%)	9 (7.4%)	16 (10.4%)
Psychologist (n)	11 (2.96%)	0 (0.0%)	0 (0.0%)	10 (17.2%)	1 (0.8%)	0 (0.0%)
Paramedic (n)	8 (2.16%)	0 (0.0%)	0 (0.0%)	2 (3.5%)	5 (4.1%)	1 (0.7%)
Physiotherapist (n)	7 (1.89%)	0 (0.0%)	0 (0.0%)	5 (8.6%)	2 (1.7%)	0 (0.0%)

Table 3. Indications and contraindications for parenteral supplementation. Data presented as number (%).

Variables	All n = 317	Czechia n = 24	Ireland n = 14	Italy n = 58	Poland n = 121	United Kingdom n = 154
Indications						
Vitamin and/or micronutrients deficiency (n)	186 (50.1%)	18 (75.0%)	2 (14.3%)	1 (1.7%)	98 (81.0%)	67 (43.5%)
Physically active individuals (n)	191 (51.5%)	11 (45.8%)	3 (21.4%)	3 (5.2%)	92 (76.0%)	82 (53.3%)
Fatigue (n)	232 (62.5%)	20 (83.3%)	6 (42.9%)	4 (6.9%)	89 (73.6%)	113 (73.4%)
Excessive stress (n)	140 (37.7%)	19 (79.2%)	3 (21.4%)	5 (8.6%)	69 (57.0%)	44 (28.6%)
Concentration loss or need of mental boost (n)	47 (12.7%)	1 (4.2%)	0 (0.0%)	1 (1.7%)	29 (24.0%)	15 (9.7%)
Symptoms of depressive disorders (n)	81 (21.8%)	14 (58.3%)	0 (0.0%)	1 (1.7%)	33 (27.3%)	33 (21.4%)
Hangover (n)	124 (33.4%)	1 (4.2%)	2 (14.4%)	4 (6.9%)	83 (68.6%)	34 (22.1%)
Immunity enhancement (n)	215 (58.0%)	23 (95.8%)	5 (35.7%)	2 (3.5%)	79 (65.3%)	106 (68.8%)
Dehydration (n)	108 (29.1%)	1 (4.2%)	5 (35.7%)	2 (3.5%)	50 (41.3%)	50 (32.5%)
Detoxification (n)	153 (41.2%)	12 (50.0%)	3 (21.4%)	5 (8.6%)	48 (39.7%)	85 (55.2%)
Anti-aging (n)	191 (51.5%)	10 (41.7%)	4 (28.6%)	48 (82.8%)	38 (31.4%)	91 (59.1%)
Beauty improvement (n)	190 (51.2%)	0 (0.0%)	10 (71.4%)	54 (93.1%)	30 (24.9%)	96 (62.3%)
Weight loss (n)	87 (23.5%)	5 (20.8%)	2 (14.3%)	1 (1.7%)	28 (23.1%)	51 (33.1%)
Improvement of libido (n)	41 (11.1%)	4 (16.7%)	0 (0.0%)	0 (0.0%)	31 (25.6%)	6 (3.9%)
For specific diseases (n)	77 (20.8%)	19 (79.2%)	3 (21.4%)	0 (0.0%)	32 (26.5%)	23 (14.9%)
Malignancy (n)	43 (11.6%)	16 (66.7%)	1 (7.1%)	0 (0.0%)	19 (15.7%)	7 (4.6%)
Contraindications						
Information on contraindications (n)	44 (11.9%)	5 (20.8%)	0 (0.0%)	4 (6.9%)	22 (18.2%)	13 (8.4%)
Hypersensitiveness (n)	19 (5.1%)	1 (4.2%)	0 (0.0%)	2 (3.5%)	10 (8.3%)	6 (3.9%)
Pregnancy (n)	17 (4.6%)	1 (4.2%)	0 (0.0%)	2 (3.5%)	11 (9.2%)	3 (1.95%)
Renal insufficiency (n)	15 (4.0%)	2 (8.3%)	0 (0.0%)	0 (0.0%)	7 (5.8%)	6 (3.9%)
Certain drugs (n)	8 (2.2%)	1 (4.2%)	0 (0.0%)	0 (0.0%)	7 (5.8%)	0 (0.0%)
Hemochromatosis (n)	8 (2.2%)	3 (12.5%)	0 (0.0%)	0 (0.0%)	3 (2.5%)	2 (1.3%)
Renal stones (n)	5 (1.4%)	4 (16.7%)	0 (0.0%)	0 (0.0%)	2 (1.7%)	0 (0.0%)

3.5. Type of Supplementation

Many facilities offered multi-ingredient intravenous supplementation, but a substantial number did not provide the exact information on the ingredients of cocktails (Table 4). Among intravenously administered supplements, the most commonly offered were vitamin C (57.4%), vitamin B12 (47.7%), and vitamin B6 (42.3%) (Table 4). The subcutaneous supplementation was offered mainly in Ireland and the United Kingdom.

The indications, contraindications, and ingredients that were not described in at least 5% of websites in each country are presented in Table S2. Parenteral supplementation was recommended as treatment for fertility problems in five facilities in Poland and in three facilities in the United Kingdom (Table S2). The other services offered in the analyzed facilities are presented in Table S3. Interestingly, the parenteral supplementation and beauty salon services were commonly provided by the same companies.

Table 4. Parenteral supplementation ingredients. Data presented as number (%).

Variables	All n = 317	Czechia n = 24	Ireland n = 14	Italy n = 58	Poland n = 121	United Kingdom n = 154
Multi-ingredient infusion (Cocktails) (n)	248 (66.9%)	8 (33.3%)	6 (42.9%)	7 (12.1%)	114 (94.2%)	113 (73.4%)
No information on exact cocktails ingredients (n)	86 (23.2%)	5 (20.8%)	1 (7.1%)	5 (8.62%)	57 (47.1%)	18 (11.7%)
Electrolytes, Macro-/Micronutrients						
Magnesium (n)	151 (40.7%)	4 (16.7%)	6 (42.9%)	3 (5.2%)	56 (46.3%)	82 (53.3%)
Zinc (n)	102 (27.5%)	0 (0.0%)	3 (21.4%)	2 (3.5%)	33 (27.3%)	64 (41.6%)
Selenium (n)	92 (24.8%)	0 (0.0%)	1 (7.1%)	2 (3.5%)	33 (27.3%)	56 (36.4%)
Multi-electrolyte fluid (n)	66 (17.8%)	2 (8.3%)	2 (14.3%)	1 (1.7%)	34 (28.1%)	27 (17.5%)
Calcium (n)	66 (17.8%)	1 (4.2%)	3 (21.4%)	0 (0.0%)	14 (11.6%)	48 (31.2%)
NaCl 0,9% (n)	44 (11.9%)	1 (4.2%)	3 (21.4%)	1 (1.7%)	11 (9.1%)	28 (18.2%)
Iron (n)	43 (11.6%)	0 (0.0%)	1 (7.1%)	0 (0.0%)	19 (15.7%)	23 (14.9%)
Chromium (n)	32 (8.6%)	0 (0.0%)	0 (0.0%)	2 (3.5%)	22 (18.2%)	8 (5.2%)
Iodine (n)	30 (8.1%)	0 (0.0%)	0 (0.0%)	0 (0.0%)	24 (19.8%)	6 (3.9%)
Molybdenum (n)	22 (5.9%)	0 (0.0%)	0 (0.0%)	0 (0.0%)	20 (16.5%)	2 (1.3%)
Copper (n)	20 (5.4%)	0 (0.0%)	0 (0.0%)	0 (0.0%)	16 (13.2%)	4 (2.6%)
Mangan (n)	14 (3.8%)	0 (0.0%)	0 (0.0%)	0 (0.0%)	11 (9.1%)	3 (2.0%)
Fluor (n)	7 (1.9%)	0 (0.0%)	0 (0.0%)	0 (0.0%)	6 (5.0%)	1 (0.7%)
Vitamins						
Vitamin A (n)	54 (14.6%)	1 (4.2%)	0 (0.0%)	0 (0.0%)	46 (38.0%)	7 (4.6%)
Vitamin B1 (n)	149 (40.2%)	3 (12.5%)	6 (42.9%)	0 (0.0%)	52 (43.0%)	88 (57.1%)
Vitamin B2 (n)	147 (39.6%)	3 (12.5%)	6 (42.9%)	0 (0.0%)	51 (42.2%)	87 (56.5%)
Vitamin B3/PP (n)	138 (37.2%)	3 (12.5%)	6 (42.9%)	0 (0.0%)	46 (38.0%)	83 (53.9%)
Vitamin B5 (n)	155 (41.8%)	3 (12.5%)	6 (42.9%)	1 (1.7%)	52 (43.0%)	93 (60.4%)
Vitamin B6 (n)	157 (42.3%)	5 (20.8%)	6 (42.9%)	0 (0.0%)	52 (43.0%)	94 (61.0%)
Vitamin B7 / H (n)	141 (38.0%)	3 (12.5%)	6 (42.9%)	1 (1.7%)	52 (43.0%)	79 (51.3%)

Table 4. *Cont.*

Variables	All n = 317	Czechia n = 24	Ireland n = 14	Italy n = 58	Poland n = 121	United Kingdom n = 154
Vitamin B9 / folic acid (n)	141 (38.0%)	3 (12.5%)	6 (42.9%)	0 (0.0%)	51 (42.2%)	81 (52.6%)
Vitamin B12 (n)	177 (47.7%)	5 (20.8%)	6 (42.9%)	2 (3.5%)	60 (49.6%)	104 (67.5%)
Vitamin C (n)	213 (57.4%)	22 (91.7%)	5 (35.7%)	5 (8.6%)	83 (68.6%)	98 (63.6%)
Vitamin D3 (n)	74 (20.0%)	2 (8.3%)	1 (7.1%)	0 (0.0%)	44 (36.4%)	27 (17.5%)
Vitamin E (n)	51 (13.8%)	1 (4.2%)	1 (7.1%)	0 (0.0%)	31 (25.6%)	18 (11.7%)
Vitamin K (n)	30 (8.1%)	0 (0.0%)	1 (7.1%)	0 (0.0%)	18 (14.9%)	11 (7.1%)
Other Supplements						
Glutathione (n)	127 (34.2%)	4 (16.7%)	5 (35.7%)	3 (5.2%)	34 (28.1%)	81 (52.6%)
Amino acids (n)	115 (31.0%)	0 (0.0%)	4 (28.6%)	2 (3.5%)	21 (17.4%)	88 (57.1%)
Alphalipoic acid (n)	52 (14.0%)	1 (4.2%)	2 (14.3%)	0 (0.0%)	37 (30.6%)	12 (7.8%)
Coenzyme Q10 (n)	39 (10.5%)	0 (0.0%)	1 (7.1%)	1 (1.7%)	17 (14.1%)	20 (13.0%)
Glucose (n)	34 (9.2%)	1 (4.2%)	1 (7.1%)	1 (1.7%)	26 (21.5%)	5 (3.3%)
Solcoseryl (n)	34 (9.2%)	0 (0.0%)	1 (7.1%)	0 (0.0%)	29 (24.0%)	4 (2.6%)
Choline (n)	12 (3.2%)	2 (8.3%)	0 (0.0%)	0 (0.0%)	5 (4.1%)	5 (3.3%)
Collagen (n)	11 (3.0%)	0 (0.0%)	0 (0.0%)	0 (0.0%)	2 (1.7%)	9 (5.8%)
Acetylcysteine (n)	9 (2.4%)	1 (4.2%)	0 (0.0%)	0 (0.0%)	8 (6.6%)	0 (0.0%)
Medications (n)	9 (2.4%)	3 (12.5%)	1 (7.1%)	0 (0.0%)	3 (2.5%)	2 (1.3%)
Ginkgo biloba (n)	5 (1.4%)	0 (0.0%)	1 (7.1%)	0 (0.0%)	1 (0.8%)	3 (2.0%)
Types of Subcutaneous Supplements						
Vitamin A (n)	17 (4.6%)	0 (0.0%)	7 (50.0%)	4 (6.9%)	0 (0%)	8 (5.2%)
Vitamin C (n)	42 (11.3%)	0 (0.0%)	5 (35.7%)	4 (6.9%)	0 (0%)	31 (20.1%)

4. Discussion

We analyzed the websites representing facilities offering parenteral supplementation in five European countries. We found that many websites lack essential information on parenteral supplementation. Simultaneously, the vitamin infusions were offered as treatment for both general ailments and serious conditions.

4.1. Main Findings

This study was the first to describe the content of websites of facilities offering parenteral supplementation. Overall, vitamin intravenous infusions were the most popular parenteral supplement offered in Czechia and the United Kingdom. The high number of facilities per 100,000 citizens in cities like Katowice, Gdańsk, and Leeds may be due to the high number of citizens living in surrounding towns.

The facilities offering parenteral supplementation are active on social media, with audiences between several hundred and several dozen thousands. Social media is a popular channel to promote alternative medicine methods and might be a valuable source for research [37–39]. Interestingly, the reviews on Google maps profile were generally positive, which may suggest that consumers had a positive experience with the facilities. However, personal belief and subjective response to vitamin infusions require further research [2]. The social media profiles of the facilities offering parenteral supplementation might be a good platform to reach the consumers.

Most of the facilities did not indicate their years of experience; thus, many companies might have a history far shorter than the median of 9 years. However, a substantial number of services are offered over a decade, which means that vitamin infusions is not a novel method; rather, it is an old method that gained attention just recently. Only Polish facilities had a median experience of less than 5 years, which suggests that the branch recently developed in this country.

Parenteral supplementation had no solid scientific evidence supporting its advertised beneficial properties [4,5]. A minority of the companies defend their standby citing scientific researches. Most of the references cited popular science books or non-systematic reviews. However, these references lacked the use of a placebo group as a comparator or did not indicate strong clinical outcomes. Interestingly, one webpage cited articles regarding the feasibility of intravenous vitamin C for sepsis—a condition that cannot be treated in ambulatory settings such as vitamin infusion facilities. None of the articles reported parenteral supplementation for common indications such as fatigue, physical endurance, or immunity enhancement.

In this study, the following two features were assessed in all facilities as qualifications for parenteral supplementation: description of initial visits and the opportunity to perform lab investigations. However, most of the analyzed facilities lacked complete information on both aspects. Because most of the ingredients were vitamins, macronutrients, or micronutrients, the status of these chemical compounds in the blood should be assessed before deciding to receive the infusion. Similarly, the initial visit is crucial to determine whether a consumer has a contraindication to the parenteral supplementation. The parenteral administration has universal contraindications, such as known hypersensitivity to any ingredient. The poorly investigated agents or the agents that are not commonly used among pregnant or breastfeeding women should be contraindicated in this group. Moreover, decompensated heart failure or end-stage chronic kidney is a contraindication to obtain excessive intravenous fluid. There are many other specific contraindications, but those mentioned above are the general contraindications. All consumers should be informed regarding them. Another concern is related to the missing information on medical professionals who work in the facilities, and who perform infusions and/or injections. Many of the companies claimed to employ dieticians, psychologists, physiotherapists, or cosmetologists, which may be associated with other offered services. However, in the analyzed countries, only physicians, nurses/midwives, or paramedics have the right to administer agents parenterally. Hence, the appropriate personnel who is certified to administer the vitamin infusions if no information is provided on the website remains unclear. Taken together, a lack of essential information creates concerns regarding the safety and reliability of the services.

Alternative treatments are often offered as management for conditions with highly subjective symptoms [40]. Indeed, most of the recommended indications were general health problems. In such settings, the placebo effect tends to be more perceptible [40]. The promotion via social media and positive comments or good ratings may support the enhanced placebo effects through social proof phenomenon [41]. The consumers may perceive that many others were satisfied by the service. For these reasons, parenteral supplement consumers may experience health improvements. Positive attitude and the desire to improve health are essential even for individuals with malignancy [42]. However, whether personal funds should be spent on an evidence-based intervention that is not gaining the placebo effect due to the use of vitamin intravenous infusions needs to be explored further. Another concern is the number of consumers that discontinued conventional treatment as they were persuaded by alternative medicine enthusiasts who shared their positive experience after receiving the supplements.

The most offered ingredients, such as vitamins, macronutrients or micronutrients, glutathione, and amino acids, are dietary supplements and are among the most popular searches in Google [17]. Many of the searched ingredients are used in hospital settings such as electrolytes, vitamin B complex, or glucose. Probably, many of the intravenous formulas used in facilities providing parenteral supplementation were purchased from companies that produce them for clinical practice. Performing a detailed analysis of the ingredients allowed in each country is beyond the scope of this study. However, the very concern is offering solutions that are not commonly used by clinicians. Further studies should focus on the source of parenteral supplements that are not also used in conventional medicine.

In this cross-sectional study, we characterized the companies offering parenteral supplementation based on their websites. This branch of complementary medicine seemed to be a wellness service rather than an evidence-based intervention. Although previous studies on parenteral supplementation were at high risk of bias, the absence of evidence does not mean that such practices were not utilized. The freedom of action on the market is an essential foundation of Western Civilization. However, this is an unjustifiable and unfair competition with conventional medicine by claiming the unproven effects of the supplementation. Moreover, many websites lack crucial information, which creates concerns about the quality and reliability of the services. The efficacy and safety of vitamin infusions services require further research.

4.2. Limitations

Our study has several limitations. We created an original data collection form from the initial analysis of a sample of 5 websites from each country. The relatively small sample of initially analyzed websites could be a source of bias in constructing the form. The analyzed countries were chosen arbitrarily, and the results cannot be extrapolated to the other countries. The arbitral choice resulted in including Ireland, which had only one city with at least 250,000 citizens, and a diverse number of found websites, e.g., 121 in Poland and only 14 in Ireland. For this reason, the characteristics of the services in these countries are not comparable. We included services offered in the cities with more than 250,000 citizens; thus, the analysis does not include services provided in smaller towns and does not represent the whole picture of each country's parenteral supplementation industry. The study only focuses on facilities that possess websites. We cannot assess the number of facilities that do not have a website and the representativeness of our sample. Furthermore, the information on services is not definitive, and consumers may obtain different information or ingredients in the facility. Finally, the study is cross-sectional, and the content of the analyzed websites may be outdated during the period data collection or may change in the future.

5. Conclusions

The market of parenteral supplementation offers numerous ingredients as treatment for general health problems and serious health conditions. Many analyzed websites lack essential information, which creates concerns regarding the quality and reliability of the services.

Supplementary Materials: The following are available online at http://www.mdpi.com/2072-6643/12/12/3614/s1. Figure S1: The workflow of the collection process of links to the websites. The first step was to choose the city (the most left upper box) to add it to the query (the most right upper box). We checked all websites suggested by Google and saved the links of websites representing parenteral supplementation services. Cz—Czech, En—English, It—Italian, Pl—Polish. Table S1: Form for collecting characteristics from websites of parental supplementation services. Table S2: Indications, contraindications, and ingredients mentioned in less than in 5% of websites in each country. Data presented as number (%). Table S3: The other services offered in the analyzed facilities.

Author Contributions: Conceptualization, M.K.; methodology, M.K., M.S., and M.K.-N.; software, M.K.; validation, M.K., M.S., and M.K.-N.; formal analysis, M.K.; investigation, M.K., M.S., A.W., P.R., and S.S.; resources, M.K., M.S., A.W., P.R., and S.S.; data curation, M.K., M.S., A.W., P.R., and S.S.; writing—original draft preparation, M.K.; writing—review and editing, M.K., M.K.-N., M.S., A.W., P.R., S.S., and P.B.; visualization, M.K.; supervision, M.K.-N. and P.B.; project administration, M.K. All authors have read and agreed to the published version of the manuscript.

Funding: This research received no external funding.

Conflicts of Interest: The authors declare no conflict of interest.

References

1. Rimmer, A. Sixty seconds on ... vitamin drips. *BMJ* **2019**, *366*, l4596. [CrossRef] [PubMed]
2. Kamiński, M.; Kręgielska-Narożna, M.; Soczewka, M.; Bogdański, P. Why Polish patients use vitamin drips services: Results of a preliminary cross-sectional survey. *Pol. Arch. Intern. Med.* **2020**. [CrossRef] [PubMed]
3. Sams, L. Science or not, IV "wellness" drips are booming. *Financial Review*, 5 April 2019.

4. Bilg, R. A Rapid Evidence Assessment on the Effectiveness of Intravenous Mega-Dose Multivitamins on Fibromyalgia, Chronic Fatigue, Cancer, and Asthma. Master's Thesis, The University of British Columbia, Vancouver, BC, Canada, 2017. [CrossRef]
5. Gavura, S. A Closer Look at Vitamin Injections. Science Based Medicine Website. Available online: https://sciencebasedmedicine.org/a-closer-look-at-vitamin--injections (accessed on 25 February 2020).
6. *England's Top Doctor Slams 'Exploitative' Party Drips*; NHS: London, UK, 2019.
7. Iqbal, N. Celebrities help the £500 vitamin jab go mainstream. *The Guardian*, 3 March 2019.
8. Vitamin Drips. Hit or Great Scam? [In Polish] Dzień Dobry TVN Website. Available online: https://dziendobry.tvn.pl/a/wlewy-witaminowe-hit-czy-wielkie-oszustwo (accessed on 25 February 2020).
9. Death of a 36-Year Old Woman in a Poznań Hospital. She Came from a Natural Medicine Clinic [In Polish]. Wprost. Available online: https://www.wprost.pl/polityka/10185708/smierc-36-latki-w-poznanskim-szpitalu-trafila-tam-z-kliniki-medycyny-naturalnej.html (accessed on 6 June 2020).
10. *Unsafe Use of Glutathione as Skin Lightening Agent*; FDA Advisory: Silver Spring, MD, USA, 2019.
11. Madsen, B.K.; Hilscher, M.; Zetner, D.; Rosenberg, J. Adverse reactions of dimethyl sulfoxide in humans: A systematic review. *F1000Research* **2019**, *7*, 1746. [CrossRef] [PubMed]
12. Lenz, J.R.; Degnan, D.; Hertig, J.B.; Stevenson, J.G. A Review of Best Practices for Intravenous Push Medication Administration. *J. Infus. Nurs.* **2017**, *40*, 354–358. [CrossRef]
13. Eysenbach, G. Infodemiology and Infoveillance. *Am. J. Prev. Med.* **2011**, *40*, S154–S158. [CrossRef]
14. Sinnenberg, L.; Buttenheim, A.M.; Padrez, K.; Mancheno, C.; Ungar, L.; Merchant, R.M. Twitter as a Tool for Health Research: A Systematic Review. *Am. J. Public Health* **2017**, *107*, e1–e8. [CrossRef]
15. Kamiński, M.; Borger, M.; Prymas, P.; Muth, A.; Stachowski, A.; Łoniewski, I.; Marlicz, W. Analysis of Answers to Queries among Anonymous Users with Gastroenterological Problems on an Internet Forum. *Int. J. Environ. Res. Public Health* **2020**, *17*, 1042. [CrossRef]
16. Baudischova, L.; Zubrova, J.; Pokladnikova, J.; Jahodar, L. The quality of information on the internet relating to top-selling dietary supplements in the Czech Republic. *Int. J. Clin. Pharm.* **2018**, *40*, 183–189. [CrossRef]
17. Kamiński, M.; Kręgielska-Narożna, M.; Bogdanski, P. Determination of the Popularity of Dietary Supplements Using Google Search Rankings. *Nutrients* **2020**, *12*, 908. [CrossRef]
18. Page, S.A.; Mannion, C.; Bell, L.H.; Verhoef, M. CAM information online: An audit of Internet information on the "Bill Henderson Protocol". *Complement. Ther. Med.* **2010**, *18*, 206–214. [CrossRef]
19. Ernst, E.; Schmidt, K. 'Alternative' cancer cures via the Internet? *Br. J. Cancer* **2002**, *87*, 479–480. [CrossRef] [PubMed]
20. Jensen, R.K.; Agersted, M.E.I.; Nielsen, H.A.; O'Neill, S. A cross-sectional study of website claims related to diagnoses and treatment of non-musculoskeletal conditions. *Chiropr. Man. Ther.* **2020**, *28*, 16. [CrossRef] [PubMed]
21. Sharma, V.; Holmes, J.H.; Sarkar, I.N. Identifying Complementary and Alternative Medicine Usage Information from Internet Resources: A Systematic Review. *Methods Inf. Med.* **2016**, *55*, 322–332. [CrossRef] [PubMed]
22. Opendatasoft. Geonames—All Cities with a Population. Available online: https://data.opendatasoft.com/explore/dataset/geonames-all-cities-with-a-population-1000%40public/table/?disjunctive.country (accessed on 2 January 2020).
23. Dijkstra, L.; Poelman, H. *Cities in Europe. The New OECD-EC Definition*; European Commission: Brussels, Belgium, 2012.
24. Witek, B.; Tabara, J.; Świderska-Kołacz, G.; Kołątaj, A. *The Reduced Glutathione Level and Glutathione Enzymes Activity in Connection with the Blood Storage*; Wydawnictwo Uniwersytetu Jana Kochanowskiego: Kielce, Poland, 2015; ISBN 978-83-7133-640-9.
25. Labrou, N. *Glutathione: Biochemistry, Mechanisms of Action & Biotechnological Implications*; Labrou, N., Flemetakis, E., Eds.; Microbiology Research Advances; Nova Science Publishers: New York, NY, USA, 2013; ISBN 978-1-62417-460-5.
26. Gutman, J.; Schettini, S.; Bounous, G. *Glutathione: Your Key to Health*; Kudo.ca Communications Inc.: Montréal, QC, Canada, 2008; ISBN 978-0-9731409-1-0.
27. Ługowski, M.; Saczko, J.; Kulbacka, J.; Banaś, T. Reactive oxygen and nitrogen species. *Pol. Merkur. Lekarski* **2011**, *31*, 313–317.

28. Vollbracht, C.; Schneider, B.; Leendert, V.; Weiss, G.; Auerbach, L.; Beuth, M.J. Intravenous vitamin C administration improves quality of life in breast cancer patients during chemo-/radiotherapy and aftercare: Results of a retrospective, multicentre, epidemiological cohort study in Germany. *In Vivo* **2011**, *25*, 983–990.
29. Yeom, C.H.; Jung, G.C.; Song, K.J. Changes of Terminal Cancer Patients' Health-related Quality of Life after High Dose Vitamin C Administration. *J. Korean Med Sci.* **2007**, *22*, 7–11. [CrossRef]
30. Mikirova, N.A.; Casciari, J.; Rogers, A.; Taylor, P. Effect of high-dose intravenous vitamin C on inflammation in cancer patients. *J. Transl. Med.* **2012**, *10*, 189. [CrossRef]
31. Mikirova, N.A.; Casciari, J.; Riordan, N.H.; Hunninghake, R. Clinical experience with intravenous administration of ascorbic acid: Achievable levels in blood for different states of inflammation and disease in cancer patients. *J. Transl. Med.* **2013**, *11*, 191. [CrossRef]
32. Fowler, A.A.; A Syed, A.; Knowlson, S.; Sculthorpe, R.; Farthing, D.; Dewilde, C.; A Farthing, C.; Larus, T.L.; Martin, E.; Brophy, D.F.; et al. Phase I safety trial of intravenous ascorbic acid in patients with severe sepsis. *J. Transl. Med.* **2014**, *12*, 32. [CrossRef]
33. Marik, P.E.; Khangoora, V.; Rivera, R.; Hooper, M.H.; Catravas, J. Hydrocortisone, Vitamin C, and Thiamine for the Treatment of Severe Sepsis and Septic Shock. *Chest* **2017**, *151*, 1229–1238. [CrossRef]
34. Biniaz, V.; Shermeh, M.S.; Ebadi, A.; Tayebi, A.; Einollahi, B. Effect of Vitamin C Supplementation on C-reactive Protein Levels in Patients Undergoing Hemodialysis: A Randomized, Double Blind, Placebo-Controlled Study. *Nephro-Urol. Mon.* **2013**, *6*, e13351. [CrossRef] [PubMed]
35. Hemilä, H. Zinc lozenges and the common cold: A meta-analysis comparing zinc acetate and zinc gluconate, and the role of zinc dosage. *JRSM Open* **2017**, *8*. [CrossRef]
36. Gaby, A.R. Intravenous nutrient therapy: The "Myers' cocktail". *Altern. Med. Rev. J. Clin. Ther.* **2002**, *7*, 389–403.
37. Delgado-López, P.D.; Corrales-García, E.M. Influence of Internet and Social Media in the Promotion of Alternative Oncology, Cancer Quackery, and the Predatory Publishing Phenomenon. *Cureus* **2018**, *10*, e2617. [CrossRef] [PubMed]
38. Hansen, M.M.; Grajales, F.J.; Martin-Sanchez, F.; Bamidis, P.D.; Miron-Shatz, T. Social Media for the Promotion of Holistic Self-Participatory Care: An Evidence Based Approach. *Yearb. Med. Inform.* **2013**, *22*, 162–168. [CrossRef]
39. George, D.R.; Rovniak, L.S.; Kraschnewski, J.L. Dangers and Opportunities for Social Media in Medicine. *Clin. Obstet. Gynecol.* **2013**, *56*, 453–462. [CrossRef] [PubMed]
40. Kaptchuk, T.J. The Placebo Effect in Alternative Medicine: Can the Performance of a Healing Ritual Have Clinical Significance? *Ann. Intern. Med.* **2002**, *136*, 817–825. [CrossRef] [PubMed]
41. Cialdini, R.B. *Influence: The Psychology of Persuasion*, Revised Edition; Morrow: New York, NY, USA, 1993; ISBN 978-0-688-12816-6.
42. Carver, C.S.; Smith, R.G.; Antoni, M.H.; Petronis, V.M.; Weiss, S.; DerHagopian, R.P. Optimistic Personality and Psychosocial Well-Being During Treatment Predict Psychosocial Well-Being Among Long-Term Survivors of Breast Cancer. *Health Psychol.* **2005**, *24*, 508–516. [CrossRef]

Article

A 12-Week, Multicenter, Randomized, Double-Blind, Placebo-Controlled Clinical Trial for Evaluation of the Efficacy and Safety of DKB114 on Reduction of Uric Acid in Serum

Yu Hwa Park [1,†], Do Hoon Kim [1,†], Jung Suk Lee [2], Hyun Il Jeong [2], Kye Wan Lee [2] and Tong Ho Kang [1,*]

1 Department of Oriental Medicine Biotechnology, Global Campus, Graduate School of Biotechnology and College of Life Sciences, Kyung Hee University, Gyeonggi 17104, Korea; best1rd@gmail.com (Y.H.P.); hoony3914@gmail.com (D.H.K.)

2 R&D Center, Dongkook Pharm. Co., Ltd., Gyeonggi 16229, Korea; ljs@dkpharm.co.kr (J.S.L.); jhi@dkpharm.co.kr (H.I.J.); lkw1@dkpharm.co.kr (K.W.L.)

* Correspondence: panjae@khu.ac.kr

† These authors contributed equally to this work.

Received: 17 November 2020; Accepted: 9 December 2020; Published: 10 December 2020

Abstract: This study sought to investigate the antihyperuricemia efficacy and safety of DKB114 (a mixture of *Chrysanthemum indicum* Linn flower extract and *Cinnamomum cassia* extract) to evaluate its potential as a dietary supplement ingredient. This clinical trial was a randomized, 12-week, double-blind, placebo-controlled study. A total of 80 subjects (40 subjects with an intake of DKB114 and 40 subjects with that of placebo) who had asymptomatic hyperuricemia (7.0–9.0 mg/dL with serum uric acid) was randomly assigned. No significant difference between the DKB114 and placebo groups was observed in the amount of uric acid in serum after six weeks of intake. However, after 12 weeks of intake, the uric acid level in serum of subjects in the DKB114 group decreased by 0.58 ± 0.86 mg/dL and was 7.37 ± 0.92 mg/dL, whereas that in the placebo group decreased by 0.02 ± 0.93 mg/dL and was 7.67 ± 0.89 mg/dL, a significant difference ($p = 0.0229$). In the analysis of C-reactive protein (CRP) change, after 12 weeks of administration, the DKB114 group showed an increase of 0.05 ± 0.27 mg/dL ($p = 0.3187$), while the placebo group showed an increase of 0.10 ± 0.21 mg/dL ($p = 0.0324$), a statistically significant difference ($p = 0.0443$). In the analysis of amount of change in apoprotein B, after 12 weeks of administration, the DKB114 group decreased by 4.75 ± 16.69 mg/dL ($p = 0.1175$), and the placebo group increased by 3.13 ± 12.64 mg/dL ($p = 0.2187$), a statistically significant difference between the administration groups ($p = 0.0189$). In the clinical pathology test, vital signs and weight measurement, and electrocardiogram test conducted for safety evaluation, no clinically significant difference was found between the ingestion groups, confirming the safety of DKB114. Therefore, it may have potential as a treatment for hyperuricemia and gout. We suggest that DKB114 as a beneficial and safe food ingredient for individuals with high serum uric acid. Trial registration (CRIS.NIH. go. Kr): KCT0002840.

Keywords: dietary supplements; asymptomatic hyperuricemia; DKB114; *Chrysanthemum indicum* Linn; *Cinnamomum cassia*; clinical trial; uric acid

---

1. Introduction

Hyperuricemia is a condition characterized by an abnormally elevated level of serum uric acid [1,2]. Uric acid is the final oxidation product of purine metabolism in humans in the absence of the hepatic enzyme uricase. Increased production and/or decreased uric acid excretion elevate serum uric acid level. The former is caused by an excessively purine-rich diet and purine metabolism overactivation, whereas the latter is caused by renal impairment and certain drugs. Hyperuricemia is diagnosed when serum uric acid level exceeds the limit of solubility (7.0 mg/dL) and increases the risk of monosodium urate or uric acid crystal deposition, which could result in acute gouty arthritis, gouty arthropathy, chronic tophaceous gout, uric acid urolithiasis, or gouty nephropathy [3–6].

Thus, long-term reduction of serum urate concentration to subsaturation level is vital in the treatment of gout [7]. Many epidemiological studies have shown that hyperuricemia and gout are associated with hypertension, cardiovascular disease, chronic kidney disease and diabetes, potentially through crystal-independent modes of action [8–11]. Clinical management of serum uric acid level often includes using a xanthine oxidase inhibitor (allopurinol) and uricosurics (probenecid and benzbromarone), which facilitate urinary excretion. However, their use can induce several adverse reactions, such as fever, skin rash, worsened renal function, and Stevens–Johnson syndrome [7,12–18].

To prevent gouty arthritis, cardiovascular disease, and renal failure, the Japanese guidelines for the management of hyperuricemia and gout recommend initiating pharmacologic urate-lowering therapy for asymptomatic hyperuricemia when serum urate level increases > 8.0 mg/dL [5]. The risk-benefit balance for using such drugs among gout-free patients with hyperuricemia is not favorable according to the guidelines for the management of hyperuricemia and gout [5,19–21]. Therefore, non-medication treatments, including a low-purine diet, exercise therapy, and natural products, are recommended for gout-free individuals with non-significantly high serum uric acid. In particular, it is commonly assumed that regular ingestion of dietary supplements is easier than dietetics or exercise therapy for individuals with non-significantly high serum uric acid and no gout pain. Although the majority of physicians have not regarded dietary supplements to be efficacious on hyperuricemia, the ingredients have been investigated for hypouricemic activity.

Our previous study demonstrated that DKB114 (a mixture of *Chrysanthemum indicum* Linn flower extract and *Cinnamomum cassia* extract) reduced serum uric acid level in normal rats and rats with PO-induced hyperuricemia and promoted excretion of uric acid in the urine, indicating that DKB114 has an anti hyperuricemic effect and may be a potent uricosuric agent. In addition, DKB114 inhibited xanthine oxidase activity and hepatic uric acid production in vitro and in vivo, as well as cellular uptake of uric acid in vitro [22].

Therefore, the present study evaluated the efficacy and safety of DKB114 intake in asymptomatic hyperuricemia (uric acid > 7.0 mg/dL) on blood uric acid reduction compared to Placebo (control foods) [23].

2. Materials and Methods

2.1. DKB114 Preparation

DKB114 (Dongkook Pharm. Co. Ltd, Suwon, Korea, Lot number is BPD170926-1) was prepared as described previously. Briefly, DKB114 was selected and prepared by mixing previously prepared *C. indicum* flower and *C. cassia* bark extracts at a weight ratio of 1:2 based on our previous report [22].

2.2. Study Design

This was a randomized, placebo-controlled, double-blinded clinical trial with a 3-month follow-up. The current protocol was registered at Neonutra Co. Ltd., Jongno, Korea (protocol number: DKP_DKB114, Version 3.3) on 14 February 2018, sponsored by Dongkook Pharm Co., Ltd. The participants were recruited from the Korean Medicine Hospital of Daejeon University, H Plus Yangji Hospital, and Bundang Jesaeng General Hospital. The study was fully conducted in accordance with the Declaration of Helsinki, and its protocol was approved by the Ethical Committee of the Institute Review Board of Daejeon University Dunsan Korean Medicine Hospital (Project identification code: DKP_DKB114; Submitted approval number: DJDSKH-17-BM-28; CRIS. NIH. go. Kr ID: KCT0002840).

Treatment was administered for 12 weeks, during which the participants were provided necessary guidance on maintaining daily diets and activities. During the clinical trial period, subjects were educated to maintain their usual diet, physical activity, and dietary intake. During the clinical trial period, subjects were educated to maintain their usual diet, physical activity, and dietary intake.

During the clinical trial testing period, the subjects were instructed to maintain the form of meals, physical activity, dietary intake was consumed as usual.

Foods related to *Cinnamomum cassia* and *Chrysanthemum indicum* Linn flower and health functional foods were not eaten. In addition, dietary survey papers were distributed to the test subjects at visit 3 and were made to fill out during the test period. The subjects were required to visit the hospital five times throughout the study. The assessments were scheduled as screening (week −2 to 0), baseline (week 0), interim (at the end of week 6), and final (at the end of week 12). The flow chart of the trial is shown in Figure 1.

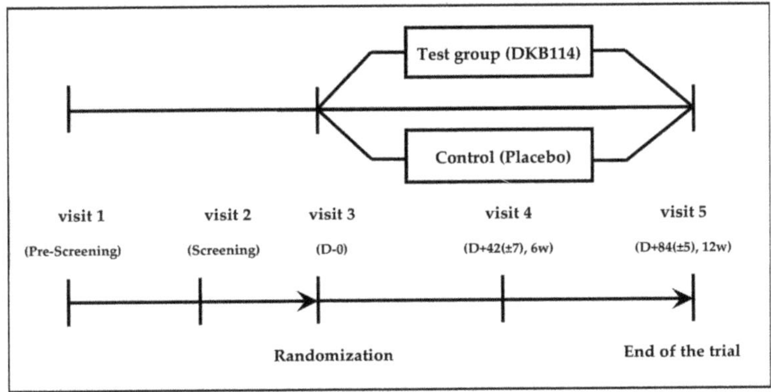

Figure 1. Flow diagram of the trial.

2.3. *Ethics Approval and Consent of the Participants*

The trial protocol (DKP_DKB114, Version 3.3, dated 22 February 2019) was approved by the ethics committee of the Korean Medicine Hospital of Daejeon University, H Plus Yangji Hospital, Bundang Jesaeng general Hospital. Before the start of the clinical trial, the participant received information on the nature, scope, and expected results of the trial and provided written consent to participate. All clinical trial subject names were kept confidential and were recorded and confirmed with the number assigned during the trial. Subjects were informed that all test data were stored on a computer and treated strictly confidentially.

2.4. *Participants*

2.4.1. Inclusion Criteria

Subjects who meet all the below criteria were qualified and enrolled in the study:

- Male or female adults aged between 20 and 75 years;
- Uric acid in serum concentration: more than 7.0 mg/dL to less than 9.0 mg/dL;
- Subjects who agree to participate in the clinical trial voluntarily and sign the informed consent form.

2.4.2. Exclusion Criteria

Subjects who have one or more of the following characteristics were not considered for the study.

- Diagnosed with gout;
- Severe cerebrovascular disease or severe heart disease;
- Mental diseases such as schizophrenia, depressive disorder, and drug dependence;
- Alcohol abuse or dependence;

- Use of uric acid inhibitor (allopurinol, febuxostat, probenecid, etc.);
- Use of thiazide diuretics 4 weeks before the start of the test;
- History of urolithiasis;
- Hypertension not controlled by drugs, with greater than 160 mmHg systolic blood pressure or 100 mmHg diastolic blood pressure;
- Uncontrolled diabetes (HbA1c greater than 8%);
- Aspartate aminotransferase (GOT) or alanine aminotransferase (GPT) level 2 times higher than the normal upper limit;
- Creatinine higher than 2 mg/dL;
- Pregnant, lactating, or planning to become pregnant within 3 months;
- Participated in other clinical trials within 2 months of the start of the clinical trial or plan to participate in other clinical trials during this trial period;
- Difficulties participating in the trial as judged by the investigator.

2.4.3. Withdrawal and Dropout

Cases in which the clinical trial was discontinued for a test subject in progress were as follows:

- Violation of inclusion criteria/exclusion criteria;
- Serious adverse event;
- Withdrawal of consent to participate;
- Use of drugs that may affect the results during the clinical period;
- Any safety reason;
- Pregnancy.

2.5. Recruitment

The participants will be recruited from the Korean Medicine Hospital of Daejeon University, H Plus Yangji Hospital, and Bundang Jesaeng General Hospital, South Korea. Recruitment for the clinical trial was advertised through posters in public places like hospitals and subways after approval from the institutional review board.

2.6. Randomization and Blinding

A randomization list was created by an independent analyst using block randomization in SAS version 9.2 (USA). Participants were given a random code and received treatment products marked with the same code. For balanced randomization between intake groups, the subjects of each group were equalized at a 1:1 ratio.

All participants and the research personnel were blinded to the assigned treatment until the end of the trial. In this clinical trial, no unblinding trials occurred.

2.7. Intervention

The test product was a 1 g purple tablet containing DKB114 functional ingredient (500 mg), crystalline cellulose, magnesium stearate, silicon dioxide, hydroxypropyl methylcellulose, glycerin fatty acid ester, titanium dioxide, lac color, and gardenia blue SB. A 1 g placebo product contained crystalline cellulose, silicon dioxide, magnesium stearate, hydroxypropyl methylcellulose, glycerin fatty acid ester, titanium dioxide, lac color, and gardenia blue SB. Participants were instructed to take 4 tablets each day, 2 in the morning and 2 at night, for a total dose of 4 g per day for 12 weeks (2 g/day as DKB114 functional ingredient). DKB114 or placebo was recommended to be administered after meals. In a rat model of calcium oxonate-induced hyperuricemia, the serum uric acid concentration decreased in the dkb114_200 mg/kg group. In addition, in the acute and chronic hyperuricemia animal models, it was confirmed that the serum uric acid concentration decreased, and the uric acid excretion concentration increased in the dkb114_200 mg/kg group. The value converted using the HED (human equivalent dose) conversion factor is 200/6.2 = 32.3 mg/kg, and when this value is calculated based on a 60 kg male, a dose of about 2.0 g is calculated. The value converted using the HED (human equivalent dose) conversion factor is 200/6.2 = 32.3 mg/kg, and when this value is calculated based on a 60 kg male, a dose of about 2.0 g is calculated.

HED (human equivalent dose) value was converted by using the conversion factor is to be 200/6.2 = 32.3 mg/kg, when calculating the value to 60 kg man is calculated based on a dose of about 2.0 g.

Based on these results, the daily intake in humans was determined as 2 g.

The DKB114 and placebo products were manufactured by COSMAX BIO Co., Ltd. (Gyeonggi, Korea) in a GMP-certified facility. The trial schedule is presented in Table 1.

Table 1. Trial schedule.

Period		**Pre-Screening**	**Screening**	**Active Treatment**		
Visit		1	2	3	4	5
Week		−3	−2	0	6	12
Window period					+/−7	+/−5
Written consent		√				
Demographic survey			√			
Medical history/drug administration		√	√	√		
Physical examination			√		√	√
Vital signs (pulse rate, blood pressure)		√	√	√	√	√
ECG			√			√
Body measurement	Height			√		
	Weight, BMI, waist circumference			√	√	√
Drinking habits questionnaire				√	√	√
Meal guidance and dietary survey				√	√	√
Clinical pathology			√			√
Pregnancy test			√			
Effectiveness evaluation	Uric acid in serum	√			√	√
	Xanthine oxidase activity in serum		√		√	√
	Blood sugar		√		√	√
	CRP		√		√	√
	Homocysteine		√		√	√
	TNF-α		√		√	√
	IL-6		√		√	√
	NO		√		√	√
	Apoprotein B		√		√	√
Suitability of subjects		√	√	√		
Randomization				√		
Test and control food prescription				√	√	
Adverse reaction assessment					√	√
Compliance assessment					√	√
Concomitant medication and combination therapy confirmation					√	√

ECG: electrocardiogram; CRP: C-reactive protein; IL-6: interleukin-6; TNF-α: tumor necrosis factor alpha; NO: nitric oxide.

2.8. Prohibited Concomitant Drugs and Therapies

The use of the following drugs and treatments could interfere with the evaluation of the safety, efficacy, and tolerability of the test product and were limited during the clinical trial.

- Uric acid inhibitors (allopurinol, febuxostat, probenecid, etc.);

- Uric acid release accelerator (urinon, narcaricin, etc.);
- Urate oxidase;
- Thiazide diuretics.

2.9. Outcome Evaluation

The following outcomes were evaluated by trained evaluators at each visit.

2.9.1. Primary Outcomes

The mean uric acid in serum was compared before ingestion (visit 1: week-3) and after ingestion (visits 4, 5: weeks 6, 12). The degree of improvement in the DKB114 group and the placebo group was analyzed and compared to evaluate whether there was a statistically significant difference.

2.9.2. Secondary Outcomes

The proportion of subjects with a uric acid serum concentration less than 7.0 mg/dL before ingestion (visit 1: week-3) and after ingestion (visits 4, 5: weeks 6, 12) was compared. Before ingestion (visit 2: week-2) and after ingestion (visits 4, 5: weeks 6, 12), the average xanthine oxidase activity in blood, blood sugar, C-reactive protein (CRP), homocysteine, TNF-α, IL-6, NO, and apoprotein B levels were compared. The degree of improvement in the DKB114 group and the placebo group was analyzed and compared to evaluate whether there was a statistically significant difference.

2.9.3. Safety Outcomes

The safety outcome variables were adverse events, vital signs (blood pressure, pulse rate, body weight, etc.), clinical test results, and electrocardiogram results.

2.10. Sample Size

Among the existing studies conducted with a design similar to this clinical trial, the number of clinical subjects was calculated using the results of Hideki et al., a clinical trial using blood uric acid, which had the same endpoint as this clinical trial [24]. As a result, the minimum number of subjects was 35 per group. Considering a dropout rate of 12%, the number of clinical trial subjects to be enrolled per group for efficacy evaluation was 40, and the total number was 80.

2.11. Statistical Analysis

Statistical analyses were performed using SAS (version 9.4, SAS Institute, Cary, NC, USA).

The efficacy evaluation analysis and safety evaluation analysis were performed at a significance level of 0.05 with a two-sided test. If the p-value was <0.05, it was judged to be significant.

Uric acid in the serum—the primary efficacy evaluation—was analyzed using a paired t-test. The degree of change between groups at each visit was evaluated for statistically significant differences by conducting ANCOVA and two-sample t-tests.

The degree of change in blood xanthine oxidase activity, blood sugar, CRP, homocysteine, TNF-a, IL-6, NO, and apoprotein B level before and after ingestion were analyzed as the secondary efficacy evaluation using paired t-test. The degree of change between groups at each visit was evaluated for statistically significant differences by conducting ANCOVA and two-sample t-tests. The proportion of subjects with a uric acid concentration in the serum <7.0 mg/dL was compared and analyzed using the chi-squared test or Fisher's exact test.

3. Results

This study was conducted at Daejeon University Dunsan Oriental Medicine Hospital, H Plus Yangji Hospital, and Bundang Jesaeng Hospital from February 2018 to July 2019. In this clinical trial, a screening evaluation was conducted on a total of 140 subjects to select suitable participants. A total of 80 subjects (40 subjects in the DKB114 group and 40 subjects in the placebo group) was randomly assigned to 60 screening eliminations. After 8 subjects withdrew consent, 4 in the test group and 4 in the control group, a total of 72 subjects completed the clinical trial (DKB114 group 36 subjects, placebo group 36 subjects).

In the per-protocol set (PP set), five people (four DKB114 groups, one placebo group) who dropped out after visit 4 were excluded. In addition, and 4 violations of more than 5 days of visit (4 placebo group) through the analysis group decision, 9 people with less than 80% compliance (3 DKB114 groups, 6 placebo group), and 1 overdose with greater than 130% secondary compliance (DKB114 group) were excluded. Thus, a total of 58 subjects (32 DKB114 groups, 26 placebo group) were included in the PP set (Figure 2)

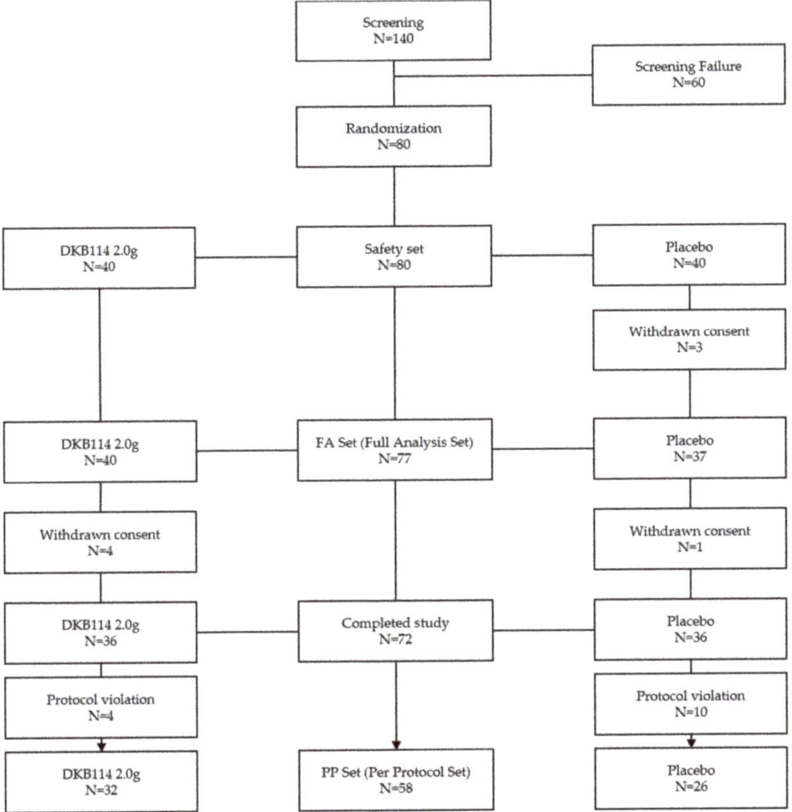

Figure 2. Disposition of clinical trial subjects.

All characteristics before ingestion, including demographic information of the subjects, were compared by ingestion group to identify factors with differences. Table 2 compares demographic information and pre-ingestion characteristics of clinical subjects.

The test group comprised 28 males (88.46%) and 4 females (12.50%), while the control group contained 23 males (88.46%) and 3 females (11.54%). There was no statistically significant difference between the intake groups. In terms of age, the DKB114 group averaged 41.84 ± 14.75 years, and the placebo group averaged 40.19 ± 12.53 years. There was no statistically significant difference between the intake groups ($p = 0.6523$). In addition, there were no statistically significant differences in smoking status, smoking amount, smoking period, and exercise status, demonstrating comparability between intake groups.

Table 2. Baseline characteristics (per-protocol set (PP) set).

		DKB114 N = 32	Placebo N = 26	Total N = 58	p-Value
Sex n (%)	Male	28 (87.50)	23 (88.46)	51 (87.93)	1.0000 ‡
	Female	4 (12.50)	3 (11.54)	7 (12.07)	
Age (years)	Mean ± SD	41.84 ± 14.75	40.19 ± 12.53	41.10 ± 13.71	0.6523 *
	Min, max	20.00, 69.00	23.00, 67.00	20.00, 69.00	
Smoking n (%)	Non-smoker	18 (56.25)	14 (53.85)	32 (55.17)	0.9367 †
	Ex-smoker (not smoking for more than 6 months)	5 (15.63)	5 (19.23)	10 (17.24)	
	Smoker	9 (28.13)	7 (26.92)	16 (27.59)	
Smoking amount	Smoker, (___) cigarettes/1 day				
	Mean ± SD	9.89 ± 7.72	10.14 ± 5.81	10.00 ± 6.73	0.9434 *
	Min, max	1.00, 20.00	3.00, 20.00	1.00, 20.00	
Smoking period	Smoker, (___) year				
	Mean ± SD	12.56 ± 6.84	11.43 ± 10.42	12.06 ± 8.29	0.7977 *
	Min, max	2.00, 21.00	2.00, 30.00	2.00, 30.00	
Exercise or not n (%)	No	12 (37.50)	10 (38.46)	22 (37.93)	0.8633 ‡
	1–2 times/week	8 (25.00)	8 (30.77)	16 (27.59)	
	3 times/week	5 (15.63)	4 (15.38)	9 (15.52)	
	4–5 times/week	4 (12.50)	1 (3.85)	5 (8.62)	
	Every day	3 (9.38)	3 (11.54)	6 (10.34)	

* p-value by two-sample t-test; † p-value by chi-squared test; ‡ p-value by Fisher's exact test.

3.1. Clinical Parameters

3.1.1. Primary Outcomes

Table 3 shows the results of the analysis of changes in serum uric acid measured at 0 weeks, 6 weeks, and 12 weeks after intake in the PP set. No significant differences between the DKB114 and placebo groups were observed in the amount of uric acid in serum after six weeks of intake. The mean 6-week uric acid in serum did not differ significantly between the DKB114 (7.47 ± 1.03 mg/dL) and placebo (7.54 ± 1.01 mg/dL) (p = 0.1734) groups. However, after 12 weeks of intake, the uric acid in serum of subjects in the DKB114 group decreased by 0.58 ± 0.86 mg/dL to 7.37 ± 0.92 mg/dL, whereas that in the placebo group decreased by 0.02 ± 0.93 mg/dL to 7.67 ± 0.89 mg/dL, a significant difference (p = 0.0229 *).

Table 3. Changes in uric acid in serum by visit (PP set).

		DKB114 N = 32		Placebo N = 26		p-Value	p-Value $
		n	Mean ± SD	n	Mean ± SD		
Uric acid in serum (mg/dL)	Baseline (visit 1)	32	7.94 ± 0.51	26	7.70 ± 0.64	0.0740 &	
	6 weeks (visit 4)	32	7.47 ± 1.03	26	7.54 ± 1.01	0.1734 &	0.2966
	Change from baseline	32	−0.47 ± 0.94	26	−0.15 ± 0.91		
	p-value **		0.0082		0.3978		
	12 weeks (visit 5)	32	7.37 ± 0.92	26	7.67 ± 0.89	0.0229 *	0.0646
	Change from baseline	32	−0.58 ± 0.86	26	−0.02 ± 0.93		
	p-value **		0.0007		0.9003		

Compared between groups, p-value by two-sample t-test; & Compared between groups: p-value by Wilcoxon rank-sum test; ** Compared within groups: p-value by paired t-test; $ Compared between groups: p-value by ANCOVA adjusted for baseline.

3.1.2. Secondary Outcomes

The proportion of subjects with a uric acid serum concentration <7.0 mg/dL before ingestion (visit 1: week-3) and after ingestion (visits 4, 5: weeks 6, 12) was compared. Before ingestion (visit 2: week-2) and after ingestion (visits 4, 5: weeks 6, 12), the average xanthine oxidase activity in blood, blood sugar, CRP, homocysteine, TNF-α, IL-6, NO, and apoprotein B levels were compared. The degree of improvement between the DKB114 group and the placebo group was analyzed and compared to evaluate whether there was a statistically significant difference.

Table 4 shows the proportion of subjects with serum uric acid concentration <7.0 mg/dL at 0 weeks, 6 weeks, and 12 weeks of intake by PP set. Such patients numbered 12 (37.5%) in the DKB114 group and 5 (19.23%) in the placebo group after 6 weeks, and there was no statistically significant difference between groups. After 12 weeks, there were 11 such patients (34.38%) in the DKB114 group and 4 (15.38%) in the placebo group. There was no statistically significant difference between the intake groups.

Table 4. Proportion of subjects with uric acid concentration in serum <7.0 mg/dL (PP set).

Percentage of Subjects with Serum Uric Acid Concentration <7.0 mg/dL		DKB114 N = 32, n (%)	Placebo N = 26, n (%)	*p*-Value †
Baseline (visit 1)	Less than 7.0 mg/dL	0 (0.00)	0 (0.00)	-
	More than 7.0 mg/dL	32 (100.00)	26 (100.00)	
6 weeks (visit 4)	Less than 7.0 mg/dL	12 (37.50)	5 (19.23)	0.1285
	More than 7.0 mg/dL	20 (62.50)	21 (80.77)	
12 weeks (visit 5)	Less than 7.0 mg/dL	11 (34.38)	4 (15.38)	0.1005
	More than 7.0 mg/dL	21 (65.63)	22 (84.62)	

† *p*-value by chi-squared test.

However, the distribution of serum uric acid levels in recruited subjects (Table 5) showed that 12 weeks after intake, the DKB114 group tended to decrease serum uric acid levels compared to the placebo group.

Table 5. Distribution of serum uric acid levels.

Uric Acid in Serum (mg/dL)	Subjects (n)			
	Visit 0		Visit 5	
	DKB114	Placebo	DKB114	Placebo
<7.0	0	0	11	4
7.0–7.5	7	13	10	6
7.6–8.0	12	4	3	8
8.1–8.5	9	4	6	5
8.6–9.0	4	5	0	2
9.0<	0	0	2	1

Table 6 shows the results of the analysis of changes in blood xanthine oxidase activity, blood glucose, C-reactive protein (CRP), homocysteine, IL-6, NO, and apoprotein B measured at 0, 6, and 12 weeks of administration. In the analysis of CRP change, after 6 weeks of administration, the DKB114 group decreased by 0.01 ± 0.15 mg/dL (p = 0.7613), and the placebo group increased by 0.17 ± 0.73 mg/dL (p = 0.2402), with no statistically significant difference between the intake groups. After 12 weeks of administration, the DKB114 group increased by 0.05 ± 0.27 mg/dL (p = 0.3187), and the placebo group increased by 0.10 ± 0.21 mg/dL (p = 0.0324), a statistically significant difference (p = 0.0443 [&]).

Table 6. Changes in serum xanthine oxidase activity, blood sugar, CRP (C-reactive protein), homocysteine, IL-6, NO, and apoprotein B during the study (PP set).

		DKB114		Placebo			
		N = 32		N = 26		*p*-Value [&]	*p*-Value [$]
		n	Mean ± SD	n	Mean ± SD		
Xanthine oxidase activity in serum (%)	Baseline (visit 2)	32	0.70 ± 0.40	26	0.84 ± 0.85	0.7486	
	6 weeks (visit 4)	32	0.74 ± 0.57	26	0.71 ± 0.93		
	Change from baseline	32	0.04 ± 0.66	26	−0.13 ± 1.35	0.4343	0.9274
	p-value **		0.7615		0.6233		
	12 weeks (visit 5)	32	0.74 ± 0.74	26	0.76 ± 0.48		
	Change from baseline	32	0.04 ± 0.71	26	−0.07 ± 1.02	0.8696	0.9418
	p-value **		0.7768		0.7109		
Blood glucose (mg/dL)	Baseline (visit 2)	32	99.34 ± 18.11	26	95.62 ± 12.87	0.6897	
	6 weeks (visit 4)	32	96.94 ± 13.16	26	96.81 ± 10.53		
	Change from baseline	32	−2.41 ± 16.34	26	1.19 ± 13.41	0.3320	0.6971
	p-value **		0.4111		0.6542		
	12 weeks (visit 5)	32	96.41 ± 15.92	26	94.96 ± 8.34		
	Change from baseline	32	−2.94 ± 15.32	26	−0.65 ± 11.10	0.4113	0.9112
	p-value **		0.2863		0.7664		
CRP (mg/dL)	Baseline (visit 2)	32	0.12 ± 0.12	26	0.14 ± 0.15	0.7279	
	6 weeks (visit 4)	32	0.12 ± 0.13	26	0.31 ± 0.75		
	Change from baseline	32	−0.01 ± 0.15	26	0.17 ± 0.73	0.2495	0.1746
	p-value **		0.7613		0.2402		
	12 weeks (visit 5)	32	0.17 ± 0.29	26	0.23 ± 0.31		
	Change from baseline	32	0.05 ± 0.27	26	0.10 ± 0.21	0.0443 [&]	0.5137
	p-value **		0.3187		0.0324		
Homocysteine (μmol/L)	Baseline (visit 2)	32	13.69 ± 8.19	26	10.78 ± 3.47	0.0990	
	6 weeks (visit 4)	32	13.21 ± 8.21	26	10.74 ± 2.32		
	Change from baseline	32	−0.48 ± 2.36	26	−0.05 ± 2.91	0.8819	0.8788
	p-value **		0.2563		0.9366		
	12 weeks (visit 5)	32	12.49 ± 5.46	26	10.95 ± 2.72		
	Change from baseline	32	−1.20 ± 3.76	26	0.16 ± 2.82	0.2841	0.7375
	p-value **		0.0808		0.7723		
TNF-α (pg/mL)	Baseline (visit 2)	32	1.00 ± 0.32	26	0.85 ± 0.31	0.0724	
	6 weeks (visit 4)	32	0.98 ± 0.35	26	0.95 ± 0.33		
	Change from baseline	32	−0.02 ± 0.27	26	0.11 ± 0.35	0.1229	0.3942
	p-value **		0.6536		0.1337		
	12 weeks (visit 5)	32	0.95 ± 0.35	26	0.97 ± 0.39		
	Change from baseline	32	−0.05 ± 0.33	26	0.13 ± 0.48	0.1009	0.4023
	p-value **		0.3974		0.1853		
IL-6 (pg/mL)	Baseline (visit 2)	32	1.75 ± 1.07	26	1.43 ± 0.82	0.1571	
	6 weeks (visit 4)	32	1.72 ± 1.27	26	1.49 ± 0.82		
	Change from baseline	32	−0.04 ± 1.10	26	0.06 ± 0.65	0.1525	0.9909
	p-value **		0.8472		0.6140		
	12 weeks (visit 5)	32	1.66 ± 1.38	26	1.75 ± 1.68		
	Change from baseline	32	−0.10 ± 1.12	26	0.33 ± 1.78	0.5368	0.4342
	p-value **		0.6256		0.3571		
NO (μmol/L)	Baseline (visit 2)	32	56.82 ± 45.34	26	49.22 ± 33.32	0.8573	
	6 weeks (visit 4)	32	49.14 ± 34.38	26	62.04 ± 52.91		
	Change from baseline	32	−7.68 ± 49.20	26	12.82 ± 49.43	0.2082	0.1683
	p-value **		0.3841		0.1979		
	12 weeks (visit 5)	32	69.44 ± 53.84	26	62.11 ± 42.51		
	Change from baseline	32	12.62 ± 62.65	26	12.89 ± 44.19	0.6901	0.6956
	p-value **		0.2631		0.1494		
Apoprotein B (mg/dL)	Baseline (visit 2)	32	108.29 ± 23.01	26	113.73 ± 32.65	0.4611 *	
	6 weeks (visit 4)	32	104.66 ± 17.31	26	112.41 ± 33.07		
	Change from baseline	32	−3.64 ± 19.55	26	−1.32 ± 9.64	1.000&	0.3540
	p-value **		0.3007		0.4917		
	12 weeks (visit5)	32	103.54 ± 21.42	26	116.85 ± 30.82		
	Change from baseline	32	−4.75 ± 16.69	26	3.13 ± 12.64	0.0518 *	0.0189 [$]
	p-value **		0.1175		0.2187		

* Compared between groups: *p*-value by two-sample *t*-test; [&] Compared between groups: *p*-value by Wilcoxon rank-sum test; ** Compared within groups: *p*-value by paired *t*-test; [$] Compared between groups; *p*-value for ANCOVA adjusted for baseline.

In the analysis of change in apoprotein B, after 6 weeks of administration, the DKB114 group decreased by 3.64 ± 19.55 mg/dL ($p = 0.3007$), and the placebo group decreased by 1.32 ± 9.64 mg/dL ($p = 0.4917$), with no statistically significant difference. After 12 weeks of administration, the DKB114 group decreased by 4.75 ± 16.69 mg/dL ($p = 0.1175$), and the placebo group increased by 3.13 ± 12.64 mg/dL ($p = 0.2187$), a statistically significant difference ($p = 0.0189^{\$}$).

In the analysis of changes in blood xanthine oxidase activity, after 6 weeks, the DKB114 group increased by 0.04 ± 0.66% ($p = 0.7615$), and the placebo group decreased by 0.13 ± 1.35% ($p = 0.6233$), with no statistically significant difference. After 12 weeks of administration, the DKB114 group increased by 0.04 ± 0.71% ($p = 0.7768$), and the placebo group decreased by 0.07 ± 1.02% ($p = 0.7109$), with no statistically significant difference.

In the analysis of blood glucose change, after 6 weeks of administration, the DKB114 group decreased by 2.41 ± 16.34 mg/dL ($p = 0.4111$), and the placebo group increased by 1.19 ± 13.41 mg/dL ($p = 0.6542$), with no statistically significant difference. After 12 weeks of administration, the DKB114 group decreased by 2.94 ± 15.32 mg/dL ($p = 0.2863$), and the placebo group decreased by 0.65 ± 11.10 mg/dL ($p = 0.7664$), with no statistically significant difference.

Regarding the amount of change in homocysteine, IL-6, and NO at each visit, there was no significant difference between the DKB114 group and the placebo group at 6 and 12 weeks of administration.

3.1.3. Safety Evaluation

The safety evaluation was carried out as the main analysis of the safety set analysis, and a total of 80 subjects (40 subjects in the DKB114 group and 40 subjects in the placebo group) who consumed the food for the clinical trial at least once after being randomized was included in the analysis. In addition, clinical pathologic examination (serum chemistry test), vital signs (heart rate, blood pressure), and weight results were analyzed (Tables 7 and 8). No serious adverse reactions occurred, and there were no dropouts due to adverse reactions.

Table 7. Serum chemistry tests (safety set).

		DKB114		**Placebo**		
		N = 32		**N = 26**		***p*-Value ***
		n	**Mean ± SD**	**n**	**Mean ± SD**	
Total cholesterol (mg/dL)	Baseline (visit 2)	40	201.38 ± 32.93	40	202.80 ± 48.16	0.8777
	12 weeks (visit 5)	36	194.03 ± 30.35	36	213.22 ± 47.11	
	Change from baseline	36	−5.94 ± 17.00	36	10.03 ± 19.72	0.0005 *
	p-value **		0.0432		0.0043	
LDL-cholesterol (mg/dL)	Baseline (visit 2)	40	121.05 ± 28.52	40	122.80 ± 40.30	0.8235
	12 weeks (visit 5)	36	112.17 ± 25.15	36	129.47 ± 39.21	
	Change from baseline	36	−8.89 ± 19.10	36	6.31 ± 21.07	0.0020 *
	p-value **		0.0084		0.0812	
HDL-cholesterol (mg/dL)	Baseline (visit 2)	40	54.26 ± 14.49	40	47.63 ± 10.87	0.0232
	12 weeks (visit 5)	36	52.68 ± 11.55	36	48.87 ± 10.45	
	Change from baseline	36	−0.39 ± 7.23	36	0.82 ± 6.35	0.4548
	p-value **		0.7506		0.4439	
Triglycerides (mg/dL)	Baseline (visit 2)	40	178.78 ± 139.42	40	243.33 ± 256.06	0.1666
	12 weeks (visit 5)	36	183.11 ± 139.99	36	226.42 ± 152.07	
	Change from baseline	36	11.44 ± 77.71	36	−7.36 ± 245.47	0.6635
	p-value **		0.3829		0.8582	
HbA1c (%)	Baseline (visit 2)	40	5.42 ± 0.33	40	5.49 ± 0.50	0.4437
	12 weeks (visit 5)	36	5.55 ± 0.32	36	5.48 ± 0.30	
	Change from baseline	36	0.12 ± 0.16	36	0.08 ± 0.14	0.3405
	p-value **		0.00001		0.0009	
AST(GOT) (IU/L)	Baseline (visit 2)	40	27.78 ± 139.42	40	26.13 ± 6.71	0.1666
	12 weeks (visit 5)	36	29.14 ± 14.67	36	33.78 ± 42.16	
	Change from baseline	36	0.83 ± 13.06	36	7.78 ± 41.84	0.3473
	p-value **		0.7041		0.2723	

Table 7. *Cont.*

		DKB114		Placebo		p-Value *
		N = 32		N = 26		
		n	Mean ± SD	n	Mean ± SD	
ALT(GPT) (IU/L)	Baseline (visit 2)	40	33.15 ± 15.06	40	29.93 ± 11.61	0.2867
	12 weeks (visit 5)	36	36.64 ± 31.00	36	34.06 ± 23.75	
	Change from baseline	36	3.44 ± 22.29	36	3.67 ± 20.20	0.9648
	p-value **		0.3602		0.2835	
BUN (mg/dL)	Baseline (visit 2)	40	13.84 ± 3.46	40	13.44 ± 3.27	0.5964
	12 weeks (visit 5)	36	14.38 ± 4.01	36	14.14 ± 3.87	
	Change from baseline	36	0.62 ± 3.68	36	0.94 ± 3.79	0.7202
	p-value **		0.3192		0.1477	
Creatinine (mg/dL)	Baseline (visit 2)	40	1.02 ± 0.16	40	1.02 ± 0.14	0.8028
	12 weeks (visit 5)	36	0.98 ± 0.14	36	1.02 ± 0.14	
	Change from baseline	36	−0.04 ± 0.10	36	0.00 ± 0.08	0.0700
	p-value **		0.0374		0.8631	

* Compared between groups: p-value by two-sample t-test; ** Compared within groups: p-value by paired t-test.

Table 8. Vital signs (heart rate, blood pressure) and weight change (safety set).

		DKB114		Placebo		p-Value *
		N = 32		N = 26		
		n	Mean ± SD	n	Mean ± SD	
Heart rate (count/min)	Baseline (visit 3)	40	75.33 ± 9.45	40	77.15 ± 9.56	0.3932
	12 weeks (visit 5)	36	77.69 ± 11.62	36	78.69 ± 13.02	
	Change from baseline	36	2.58 ± 7.54	36	1.14 ± 11.96	0.5422
	p-value **		0.0474		0.5713	
Systolic pressure (mmHg)	Baseline (visit 3)	40	128.80 ± 11.82	40	125.58 ± 10.38	0.1987
	12 weeks (visit 5)	36	128.22 ± 10.21	36	127.06 ± 10.67	
	Change from baseline	36	0.14 ± 11.57	36	2.14 ± 10.49	0.4448
	p-value **		0.9430		0.2292	
Diastolic pressure (mmHg)	Baseline (visit 3)	40	80.40 ± 10.17	40	78.90 ± 7.97	0.4650
	12 weeks (visit 5)	36	79.47 ± 8.72	36	81.31 ± 8.14	
	Change from baseline	36	0.47 ± 6.98	36	2.39 ± 10.12	0.3532
	p-value **		0.6873		0.1655	
Weight (kg)	Baseline (visit 3)	40	82.56 ± 15.27	40	79.24 ± 12.47	0.2904
	12 weeks (visit 5)	36	80.91 ± 12.92	36	80.21 ± 13.18	
	Change from baseline	36	−0.38 ± 3.98	36	0.52 ± 2.38	0.2518
	p-value **		0.5724		0.2016	

* Compared between groups: p-value by two-sample t-test; ** compared within groups: p-value by paired t-test.

In the analysis of changes in total cholesterol during the serum chemistry test, after 12 weeks of intake, the DKB114 group decreased by 5.94 ± 17.00 mg/dL (p = 0.0432), and the placebo group increased by 10.03 ± 19.72 mg/dL (p = 0.0043), a significant difference (p = 0.0005). In the analysis of LDL cholesterol change, after 12 weeks of administration, the DKB114 group decreased by 8.89 ± 19.10 mg/dL (p = 0.0084), and the placebo group increased by 6.31 ± 21.07 mg/dL (p = 0.0812), a significant difference (p = 0.0020). All values were within the normal ranges, and there was no statistically significant difference between the administration groups in other serum chemical tests.

In the analysis of vital signs (pulse, blood pressure) and weight change, there were no statistically significant differences between intake groups at 6 weeks and 12 weeks.

4. Discussion

Hyperuricemia is diagnosed when serum uric acid level exceeds the limit of solubility (7.0 mg/dL), which increases the risk of monosodium urate or uric acid crystal deposition and can result in acute gouty arthritis, gouty arthropathy, chronic tophaceous gout, uric acid urolithiasis, or gouty nephropathy (6). In addition, hyperuricemia is a risk factor for cardiovascular diseases (25). Treatment of asymptomatic hyperuricemia is not necessary for most patients unless they have a very high level of uric acid or are otherwise at risk of complications, such as those with a personal or strong family history of gout, urolithiasis, or uric acid nephropathy. However, no direct role of hyperuricemia in the pathogenesis or outcome of these conditions was confirmed.

There is a need to develop a functional ingredient that can prevent gout and related complications caused by uric acid in the blood. In a previous study, we performed in vitro and in vivo tests using DKB-114, a natural functional ingredient. DKB114 markedly reduced serum uric acid level in normal rats and rats with PO-induced hyperuricemia while increasing renal uric acid excretion. Furthermore, it inhibited the activity of xanthine oxidase (XOD) in vitro and in the liver, in addition to reducing hepatic uric acid production. DKB114 decreased cellular uric acid uptake in oocytes, and HEK293 cells expressing human urate transporter (hURAT)1 and decreased the protein expression levels of urate transporters, URAT1, and glucose transporter, GLUT9, associated with reabsorption of uric acid in the kidney. DKB114 exerts antihyperuricemic effects and uricosuric effects, which are accompanied by a reduction in the production of uric acid and promotion of uric acid excretion via inhibition of xanthine oxidase activity and reabsorption of uric acid (22).

The present study aimed to investigate the antihyperuricemia efficacy and safety of DKB114 to evaluate its potential as a functional food ingredient. The present study performed a randomized controlled study and investigated the effect of oral administration of DKB114 on serum uric acid levels in 7.0–9.0 mg/dL with insignificantly high serum uric acid.

As an efficacy evaluation index, uric acid concentration in serum, the proportion of subjects with serum uric acid concentration <7.0 mg/dL, blood xanthine oxidase activity, blood glucose, CRP, homocysteine, TNF-α, IL-6, NO, and apoprotein B were measured. Among them, uric acid concentration in serum, CRP, and apoprotein B were clinically significantly improved in the DKB114 intake group compared to the placebo group. After 12 weeks of intake, the serum uric acid concentration in the DKB114 group was significantly decreased compared to before intake, but the placebo group had no statistical significance compared to before ingestion; as a result, a statistically significant difference was found between intake groups ($p = 0.0229$).

In the analysis of CRP change, after 12 weeks of intake, the DKB114 group increased by 0.05 ± 0.27 mg/dL ($p = 0.3187$), and the placebo group increased by 0.10 ± 0.21 mg/dL ($p = 0.0324$), a significant difference ($p = 0.0443$). C-reactive protein (CRP) is a protein that reacts with c-polysaccharide, a surface antigen of Streptococcus pneumoniae, and is one of the acute phase reactants whose concentration changes in a nonspecific response to inflammation or tissue damage. Through the intake of DKB114, CRP significantly changed compared to the placebo group, and DKB114 can be expected to act positively on the inflammatory response as well as uric acid reduction.

At 12 weeks after ingestion of apoprotein B, the level in the DKB114 group decreased, while that of the placebo group increased, showing a statistically significant difference between groups ($p = 0.0189$). Cardiovascular disease-related apoprotein B decreased through DKB114 intake but increased in the placebo group. There was no statistically significant difference between the groups in the proportion of subjects with serum uric acid concentration <7.0 mg/dL. However, after 12 weeks of ingestion, 11 patients in the DKB114 group and 4 patients in the placebo group showed a tendency to improve.

DKB114 showed no statistically significant difference in adverse reactions between groups, all of which were mild. Thus, it was judged that there was no effect of the food on the clinical results.

In the clinical pathology test, vital signs and weight measurement, and electrocardiogram test conducted for safety evaluation, no clinically significant difference was found between the ingestion groups, confirming the safety of DKB114. These findings indicate its potential as a treatment for hyperuricemia and gout. However, the group of subjects recruited (DKB114, controls) could be small to draw conclusions about the efficacy of DKB114. In the process of selecting only those who meet the per protocol (PP) set criteria among recruited subjects, the number of subjects in the controls group decreased. Based on the results of the current clinical trial, additional trials are being planned by increasing the number of subjects.

5. Conclusions

For the treatment of asymptomatic hyperuricemia in Japan, it is recommended to achieve blood uric acid concentration <7.0 mg/dL more and less than 9.0 mg/dL, and medicines are recommended for those with blood uric acid concentration ±9.0 mg/dL [5]. The United States and Europe do not specifically recommend drug treatment for asymptomatic hyperuricemia [19,20]. However, if the blood uric acid level is more than 9.0 mg/dL, the possibility of kidney damage increases, so uric acid-lowering medication is considered according to the patient [25]. Reducing uric acid production and increasing uric acid excretion may be a useful therapeutic approach for hyperuricemia treatment [26]. Currently, xanthine oxidase inhibitors such as allopurinol and febuxostat and uricosuric agents such as benzbromarone and probenecid are used for clinical hyperuricemia treatment [26]. However, these drugs are poorly tolerated and induce side effects, such as drug allergies, gastrointestinal symptoms, kidney disease, hypersensitivity syndrome, and hepatotoxicity [27–30]. Thus, new therapeutic strategies with minimal side effects are needed. An increasing interest in natural products, such as herbal medicines, has elucidated new approaches that can overcome these limitations, suggesting natural products as a promising pool of candidates for drugs and functional foods for hyperuricemia management [30].

We suggest DKB114 as a beneficial and safe food ingredient for individuals with high serum uric acid.

Author Contributions: Conceptualization, D.H.K., Y.H.P., J.S.L., H.I.J. and K.W.L.; writing—review and editing, D.H.K., Y.H.P. and T.H.K. All authors have read and agreed to the published version of the manuscript.

Funding: This research was supported by the INNOPOLIS Foundation (2017-01-DD-031), funded by the Korean government (Ministry of Science and ICT).

Conflicts of Interest: The authors declare that they have no conflicts of interest.

References

1. Brook, R.A.; Forsythe, A.; Smeeding, J.E.; Lawrence, E.N. Chronic gout: Epidemiology, disease progression, treatment and disease burden. *Curr. Med. Res. Opin.* **2010**, *26*, 2813–2821. [CrossRef]
2. Caroline, L.B.; Pinky, D.; Rachel, G.; Peter, L.; Pike, A.; Ian, S.R.; Ciara, V. Physiology of Hyperuricemia and Urate-Lowering Treatments. *Front. Med.* **2018**, *5*, 160.
3. Shin, Y.T.; Kim, K.K.; Hwang, I.C. Clinical Implication of Plasma Uric acid level. *Korean J. Fam. Med.* **2009**, *30*, 670–680.
4. John, K.; Daniel, I.F.; Richard, J.J. Does asymptomatic hyperuricaemia contribute to the development of renal and cardiovascular disease? An old controversy renewed. *Nephrology* **2004**, *9*, 394–399.
5. Hisashi, Y. Essence of the Revised Guideline for the management of hyperuricemia and gout. *JMAJ* **2012**, *55*, 324–329.
6. Wortmann, R.L. Gout and hyperuricemia. *Curr. Opin. Rheumatol.* **2002**, *14*, 281–286. [CrossRef] [PubMed]
7. Song, J.S. New Classification Criteria and Guideline for Management of Gout. *Korean J. Med.* **2018**, *93*, 344. [CrossRef]
8. Choi, H.K.; Karen, A.; Elizabeth, W.K.; Gary, C. Obesity, Weight Change, Hypertension, Diuretic Use, and Risk of Gout in Men: The health professionals follow-up study. *Arch. Intern. Med.* **2005**, *165*, 742–748. [CrossRef]
9. Roubenoff, R.; Klag, M.J.; Mead, L.A.; Liang, K.Y.; Seidler, A.J.; Hochberg, M.C. Incidence and risk factors for gout in white men. *JAMA* **1991**, *266*, 3004–3007. [CrossRef]
10. Manuel, M.A.; Steele, T.H. Changes in renal urate handling after prolonged thiazide treatment. *Am. J. Med.* **1974**, *57*, 741–746. [CrossRef]
11. Roubenoff, R. Gout and hyperuricemia. *Rheum. Dis. Clin. N. Am.* **1990**, *16*, 539–550.
12. Roujeau, J.C.; Kelly, J.P.; Naldi, L.; Rzany, B.; Stern, R.S.; Anderson, T.; Auquier, A.; Bastuji-Garin, S.; Correia, O.; Locati, F. Medication use and the risk of Stevens-Johnson syndrome or toxic epidermal necrolysis. *N. Engl. J. Med.* **1995**, *333*, 1600–1607. [CrossRef] [PubMed]

13. Hande, K.R.; Noone, R.M.; Stone, W.J. Severe allopurinol toxicity. Description and guidelines for prevention in patients with renal insufficiency. *Am. J. Med.* **1984**, *76*, 47–56. [CrossRef]
14. Karen, P.; Peter, J.G.; Nicola, D. Efficacy and Tolerability of Probenecid as Urate-lowering Therapy in Gout; clinical experience in high-prevalence population. *J. Rheumatol.* **2013**, *40*, 872–876.
15. Dalbeth, N.; Kumar, S.; Stamp, L.; Gow, P. Dose adjustment of allopurinol according to creatinine clearance does not provide adequate control of hyperuricemia in patients with gout. *J. Rheumatol.* **2006**, *33*, 1646–1650.
16. Sunil, K.; Jennifer, N.G.; Peter, G. Benzbromarone therapy in management of refractory gout. *N. Z. Med. J.* **2005**, *118*, U1528.
17. Thompson, G.R.; Duff, I.F.; Robinson, W.D.; Mikkelsen, W.M. Long term uricosuric therapy in gout. *Arthritis. Rheum.* **1962**, *5*, 384–396. [CrossRef]
18. Yu, T.F.; Gutman, A.B. Effect of allopurinol (4-hydroxypyrazolo (3,4-d) pyrimidine) on serum and urinary uric acid in primary and secondary gout. *Am. J. Med.* **1964**, *37*, 885–898. [CrossRef]
19. Richette, P.; Doherty, M.; Pascual, E.; Barskova, V.; Becce, F.; Castañeda-Sanabria, J.; Coyfish, M.; Guillo, S.; Jansen, T.L.; Janssens, H.; et al. 2016 updated EULAR evidence-based recommendations for the management of gout. *Ann. Rheum. Dis.* **2017**, *76*, 29. [CrossRef]
20. Khanna, D.; Fitzgerald, J.D.; Khanna, P.P.; Bae, S.; Singh, M.K.; Neogi, T.; Kaldas, M. 2012 American college of Rheumatology guidelines for management of gout. Part 1: Systematic nonpharmacologic and pharmacologic therapeutic approaches to hyperuricemia. *Arthritis Care Res.* **2012**, *64*, 1431–1446. [CrossRef]
21. Michelle, H.; Alison, C.; Stewart, C.; Graham, D.; Michael, D.; Harry, F.; Wendy, J.; Kelsey, M.J.; Christian, D.M.; Thomas, M.M.; et al. The British Society for Rheumatolgy Guideline for the management of Gout. *Rheumatology* **2017**, *56*, 1056.
22. Lee, Y.S.; Kim, S.H.; Yuk, H.G.; Kim, D.S. DKB114, A Mixture of *Chrysanthemum indicum* Linne Flower and *Cinnamomum cassia* (L.) J. Presl Bark Extracts, Improves Hyperuricemia through Inhibition of Xanthine Oxidase Activity and Increasing Urine Excretion. *Nutrients* **2018**, *10*, 1381. [CrossRef] [PubMed]
23. Erhan, D.H.; Ayse, P.D.; Dennis, J.L. Asymptomatic hyperuricemia: To treat or not to treat. *Clevel. Clin. J. Med.* **2002**, *69*, 594–597.
24. Hideki, H.; Toshiro, O.; Tetsuro, Y.; Leo, A.; Kazuhisa, O. Effects of a fermented barley extract on subjects with slightly high serum uric acid or mild hyperuricemia. *Biosci. Biotechnol. Biochem.* **2010**, *74*, 828–834.
25. Song, J.S. Recent advances in management of gout. *J. Korean Med. Assoc.* **2016**, *59*, 379. [CrossRef]
26. Ishibuchi, S.; Morimoto, H.; Oe, T.; Ikebe, T.; Inoue, H.; Fukunari, A.; Kamezawa, M.; Yamada, I.; Naka, Y. Synthesis and structure-activity relationships of 1-phenylpyrazoles as xanthine oxidase inhibitors. *Bioorg. Med. Chem. Lett.* **2001**, *11*, 879–882. [CrossRef]
27. Azevedo, V.F.; Buiar, P.G.; Giovanella, L.H.; Severo, C.R.; Carvalho, M. Allopurinol, benzbromarone, or a combination in treating patients with gout: Analysis of a series of outpatients. *Int. J. Rheumatol.* **2014**, *2014*, 263720. [CrossRef]
28. Bardin, T. Current management of gout in patients unresponsive or allergic to allopurinol. *Jt. Bone Spine* **2004**, *71*, 481–485. [CrossRef]
29. Pacher, P.; Nivorozhkin, A.; Szabo, C. Therapeutic effects of xanthine oxidase inhibitors: Renaissance half a century after the discovery of allopurinol. *Pharmacol. Rev.* **2006**, *58*, 87–114. [CrossRef]
30. Hao, S.; Zhang, C.; Song, H. Natural Products Improving Hyperuricemia with Hepatorenal Dual Effects. *Evid. Based Complement. Altern. Med.* **2016**, *2016*, 7390504. [CrossRef]

 *nutrients*

Review

Scientific Evidence Supporting the Beneficial Effects of Isoflavones on Human Health

Saioa Gómez-Zorita [1,2,3], Maitane González-Arceo [1], Alfredo Fernández-Quintela [1,2,3], Itziar Eseberri [1,2,3,*], Jenifer Trepiana [1,2,3,*] and María Puy Portillo [1,2,3]

1 Nutrition and Obesity Group, Department of Pharmacy and Food Science, University of the Basque Country (UPV/EHU) and Lucio Lascaray Research Institute, 01006 Vitoria, Spain; saioa.gomez@ehu.eus (S.G.-Z.); maitane.gonzalez@ehu.eus (M.G.-A.); alfredo.fernandez@ehu.eus (A.F.-Q.); mariapuy.portillo@ehu.eus (M.P.P.)
2 CIBEROBN Physiopathology of Obesity and Nutrition, Institute of Health Carlos III, 01006 Vitoria, Spain
3 Bioaraba Health Research Institute, 01002 Vitoria, Spain
* Correspondence: itziar.eseberri@ehu.eus (I.E.); jenifer.trepiana@ehu.eus (J.T.)

Received: 30 November 2020; Accepted: 15 December 2020; Published: 17 December 2020

Abstract: Isoflavones are phenolic compounds with a chemical structure similar to that of estradiol. They are present in several vegetables, mainly in legumes such as soy, white and red clover, alfalfa and beans. The most significant food source of isoflavones in humans is soy-derived products. Isoflavones could be used as an alternative therapy for pathologies dependent on hormonal disorders such as breast and prostate cancer, cardiovascular diseases, as well as to minimize menopausal symptoms. According to the results gathered in the present review, it can be stated that there is scientific evidence showing the beneficial effect of isoflavones on bone health and thus in the prevention and treatment of osteoporosis on postmenopausal women, although the results do not seem entirely conclusive as there are discrepancies among the studies, probably related to their experimental designs. For this reason, the results should be interpreted with caution, and more randomized clinical trials are required. By contrast, it seems that soy isoflavones do not lead to a meaningful protective effect on cardiovascular risk. Regarding cancer, scientific evidence suggests that isoflavones could be useful in reducing the risk of suffering some types of cancer, such as breast and endometrial cancer, but further studies are needed to confirm these results. Finally, isoflavones could be useful in reducing hot flushes associated with menopause. However, a limitation in this field is that there is still a great heterogeneity among studies. Lastly, with regard to isoflavone consumption safety, it seems that they are safe and that the most common adverse effect is mild and occurs at the gastrointestinal level.

Keywords: isoflavones; flavonoids; phytoestrogens; soy; bone health; cardiovascular risk; cancer; menopausal symptoms

1. Introduction

Phenolic compounds are secondary metabolites, which are produced by plants as a defense mechanism against infection, water stress, cold stress, ultraviolet radiation and high visible light, among others [1,2]. Phenolic compounds, characterized by the presence of a hydroxyl group attached to at least one aromatic ring, can be classified as flavonoids and non-flavonoids. Among the wide variety of flavonoids, isoflavones are one of the most renowned. Due to its structural similarity to the estrogen-like compound 17β-estradiol (Figure 1), they are referred to as phytoestrogens. Consequently, isoflavones can have estrogenic or anti-estrogenic effects [3].

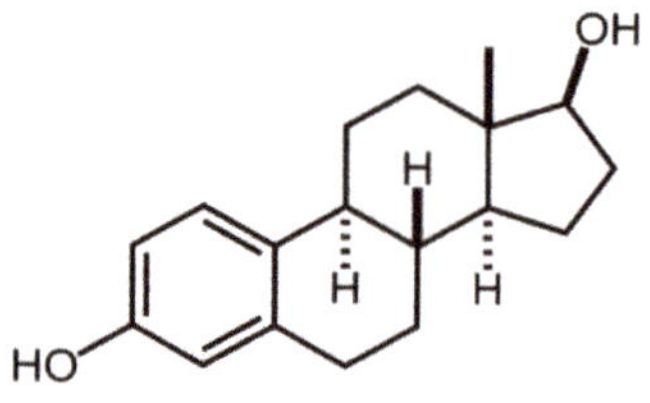

Figure 1. Chemical structure of 17β-estradiol. Modified from Wang et al., 2006 [4].

Isoflavones have received attention due to their putative healthy properties. In this sense, it has been described that both isoflavones and other estrogenic molecules could mediate their beneficial effects due to two different mechanisms: the classical estrogen receptor (ER)-mediated signaling pathway and the activation of intracellular pathways such as protein tyrosine kinase, phospholipase C and mitogen-activated protein kinase (MAPK) [5,6]. In the aforementioned mechanism, it has been reported that isoflavones bind to both ERα and ERβ, although they present a higher affinity towards ERβ receptors [7]. Moreover, even though the binding affinity of isoflavones to the ER receptors is less than that of 17β-estradiol, they can also act as estrogenic compounds when the endogenous estradiol is not available [8]. Thus, its consumption, according to epidemiological and clinical studies, has been postulated to be related to a decrease in the risk of different diseases. The aim of the present review is to gather the scientific evidence existing nowadays on the main beneficial effects of isoflavones on health: bone health, cardiovascular risk, cancer and menopausal symptoms. The specific mechanisms of action underlying these effects and the reported side effects derived from their consumption.

2. Oral Intake and Occurrence

Isoflavones are found in several vegetables, mainly in legumes (*Fabaceae* family) such as soy, white and red clover, alfalfa and beans [9,10]. The most significant food source of isoflavones in humans is soy-derived products, soybeans, soy flour, soy flakes, soy beverages and fermented soy products such as miso and tempeh, among others [9]. Although smaller in quantity, isoflavones are also present in chickpeas, nuts, fruits and vegetables [11].

The main isoflavones found in foodstuffs are daidzein, genistein and glycitein, as well as biochanin A and formononetin [9,12]. Some authors label equol as an isoflavone; however, since it is a daidzein-derived bacterial metabolite and not a naturally occurring isoflavone, this classification is not correct [10]. Nevertheless, it is important to consider this metabolite since it is key in the biological activity of isoflavones [13]. As with other phenolic compounds, isoflavones can be presented as free forms (aglycones) or conjugated with carbohydrates (glycosides). The latter is the most common form in foodstuffs, with the exception of fermented soy products, such as miso and tempeh. In miso and tempeh, the aglycone form is the most abundant one [9]. In soybeans, the three most abundant isoflavones are present in four chemical forms: aglycone, glycoside, acetylglycoside and malonylglucoside (Table 1 and Figure 2). The aglycone and conjugate forms of genistein represent up to 60% of total isoflavones, this being up to 30% in the case of daidzein. The main forms of conjugates in soybeans are the malonyl derivatives [14].

Table 1. Presence of four chemical forms of the three isoflavones found in soybeans.

Isoflavone Content and Chemical Forms in Soybeans	
Aglycones	Daidzein, genistein, glycitein
Glycosides	Daidzin, genistin, glycitin
Acetylglycosides	Acetyldaidzin, acetylgenistin, acetylglycitin
Malonylglycosides	Malonyldaidzin, malonylgenistin, malonylglycitin

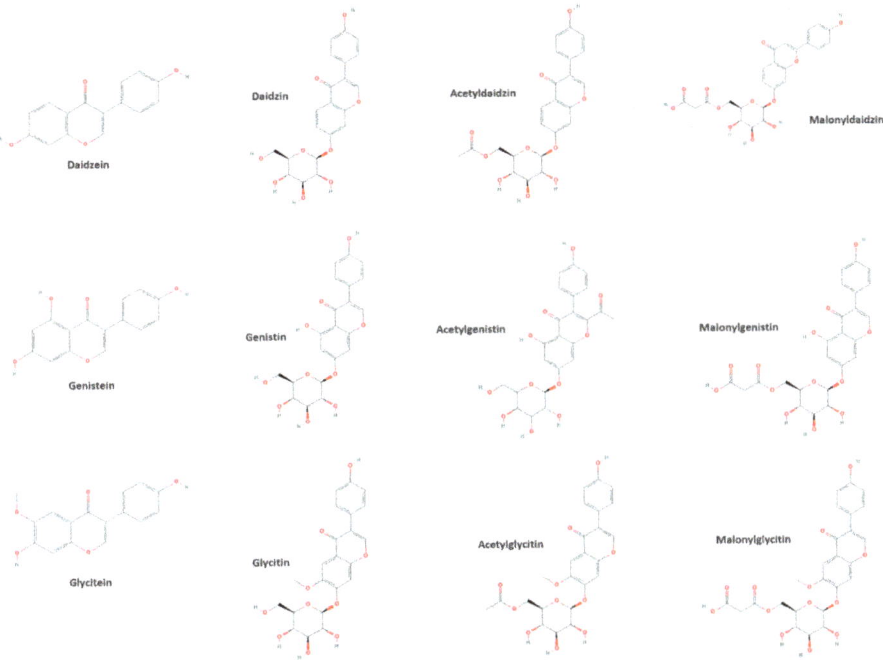

Figure 2. The four chemical forms of the three most abundant isoflavones.

With regard to isoflavone content in foodstuffs, it has been observed that in soybeans, the amount of these phenolic compounds ranges from 1.2 to 4.2 mg per g on dry weight, whereas in red clover, the amount varies between 10 and 25 mg per g on dry weight [15,16]. Even though red clover has a greater amount of isoflavones, it is not an edible plant, and it is consumed as a food complement. Thus, soybean can be considered as the best food source. Isoflavone content in soy-derived products is usually lower than in soybeans [17]. Importantly, the soy variety, environmental conditions of the culture and the processing of soybeans notably influence their isoflavone content [18]. Furthermore, isoflavones are associated with proteins; thus, alcohol extraction and processing diminish isoflavone content in foodstuffs such as fermented soybean, tofu or soy beverages. By contrast, the total isoflavone amount is high in soy flour, protein isolate or edamame [19].

Focusing on the overall intake of isoflavones in humans, several studies have reported varied results. In general terms, it has been observed that in Asian countries where there is high consumption of soy and its derived foodstuffs, isoflavone intake is high, ranging from 15 to 60 mg/day [20,21], whereas, in western countries, it is notably lower, around 1–2 mg/day [22–24].

3. Bioavailability

As previously mentioned, soybeans and their derived foodstuffs are the main isoflavone sources in the human diet, at least if we do not consider red clover extract supplements, which are being widely used by menopausal women as "nutraceutical therapy" [13]. For this reason, the bioavailability of isoflavones has been extensively characterized in animal models and humans [25–27].

Isoflavone glycosides have to be hydrolyzed to the aglycone form prior to its absorption by passive diffusion in the upper small intestine [28]. The enzymes responsible for the hydrolysis of isoflavone glycosides are glucosidases, which can be produced by the intestinal mucosa or the microbiota [29]. The enzyme lactase-phlorizin hydrolase (LPH) deglycosylases phenolic compounds in the intestinal lumen, and after that, the aglycone form enters the epithelial cells by passive diffusion.

Moreover, sodium-dependent glucose transporter 1 (SGLT1) allows phenolic compound-glycosylates to directly enter into epithelial cells, where they are hydrolyzed by cytosolic glucosidases [30,31]. Interestingly, in a study conducted by Tamura et al., the authors observed that in individuals without the LPH enzyme, the concentration of isoflavones and derived metabolites was similar to that found in individuals without an enzyme deficiency. This finding suggests that bacterial deglycosylation contributes significantly to isoflavone absorption [32]. Daidzein, genistein and glycitein occur naturally in the aglycone form, and thus their absorption is faster than that of others.

It has been described that after the oral intake of isoflavones, their peak plasma concentration in humans occurs at around seven hours [33]. As in the case of other phenolic compounds, genistein and daidzein undergo phase II xenobiotic metabolism, mainly glucuronidation and sulfation reactions at 4′and/or 7′ positions of the isoflavone ring [34,35]. Produced metabolites appear in plasma at variable concentrations and can enter enterohepatic circulation [28]. In a study reported by Ko et al., the authors quantified the plasma concentrations of daidzein, genistein, glycitein and equol in Korean men and women [36]. After the consumption of around 17 mg of isoflavones per day, the median plasma concentrations of genistein, glycitein, daidzein and equol were 245.3 ng/mL, 9.8 ng/mL, 86.8 ng/mL and 12.7 ng/mL, respectively.

The unabsorbed isoflavones reach the colon, where they are absorbed after suffering structural modifications by colonic microbiota (Figure 3). Daidzein is first converted into dihydrodaidzein, which is the precursor of both equol and O-demethylangolensin (O-DMA) [37]. Equol has been considered as the most interesting colonic metabolite due to its beneficial biological activity, which differs from that of its precursor, daidzein [13,38,39]. In addition, genistein and glycitein are transformed into dihydrogenistein and dihydroglycitein, respectively, before being further converted into other metabolites. Nevertheless, the metabolism of formononetin and biochanin A, the main isoflavones in red clover, has been less studied due to their limited presence in foodstuffs. Anywise, it has been observed that both isoflavones are demethylated by microbiota, formononetin to daidzein and biochanin A to genistein. After this, each metabolite undergoes its own metabolic pathway [40]. In addition, it is important to highlight that interindividual differences in gut bacterial populations are responsible for the diverse effects of isoflavones in humans. While equol production seems to be similar in animals, it is very variable in humans [41], and therefore, beneficial effects referred to equol have been observed only in individuals with a specific microbiota composition [38].

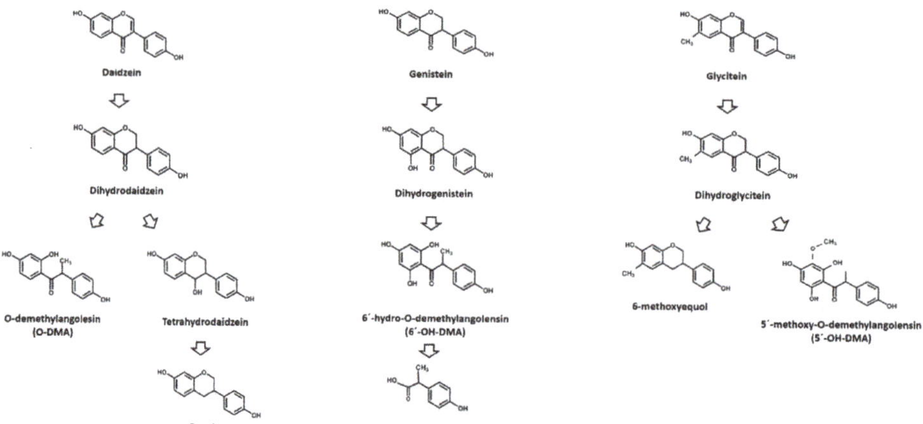

Figure 3. Bacterial metabolites of isoflavones forming in the gut. Modified from Mace et al. and Rossi et al. [12,42].

Isoflavone excretion occurs mainly through urine and feces, primarily in the conjugate forms, and up to 95% takes place in 24 h [43]. In addition, several studies have observed that the urine excretion of daidzein metabolites is notably higher than that of genistein metabolites [44–46]. In a study conducted by Karr et al., the authors analyzed the urinary excretion of genistein, daidzein, equol and O-DMA in fourteen young men and women [47]. They observed that it was dose-dependent at low to moderate soybean consumption.

4. Biological Activity

Isoflavones could be used as an alternative therapy for pathologies dependent on hormonal disorders such as breast and prostate cancer, cardiovascular diseases, as well as to minimize menopausal symptoms. For the development of this section, the meta-analyses and systematic reviews published between 2015 and 2020 were taken into account.

4.1. Bone Health Maintenance

During menopause, there is a loss of bone density that can cause osteoporosis. In this sense, soy isoflavones have been proposed as beneficial because, in theory, they may contribute to the maintenance of good bone health (mass, mineral density and bone structure) in women who are at this stage in their life. Regarding this topic, three systematic reviews and three meta-analyses were analyzed (Table 2).

Table 2. Effects of isoflavones in bone health maintenance.

Authors	Number of Studies Included	Type of Studies Included	Number of Participants and Gender/Age/Characteristics	Compound and Doses	Observed Effects
Meta-analysis					
Lambert et al., 2017 [48]	26	Randomized Clinical trials	2652 estrogen-deficient women	Isoflavones (different forms) Intervention period: ≥3 months	Moderate attenuation of bone loss, primarily at the level of the lumbar spine and the femoral neck
Akhlaghi et al., 2019 [49]	52	Controlled trials	5313 patients	Soy isoflavones 40–300 mg/day Intervention period: 1 month–3 years	Prevention of osteoporosis-related bone loss in any weight status or treatment duration
Sansai et al., 2020 [50]	63	Controlled trials	6427 postmenopausal women	Isoflavones (different forms) Intervention period: 1–36 months	Isoflavone interventions, genistein (54 mg/day) and ipriflavone (600 mg/day) in particular hold great promise in the prevention and treatment of bone mineral density

Table 2. *Cont.*

Authors	Number of Studies Included	Type of Studies Included	Number of Participants and Gender/Age/Characteristics	Compound and Doses	Observed Effects
Systematic reviews					
Abdi et al., 2016 [51]	23	Clinical trials	3494 participants	Isoflavones Intervention duration: 7 weeks–3 years	Probably they have beneficial effects on bone health in menopausal women but there are controversial reports about changes in bone mineral density
Perna et al., 2016 [52]	9	Clinical trials	1379 menopausal and postmenopausal women	Soy isoflavones (20–80 mg) and equol (10 mg)	May be protective in osteoporosis
Chen et al., 2019 [53]	3	1Meta-analysis 1Systematic review and 1clinical trial	3663 menopausal and postmenopausal women	Soy isoflavones	Attenuation of lumbar spine bone mineral density

In the systematic review reported by Perna et al. (2016), nine studies addressed to menopausal women and focused on bone health were analyzed [52]. The authors indicated that no consensus was found regarding the protective effects of soy isoflavones (20–80 mg) and equol (10 mg) on bone resorption. Abdi et al. (2016) published another systematic review of 23 randomized controlled trials to determine the effect of soy isoflavone extracts on bone mineral density in postmenopausal women [51]. The conclusion was that isoflavones exerted little influence over bone mineral density and thus over bone health during menopause, although not all the studies observed the same effect. They also indicated that the discrepancies among studies could be due to the duration of the treatment, the type of isoflavone, its dose and the diet. Regarding the effect of specific isoflavones, they suggested that genistein, alone or in combination with daidzein, improved bone density and bone turnover in women after menopause. However, the authors stated that they could not reach definitive conclusions due to, among other reasons, the lack of inclusion in the review of randomized controlled clinical trials focused on the treatment of osteoporosis in early menopause. In a more recent systematic review, including two meta-analyses, as well as one multicenter and randomized controlled trial, Chen and his team concluded that isoflavones reduced lumbar spine bone mineral density loss [53]. It is important to point out that the spine has a high proportion of trabecular bone and that this is probably the reason why the spine is thought to be more sensitive to isoflavones.

With regard to the published meta-analysis, the studies included were restricted to randomized controlled trials. The work reported by Lambert et al. (2017) included 26 trials with 2652 participants [48]. It supports the evidence that isoflavones can moderately attenuate bone resorption in women with low estrogen levels, primarily at the level of the lumbar spine and the femoral neck. They also note that the effects of isoflavones are greater when administered as aglycones. This fact was not taken into account in previous meta-analyses, and thus it may justify the discrepancies amongst them. In this line, another meta-analysis, which included 52 controlled trials and 5313 patients, shows that soy isoflavones have beneficial effects on bone mineral density in the femur neck, lumbar spine, and hip, regardless of body weight or ethnicity [49]. The improvement was greater in treatments lasting more than one year and in subjects with normal weight, probably because subjects with excessive body weight have a lesser risk of bone loss. In contrast, the effects on bone resorption biomarkers were more favorable in overweight or obese women than in women with a healthy weight. Sansai et al. (2020) carried out their meta-analysis of 63 controlled trials, including 6427 postmenopausal women and concluded that isoflavones improve bone density at the lumbar spine, femoral neck and the distal radius in menopausal women [50]. These positive effects were associated with 54 mg/day of genistein and 600 mg/day of synthetic isoflavone ipriflavone.

In view of these results, we can conclude that there is scientific evidence that supports the beneficial effect of isoflavones on bone health and thus in the prevention and treatment of osteoporosis in postmenopausal women. However, the results do not seem entirely conclusive, for there are discrepancies among the studies, probably related to their experimental designs. For this reason, the results should be interpreted with caution, and more randomized clinical trials are required.

Although the mechanisms of action of isoflavones are not completely understood, it seems that isoflavones not only reduce the rate of bone resorption but also increase the rate of bone formation. The enhanced bone formation is due to the stimulation of osteoblastic activity mainly through (a) the activation of estrogen receptors because they bind to nuclear estrogen receptors and exhibit estrogenic activity due to its similarity to 17β-estradiol, and (b) the promotion of insulin-like growth factor-I (IGF-I) production [54,55] (Figure 4).

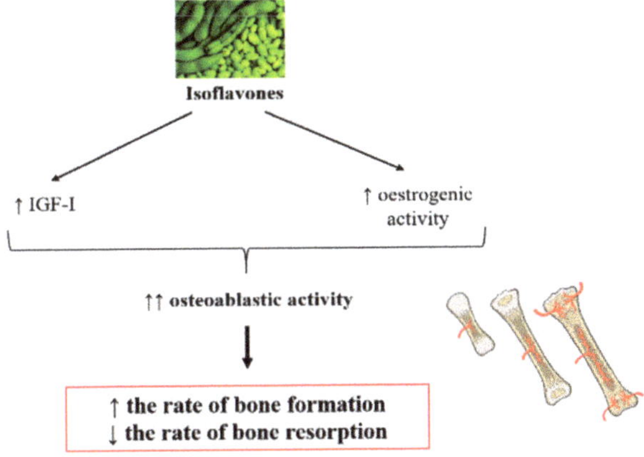

Figure 4. Potential mechanism of action of isoflavones on bone metabolism.

4.2. Cardiovascular Risk

In some Asian countries, isoflavone intake, usually derived from soy consumption, can be associated with a lower prevalence of cardiovascular diseases (CVDs). In this revision, three systematic reviews and five meta-analyses were included to explore the association of isoflavones and markers related to CVDs (Table 3).

Table 3. Effects of isoflavones in cardiovascular disease-related markers.

Authors	Number of Studies Included	Type of Studies Included	Number of Participants and Gender/Age/Characteristics	Compound and Doses	Observed Effects
Meta-analysis					
Kim and Je, 2017 [56]	13	Prospective studies	338,541 participants Age ranging from 40–84 years Follow-up period from 4 to 28 years	Intake data not provided	No association with mortality from CVD
Yan et al., 2017 [57]	17	Prospective cohort and case-control studies	508,841 participants, 17,269 with CVD events (stroke, coronary heart disease, ischemic stroke) Follow-up period from 6.3 to 16 years	Isoflavones 0.025–53.6 mg/day	No associations between soy isoflavones consumption and risk of cardiovascular disease, stroke, and coronary heart disease
Abshirini et al., 2020 [58]	5	Randomized clinical trials	548 participants (272 case and 276 controls) Age not available	Isoflavones 49.3–118 mg/day Intervention period: 1–12 months	Non-significant change in flow-mediated dilation (parameter of endothelial function)
Man et al., 2020 [59]	8	Various designs (double-blind, placebo-controlled, parallel design, crossover design)	485 participants (276 women and 209 men) Age ranging from 35–75 years	Isoflavones 80–118 mg/day and 10 mg/day of S-equol Intervention period: 1 day–12 weeks	Positive effect of soy isoflavones on arterial stiffness
Systematic reviews					
Perna et al., 2016 [52]	12	Randomized clinical trials	139–1268 menopausal and postmenopausal women	Isoflavones 20 to 100 mg/day Intervention period: 8 weeks–2 years	Reduction in total cholesterol and triglyceride plasma concentrations Reduction in nitric oxide and malonaldehyde
Chalvon-Demersay et al., 2017 [60]	17 3	Randomized clinical trials; nutritional intervention	337 healthy, diabetic or hypercholesterolemic individuals Age: 18–74 years 406 healthy, postmenopausal or obese participants Age: 50–79 years	Isoflavones 3 to 102 mg/day Intervention period: 4–208 weeks Isoflavones 60 to 135 mg/day Intervention period: 3–12 months	Reduction in total cholesterol and LDL cholesterol Changes in systolic or diastolic blood pressure (increase and decrease, depending on the study)
Rienks et al., 2017 [61]	3	Prospective studies	68,748 individuals Age: 40–70 years	Follow-up period: up to 10 years	Decreased risk of acute coronary syndrome or coronary heart disease No association with ischemic stroke

CVD: cardiovascular disease, LDL: low-density lipoprotein.

In their systematic revision, which included 1307 menopausal (n = 139) and post-menopausal (n = 1268) women, Perna et al. (2016) suggested that a daily intake of soy isoflavones, ranging from 20 to 100 mg/day, reduced total cholesterol and triglyceride plasma concentrations, as well as some markers of oxidative stress (nitric oxide and malonaldehyde), thus reducing cardiovascular risk [52]. Further, Chalvon-Demersay et al. aimed to compare the effects of animal and plant-sourced proteins on lipemia and blood pressure [60]. In this systematic review, 123 studies were included, with a total number of 516,330 participants. The authors found, in a small number of studies (17 studies including 337 healthy, diabetic or hypercholesterolemic individuals), that a soy protein-based diet (rich in isoflavones) as compared to animal protein-based diets (meat, milk, casein, whey) resulted in reduced total cholesterol or LDL-cholesterol concentrations, increased HDL cholesterol concentrations, improved LDL:HDL cholesterol ratio, and decreased triglycerides. Interestingly, soy protein isolate (void of isoflavones through alcoholic extraction during the protein solation phase) did not show these effects. Regarding blood pressure, inconsistent results were reported in three further interventional studies (n = 406 healthy, postmenopausal or obese participants aged 50–79 years) either in diastolic or systolic blood pressure, after a supplementation with soy-protein.

In another systematic review of three observational studies (including 68,748 participants), low evidence of isoflavones was shown in the prevention of cardiovascular diseases [61]. The authors concluded that evidence for the role of isoflavones and soy products in the prevention of cardiovascular diseases was scarce and inconclusive.

Two meta-analyses, which covered prospective cohort studies, showed that soy isoflavone consumption was not associated significantly with mortality linked to cardiovascular events [56,62]. Indeed, Kim and Je (2017) indicated that a high intake of flavonoids is associated with a reduced risk of mortality from cardiovascular diseases in men and women [56]. However, when a subgroup analysis by a class of flavonoids was carried out, they concluded that these inverse associations were significant for all categories of flavonoids except for isoflavones and flavonols.

Additionally, Yan and co-workers, in their meta-analysis with a total of ten prospective cohorts and seven case-control studies (including 508,841 participants and 17,269 cardiovascular disease events), observed that the consumption of soy foods, but not that of isoflavones, was associated with a lower risk of total cardiovascular disease (including coronary heart disease and stroke) [57]. This fact suggests that other components, such as fiber in the soy foods, may account for these negative associations. In like manner, a meta-analysis by Simental-Mendía et al. (2018) showed no effect of soy isoflavones on plasma concentrations of lipoprotein a (Lpa), a low-density lipoprotein associated with increased cardiovascular risk due to its pro-thrombotic and atherogenic properties [63].

Lastly, Abshirini et al. (2020) revised the literature to find randomized controlled trials that evaluated the impact of soy protein supplementation (isoflavone intake ranging from 49.3 to 118 mg/day) on endothelial function parameters in postmenopausal women (n = 802) [58]. Apparently, soy isoflavones modified adhesion molecules by binding to vascular endothelium and mimicking the effect of a modulator of estrogen receptors [64]. Furthermore, soy protein has been linked to an improvement in nitric oxide production and contributes to increased arterial compliance [65]. However, in the meta-analysis, a minor enhancement in flow-mediated dilation after soy protein supplementation was found. On the contrary, a systematic review and meta-analysis reported by Man et al. (2020) revealed that supplementation of soy isoflavones had a positive effect in reducing arterial stiffness, also known as the loss of arterial elasticity [59]. The physical stiffening of arteries has major health implications for its connection to various adverse cardiovascular and other health outcomes such as coronary heart disease, stroke, hypertension or heart failure [66]. However, the authors acknowledged several limitations to their study, such as the small sample size in the meta-analysis (middle-aged 276 women and 209 men), and recognized that a large number of subjects as well as a lengthier intervention are needed to investigate whether supplementation of soy isoflavones actually improves these outcomes.

Taken as a whole, all these results suggest that soy isoflavones do not lead to a meaningful protective effect on cardiovascular risk. It follows that additional high-quality, large-scale randomized controlled trials are needed.

As far as the mechanisms of action involved in the decrease in triglyceride concentration are concerned, it has been reported that hepatic lipase and/or lipoprotein lipase activity may be increased as a result of isoflavone presence. Indeed, some authors speculate that hepatic and lipoprotein lipase activity may have been altered after variable isoflavone intake from soybean, although no experimental data were provided [67]. Since hepatic lipase and lipoprotein lipase hydrolyze triglycerides into their constituents glycerol and fatty acids, their activation results in the delivery of the fatty acids to tissues such as muscle and adipose tissue.

4.3. Cancer

Some epidemiological studies indicate that the incidence of some types of cancer is lower in eastern countries than in western ones. This fact does not seem to be influenced by genetics since migrating from eastern to western countries appears to cause the loss of this protective effect. In this section, 16 meta-analyses have been revised: Nachvack et al. (2019) conducted a meta-analysis where 23 prospective studies with a total of 330,826 participants were included. The authors observed that soy isoflavone consumption was inversely associated with cancer deaths. Moreover, a higher intake of soy isoflavones was associated with a lower risk of mortality from gastric, colorectal, and lung cancers. Indeed, they reported that an increase of 10 mg/day of soy isoflavones consumption was associated with a 7% lower risk of cancer mortality [62].

Regarding breast cancer (Table 4), the same increase in soy isoflavone intake (10 mg/day) was related to a 9% lower risk of breast cancer mortality. However, it is important to point out that the beneficial effects of soy isoflavone consumption have been exhibited by women with estrogen receptor-negative breast cancer, but not by women with receptor-positive breast cancer, who present a better prognosis [62]. In this sense, Micek et al. (2020) carried out a meta-analysis with 15 cohort studies with the aim of exploring the association between isoflavone intake (<62.7 mg/day) and breast cancer mortality and its recurrence [68]. The authors found a significant inverse association between isoflavone intake and both overall mortality and breast cancer recurrence. These two associations were significant for postmenopausal participants. In addition, Qiu et al. (2018) analyzed 12 studies with 37,275 women with breast cancer and reported that soy isoflavone consumption at a pre-diagnosis stage might have a small effect on the survival of postmenopausal women with breast cancer [69]. However, the authors pointed out several limitations of this study, such as the important source of heterogeneity in their study. In fact, the number of isoflavones present was very variable among the different soy food types. Therefore, it is necessary to support these results with additional studies. Future research in this field should quantify isoflavones as accurately as possible, analyzing larger cohorts and lengthening the follow-up stage. Moreover, the number of publications grouping women according to pre- or post-diagnosis intake criteria is small.

Table 4. Effects of isoflavones in different cancer types.

Authors	Number of Studies Included (Meta-Analysis)	Type of Studies Included	Number of Participants and Gender/Age Characteristics	Compound and Doses	Observed Effects
Nachvack et al., 2019 [62]	23	Prospective study	330,826 (12 studies in both genders and 11 in women)	Soy/soy products (10 mg/day)	Inverse association with cancer deaths. 7% lower risk of gastric, colorectal, and lung cancer mortality
Breast cancer					
Nachvack et al., 2019 [62]	23	Prospective study	330,826 (12 studies in both genders and 11 in women)	Soy/soy products (10 mg/day)	9% lower risk of estrogen receptor-negative breast cancer mortality
Micek et al., 2020 [68]	15	Cohort study	49,659	Isoflavone intake (0.0036–62.7 mg/day)	Inverse association between isoflavone intake and both overall mortality and breast cancer recurrence
Qiu et al., 2018 [69]	12	Prospective cohort study	37,275 women	Isoflavones (the amount varies greatly among different soy foods)	Pre-diagnosis, soy isoflavone consumption has a poor effect on survival of postmenopausal women
Zhao et al., 2019 [70]	16	Prospective cohort study	648,913 (11,169 breast cancer cases)	High dietary intake of soy foods (dose non-defined)	Significant reduction of breast cancer risk
Rienks et al., 2017 [61]	10	Case-control study	Sample sizes (from 100 to 15,688 participants)	Daidzein, genistein, and equol (dose non-defined)	Daidzein (34%) and genistein (28%) were associated with a lower risk of breast cancer
Endometrial cancer					
Zhong et al., 2016 [71]	13	Prospective cohort study (3) Case-control study (10)	178,947 (7067 cases and 171,880 controls)	Soy products (0.05–130 g/day) and total isoflavones (0.28 to 63 mg/day).	19% reduction in endometrial cancer risk. Higher reduction in Asian women
Liu et al., 2016 [72]	23	Randomized controlled trials	2167	More than 54 mg isoflavone/day	Reduction of the endometrial thickness in North American women (for 0.23 mm). Opposite effect in Asian women
Grosso et al., 2017 [73]	8	Prospective studies (3) Case-control studies (5)	Non-defined	Isoflavone consumption (>45 mg/day among the Asian population, and >1 mg/day among the non-Asian population)	Potential reduction risk associated with isoflavone consumption
Ovarian cancer					
Hua et al., 2016 [74]	12	Prospective cohort study (5) Case-control study (7)	6275 cases and 393,776 controls	Isoflavone intake (0.01–41 mg/day)	33% reduction in ovarian cancer risk

Table 4. *Cont.*

Authors	Number of Studies Included (Meta-Analysis)	Type of Studies Included	Number of Participants and Gender/Age Characteristics	Compound and Doses	Observed Effects
Prostate cancer					
Pérez-Cornago et al., 2018 [75]	7	Cohort prospective study (2 studies from Japan and 5 studies from Europe)	241 cases and 503 controls (from Japanese studies), and 2828 cases and 5593 controls (from European studies) 60–69 years-old	Circulating isoflavone concentrations (nmol/L): Daidzein (Japanese: 115–166; European: 2.84–3.96) Genistein (Japanese: 277–454; European: 4.84–5.97), and equol (Japanese: 10.3–24; European: 0.25–0.65)	Genistein, daidzein and equol did not affect prostate cancer risk in both Japanese and European men
Applegate et al., 2018 [76]	30	Case-control study (15) Cohort study (8) Nested case-control study (7)	266,699 (21,612 patients with prostate cancer)	Soy foods (<90 mg/day)	Isoflavones were not associated with a reduction of prostate cancer risk Metabolites (genistein and daidzein) consumption was associated with a reduction of prostate cancer risk
Rienks et al., 2017 [61]	8	Case-control study	Sample sizes (10–15,688 participants)	Daidzein, genistein, and equol (dose not defined)	19% reduction in prostate cancer risk was found at high concentrations of daidzein, but not with genistein or equol
Zhang et al., 2017 [77]	23	Case-control study (21) Cohort study (2)	Participants: 11,346 cases and 140,177 controls	Daidzein, genistein, and equol (dose not defined)	Daidzein and genistein intakes were associated with a reduction of prostate cancer risk (no effect with equol)
Colorectal cancer					
He et al., 2016 [78]	18	Case-control study (9) Cohort study (9)	559,486 (among them 16,917 colorectal cancer cases)	Foods rich in isoflavones (dose not defined)	Reduction of colorectal cancer risk. This association is stronger among postmenopausal women than premenopausal women
Yu et al., 2016 [79]	17	Case-control study (13) Prospective cohort study (4)	272,296 participants	Soy foods (30 mg/day–170 g/day) and isoflavones (0.014–60 mg/day)	23% reduction in colorectal cancer risk. Potential protective effect in the Asian population (their consumption is higher than in the Western population)
Jiang et al., 2016 [80]	17	Case-control study (9) Cohort study (8)	317,599 participants	Isoflavones (0.025–74 mg/day).	Inverse association between isoflavone consumption and colorectal cancer risk in case-control studies, but not in cohort studies. 8% reduction in colorectal neoplasm risk for every 20 mg/day increase in isoflavone intake (Asian population) and for every 0.1 mg/day increase (Western population)
Gastric cancer					
You et al., 2018 [81]	12	Cohort study (6) Case-control study (6)	596,553 participants	Isoflavones (high dose: 0.6–75.5 mg/day; and low dose: 0.01–20.1 mg/day).	No association between isoflavone consumption and gastric cancer risk with the highest versus the lowest categories of dietary isoflavone intake

Another meta-analysis, conducted by Zhao et al. (2019), included 16 prospective cohort studies involving 11,169 breast cancer cases and 648,913 participants [70]. The authors studied all possible correlations between isoflavone consumption and the risk of breast cancer. While a moderate intake of soy isoflavones did not significantly affect breast cancer risk, a significant reduction was shown when the intake was high. However, some limitations were found in these prospective studies, such as the small sample size of the cohorts and the unspecific definition of high, moderate and low isoflavone intake. Moreover, Rienks et al. (2017) conducted a meta-analysis with ten observational studies to assess the association between the isoflavone compounds daidzein, genistein and equol, and breast cancer [61]. The authors indicated that high concentrations of daidzein and genistein (no specific values concerning these concentrations were available) were related to 34% and 28% reduction in breast cancer risk, respectively, whereas no association with equol was found. In conclusion, this study reported some evidence that suggests the beneficial role that daidzein and genistein play in preventing breast cancer risk. Furthermore, among the four studies of breast cancer recurrence, three studies found that isoflavone intake decreased the risk of breast cancer recurrence, although further studies are required to confirm this finding [82].

Endometrial cancer (Table 4), as well as breast cancer, is an estrogen-dependent cancer type related to high circulating estradiol concentrations. In this line, three meta-analyses were conducted to study isoflavone consumption and endometrial cancer risk. Zhong et al. (2018) revised 13 epidemiologic studies involving 178,947 participants: 7067 cases and 171,880 controls; among them, three were prospective cohort studies, and ten were population-based case-control studies [71]. The authors reported that isoflavone intake from soy products and legumes was associated with a 19% reduction in endometrial cancer risk. However, the magnitude of risk reduction in Asian women was slightly larger than in non-Asian women (22% vs. 18%). Diversely, Liu et al. (2016) analyzed the effect of oral isoflavone supplementation on endometrial thickness in pre- and post-menopausal women, which is a biomarker for the proliferative effects of estrogens associated with an increased endometrial risk [72]. For this purpose, 23 studies with 2167 patients were included in this meta-analysis, which showed that a daily dose of more than 54 mg could decrease the endometrial thickness in North American women, whereas the contrary effect was observed in the Asian sample. The authors suggested that this difference could be due to different genetic backgrounds and dietary patterns among Asian and Western populations. Grosso et al. (2017) carried out a meta-analysis of 143 case-control studies, where isoflavone intake was associated with endometrial, ovarian, and breast cancers [73]. Regarding endometrial cancer, three prospective and five case-control studies were analyzed, which suggested a lower endometrial cancer risk associated with isoflavone consumption. However, more studies are required to provide stronger evidence.

Hua et al. (2016) conducted a meta-analysis aimed at ovarian cancer, which affects women exclusively [74]. In this study, five prospective cohort studies and seven case-control studies, with 6275 cases and 393,776 controls, were included. The results indicated that the isoflavone intake decreased the ovarian cancer risk by 33%. Therefore, the authors concluded that the intake of dietary isoflavones paid a protective role against ovarian cancer.

With regard to prostate cancer, it has been proposed that it might be affected by isoflavone consumption due to its effect on hormone metabolism. Four meta-analyses have been reported to address this issue. Pérez-Cornago et al. (2018) published a meta-analysis of seven prospective cohort studies (two studies from Japan with 241 cases and 503 controls, and five studies from Europe with 2828 cases and 5593 controls) [75]. In Japanese men, pre-diagnostic circulating levels of genistein, daidzein and equol did not significantly affect prostate cancer risk. In the same line, in European men, isoflavone concentrations did not affect the risk of developing prostate cancer. In the meta-analysis reported by Applegate et al. (2018), 30 articles were included for analysis; 15 of them were case-control studies, eight cohort studies and seven nested case-control studies [76]. This meta-analysis included 266,699 participants (21,612 patients with prostate cancer among them). The authors reported that neither soy food intake nor circulating isoflavones were associated with a reduced risk of prostate cancer. However, the intake of specific isoflavones, genistein and daidzein, was indeed significantly associated with a reduced risk of this type of cancer. The authors suggested that further studies grouping subjects by isoflavone intake or by the circulating isoflavone levels might enhance this association. Rienks et al. (2017) also analyzed eight studies of prostate cancer, and they found an association between isoflavone intake and prostate cancer risk [61]. A significant 19% reduction of the risk was found at high concentrations of daidzein, but not at genistein or equol concentrations. Lastly, Zhang et al. (2017) conducted a meta-analysis with 21 case-control studies and two cohort studies, with a total number of 11,346 cases and 140,177 controls [77]. They observed that daidzein and genistein intakes were associated with a significant reduction of prostate cancer risk. Nevertheless, this association was not shown when equol or isoflavones were consumed. Moreover, the authors suggested that ethnicity could have had influenced this association, but studies with larger sample sizes are needed to confirm this hypothesis.

Due to the high prevalence of colorectal cancer, three meta-analyses addressing this issue are included in the present review. He et al. (2016) conducted a meta-analysis of 18 studies involving 16,917 colorectal cancer cases in 559,486 participants [78]. The results showed that a higher regular intake of foods rich in isoflavones might potentially decrease colorectal cancer incidence. Moreover, this association was more prominent among postmenopausal women than premenopausal women. However, the authors suggested that more prospective cohort studies are needed to further investigate this association and that the quantity of isoflavone intake should be accurately measured because flavonoid contents in food can vary significantly. In the same line, Yu et al. (2016) published a meta-analysis of 13 case-control and four prospective cohort studies, where the results revealed that soy isoflavone consumption reduced the risk of developing colorectal cancer by 23% [79]. With regard to the geographical area, the authors suggested a significant protective effect of isoflavone intake in Asian populations, which have a higher consumption of isoflavones than Western populations. For their meta-analysis, Jiang et al. (2016) included 17 studies (nine case-control studies and eight cohort studies) [80]. The authors found a statistical inverse association between isoflavone consumption and colorectal cancer risk in case-control studies, although not in cohort studies. Furthermore, the dose-response analysis revealed an 8% reduction in colorectal neoplasm risk for every 20 mg/day increase in isoflavone intake in Asian populations, and for every 0.1 mg/day increase in Western populations. The difference for isoflavones required to find an effect can be due to variations in the amount of isoflavone intake between both populations. Thus, whereas Asian populations have an average isoflavone consumption of >30 mg/day, Western populations consume <1 mg/day [83].

Lastly, the association between dietary isoflavone intake and the risk of developing gastric cancer has also been analyzed. You et al. (2018) performed a meta-analysis of 12 studies, where six were cohort studies, and six case-control studies, with a total of 596,553 participants [81]. The results showed no significant association between isoflavone consumption and gastric cancer risk, with the highest (0.6–75.5 mg/day) *versus* the lowest (0.01–20.1 mg/day) categories of dietary isoflavone intake. In this sense, the authors suggested that the composition of the gut microflora may influence isoflavone absorption and metabolism and could affect the production of isoflavone metabolites, which, in turn, could mediate their biological activity.

Overall, evidence regarding the use of isoflavones in cancer prevention suggests that they may be useful in reducing the risk of suffering from some types of cancer, such as breast and endometrial cancer, two of the types for which the association between isoflavone intake and cancer risk has been studied more in-depth. However, all the authors agree that further studies are necessary so as to confirm these results. Indeed, the dose levels of dietary isoflavones require a better definition, and the measurements of isoflavone consumption should be provided in quantifiable terms.

With regard to the mechanisms by which isoflavones act to reduce cancer risk, it has been suggested that the effects are due to the similarity of the isoflavone molecule to estradiol. This chemical feature confers estrogenic or antiestrogenic effects to isoflavones, depending on the cell type and the binding to α- or β-estrogen receptors. In this regard, it has been reported that isoflavones have a higher affinity to the β-estrogen receptor [7], which, contrary to the α-estrogen receptor, inhibits cell proliferation and stimulates apoptosis [84]. Furthermore, isoflavones are able to inhibit aromatase activity, the enzyme that converts androgen to estrogen. It is also important to mention that isoflavones could act as potential anticancer compounds due to their antioxidant role in malignant cell proliferation and differentiation (Figure 5).

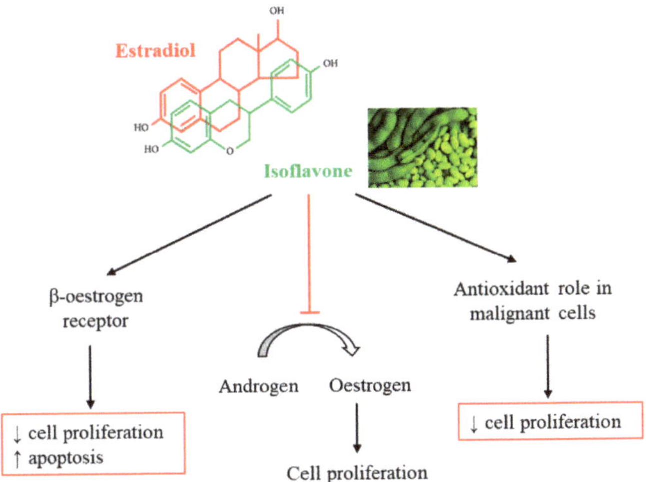

Figure 5. Possible mechanism of action of isoflavones for cancer prevention.

4.4. Menopausal Symptoms

Menopause is characterized by a decrease in estrogen levels and is often accompanied by a range of symptoms. Among these, vasomotor symptoms comprising hot flushes and night sweats are the most common and bothersome, which, in turn, have a negative impact on women's quality of life. Hormone replacement therapy has been proven to be effective in reducing vasomotor symptoms, although isoflavones have gained popularity as an alternative treatment to hormone replacement therapy to relieve menopausal symptoms. This is mainly due to their potential adverse effects, which range from increased coronary heart disease to stroke and cancer [53,85]. Concerning this issue, six meta-analyses and two systematic reviews are included in the present review (Table 5).

The six meta-analyses show a slight improvement in the frequency and intensity of hot flushes. In the meta-analysis reported by Chen et al. (2015), the authors examined the efficacy of phytoestrogens in reducing hot flushes [86]. The meta-analysis included 15 studies in which the number of participants ranged from 30 to 252. The mean age of women ranged from 49 to 58.3 and 48 to 60.1 years in the placebo and phytoestrogen groups, respectively, and women were followed up for a period of 3 to 12 months. Isoflavones were the phytoestrogens used in most of the studies; therefore, the meta-analysis only included outcomes regarding isoflavones, while outcomes with other phytoestrogens were not included. Ten studies reported a significant reduction in hot flush frequency in the phytoestrogen group when compared with the placebo group. Moreover, five studies reported side-effect data, and after the analysis, no significant differences were observed between the two groups.

Table 5. Effects of isoflavones in menopausal symptoms.

Authors	Number of Studies Included	Type of Studies Included	Number of Participants and Gender/Age/Characteristics	Compound and Doses	Observed Effects
Meta-analysis					
Chen et al., 2015 [86]	15	RCT	30–252 perimenopausal or postmenopausal women/report (1753 in total) 49–58.3 years (placebo group) 48–60.1 years (phytoestrogen group)	Isoflavones 5–100 mg/day Intervention period: 3–12 months	Reduction of hot flush frequency (vs. placebo)
Li et al., 2015 [87]	16	RCT	24–236 women/report (median 90) 40–65 years	Soy isoflavones 30–200 mg/day Intervention period: 4 weeks–2 years (median 12 weeks)	Slight and slow attenuation of hot flushes (vs. estradiol)
Li et al., 2016 [88]	39	RCT	24–620 women/report (median 200) Age not available	SSRIs/SNRIs: 7.5–200 mg/day Gabapentin: 300–1800 mg/day Clonidine: 0.1–0.4 mg/day Soy isoflavones: 30–200 mg/day Intervention period: 2–96 weeks (average 12 weeks)	Slight and slow attenuation of hot flushes (vs. non-hormonal drugs)
Daily et al., 2019 [85]	5	RCT	728 menopausal women (total subjects) 50.5–58.8 years (mean)	Soy isoflavones: 33–200 mg/day and 6 g soy extract/day Equol: 10 mg/day Intervention period: not available	Equol or isoflavone in equol-producers more effective than placebo
Sarri et al., 2017 [89]	32	RCT	4165 menopausal women (total subjects) 45+ years	Isoflavones and black cohosh (Doses not available)	Reduction of VSM (hot flushes and night sweats) compared to placebo No beneficial effect (vs. pharmacological treatment)
Franco et al., 2016 [90]	17	RCT	30–252 women/trial 40–69 years	Dietary soy isoflavones: 42–90 mg/day Supplements and extracts of soy isoflavones: 10–100 mg/day Red clover: 40–160 mg/day Intervention period: 12–48 weeks	Reduction of hot flush frequency by means of dietary isoflavones and supplements) Reduction of night sweat frequency by red clover

Table 5. *Cont.*

Authors	Number of Studies Included	Type of Studies Included	Number of Participants and Gender/Age/Characteristics	Compound and Doses	Observed Effects
Systematic reviews					
Chen et al., 2019 [86]	15	RCT (9) Prospective study (2) Systematic review (2) Randomized crossover trial (1) Meta-analysis (2)	51–403 menopausal and postmenopausal women	Soy (soy nut, soy protein, soy extracts) Natural isoflavones Synthetic isoflavones	Beneficial effects of isoflavones (vs. placebo) Synthetic or combination of isoflavones more effective than natural soy HRT more effective than soy or its extracts Isoflavone in equol-producers or equol supplementation more effective than placebo.
Perna et al., 2016 [52]	7	RCT	40–403 menopausal and postmenopausal women	Isoflavones 50–120 mg/day Intervention period: 8 weeks–2 years	Reduction of hot flush frequency

HRT: hormone replacement therapy, RCT: randomized controlled trial, VSM: vasomotor symptoms.

Similarly, Li et al. (2015) concluded that soy isoflavones have both a slight and slow effect in attenuating menopausal hot flushes [87]. A total of 16 studies were included in their meta-analysis with a range of 24 to 236 subjects. Study duration ranged from four weeks to two years, with a median of 12 weeks. The dose of isoflavones ranged from 30 to 200 mg/day. Despite the positive effect of isoflavones, the authors emphasized that the effects took a long time to appear. Estradiol, which has proven efficacy on hot flushes, needs 3.09 weeks to achieve half of its peak activity, while isoflavones require 13.4 weeks, suggesting that 12 weeks of treatment is not enough for soy isoflavones to exert a beneficial effect. The same results were reported in another meta-analysis published by the same authors a year later (Li et al., 2016), when they compared the efficacy of several non-hormonal drugs on hot flushes, including selective serotonin reuptake inhibitors (SSRIs), serotonin–norepinephrine reuptake inhibitors (SNRIs), gabapentin, clonidine and soy isoflavones [88]. Thirty-nine studies were included with a sample size of 200 women per trial, in which the intervention period ranged from two to 96 weeks (median 12 weeks). As in the previous study, the onset of soy isoflavones was very slow in comparison with other non-hormonal treatments. Specifically, the time to achieve half of the efficacy was 11.6 weeks for soy isoflavones, while for the rest of the compounds, the onset time was near 0 weeks. They also explained that the differences between isoflavones and the rest of the drugs might be because the latter act by modulating central neurotransmitters and, as a consequence, by regulating the central thermoregulatory centers in the hypothalamus. In contrast, it seems that isoflavones do not act on central neurotransmitters directly. This difference could explain the slow effect of isoflavones in alleviating hot flushes.

Daily et al. (2019) studied whether equol supplementation could benefit equol nonproducer subjects in lowering the incidence and severity of hot flushes [85]. Equol is produced from the isoflavone daidzein by intestinal bacteria, mainly in the large intestine. Their meta-analysis included five studies with a total of 728 menopausal women between 50.5 and 58.5 years old and concluded that women who are not able to produce equol could benefit from equol supplementation. On the contrary, women who were already equol producers did not obtain any additional benefit from supplements of equol or isoflavones.

Another meta-analysis explored the effects of several pharmacological and non-pharmacological treatments in the relief of vasomotor symptoms, taking into account not only hot flushes but also night sweats [89]. Non-pharmacological treatments included isoflavones, which appeared to be better than placebo in the relief of vasomotor symptoms, although not significantly better than the treatment with estradiol and progesterone. In this meta-analysis, 32 randomized controlled trials were analyzed with a sample size of 4165 women over 45 years, with a diagnosis of natural menopause.

Franco et al. (2016) also studied whether there was an association between different types of interventions and the reduction in hot flushes and night sweats [90]. They performed a separated meta-analysis for each of the interventions, including dietary soy isoflavones (including four studies), supplements and extracts of soy isoflavones (including six studies) and red clover isoflavones (including seven studies). They found that the use of dietary isoflavones or supplements and extracts was associated with an improvement in hot flushes, but not with a reduction in the frequency of night sweats. In contrast, red clover was not associated with changes in the hot flush frequency. Only one study examined the effects of red clover on night sweats, and a significant association was found, which showed a decreased frequency in night sweats.

The two systematic reviews conclude that isoflavones have an important role in reducing hot flushes [52,53], although some observations need to be pointed out. Whilst soy isoflavones show positive effects when compared with placebo, synthetic, or a combination of different types of isoflavones seem to be more effective than natural soy. When a group receiving hormone replacement therapy was included in the study, significant differences between the effects of hormone replacement therapy and soy were found, being the effect of the former superior to that of isoflavones. Two of the studies also revealed that women who were able to produce equol or who had received equol supplementation could have had a greater benefit from isoflavones [53]. Perna et al. (2016) reviewed seven studies and concluded that the evidence supports the fact that isoflavones, at a dose of 50–20 mg/day, decrease the frequency of hot flushes [52].

In general, evidence regarding the use of isoflavones on vasomotor symptoms suggests that they could be useful in reducing hot flushes. However, there is still a great heterogeneity among studies, which makes the current data inconclusive. Regarding the mechanisms by which isoflavones act to alleviate vasomotor symptoms, is has been proposed that they bind to estrogen receptors and activate endothelial nitric oxide synthase (eNOS) transcription, leading to eNOS synthesis and nitric oxide production, the main mediator of vasodilatation, which allows cutaneous heat dissipation [91] (Figure 6).

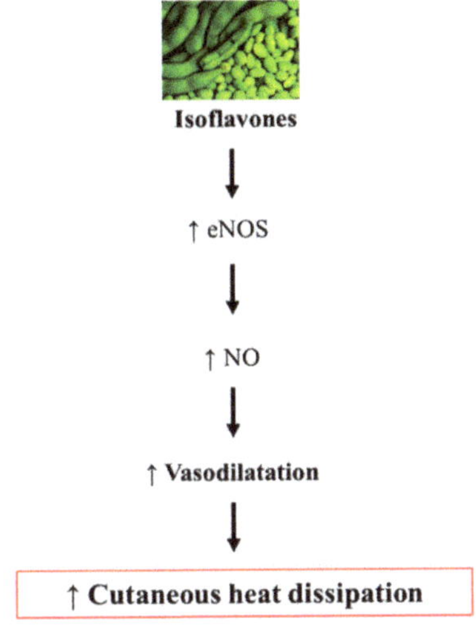

eNOS: endothelial nitric oxide synthase
NO: nitric oxide

Figure 6. Possible mechanism of action of isoflavones for menopausal symptoms.

5. Side Effects and Safety

Potential health adverse effects in the intake of isoflavones have been postulated due to their estrogenic activity, especially in relation to breast cancer. However, it is important to mention that this connection has been established based on pre-clinical studies, whereas data derived from clinical trials have not demonstrated such association [92]. In fact, in 2015, in relation to the possible adverse effects on the mammary gland, uterus and thyroid, the European Food Safety Agency (EFSA) offered a scientific opinion with regards to the assessment of potential increased risk in peri- and postmenopausal women who took food supplements that contained isolated isoflavones. After an extensive bibliographical review, the expert panel concluded that the results of observational studies do not support the hypothesis of an increased risk of breast cancer after isoflavone-rich food supplement intake [93]. Moreover, an increase in thyroid hormone levels after isoflavone intake by food supplements has not been observed, and no side effects were confirmed with regard to the endometrial thickness or histopathological changes in the uterus. Thus, the EFSA scientific committee's opinion agreed that isoflavone intake was safe, at least in the doses that the analyzed studies have used. In any case, it is important to bear in mind that a great number of expert scientists and doctors think that breast cancer survivors should not increase their isoflavone intake [94], although others consider that their consumption is safe and beneficial [95]. In general terms, caution should be exercised with the use of high doses of isoflavones in women with a family history of breast cancer [49].

Finally, it is important to mention that despite the fact that potential isoflavone adverse effects have been widely studied, little is known about the side effect of the bacterial metabolite equol. As it has been stated in the Bioavailability section, microbiota composition in humans makes equol production very variable, which yields high interindividual differences. Reproductive and developmental toxicity studies have stated that 1 g and 2 g of equol per day is safe [96], although further studies are required [92].

6. Concluding Remarks

In view of the results gathered in the present review, we can conclude that there is scientific evidence that reveals the beneficial effect of isoflavones on bone health and thus in the prevention and treatment of osteoporosis in postmenopausal women. Nevertheless, the results do not seem entirely conclusive, as there are discrepancies among the studies included in the systematic reviews and in the meta-analyses, probably related to their experimental designs. For this reason, the results should be interpreted with caution, and more randomized clinical trials are imperative. By contrast, it seems that soy isoflavones do not lead to a meaningful protective effect on cardiovascular risk. Regarding cancer, scientific evidence suggests that isoflavones could be useful in reducing the risk of suffering from some types of cancer, such as breast and endometrial cancer, although further studies are needed to confirm these results. In the end, isoflavones could be useful in reducing hot flushes associated with menopause. However, a limitation in this field is that there is still a great heterogeneity among studies. Altogether, these results show that further studies are needed in order to find stronger conclusions.

Considering the isoflavone chemical structure, potential health adverse effects of these compounds related to their estrogenic activity have been postulated, despite the scientific opinion offered by EFSA indicating that isoflavone intake was safe.

Author Contributions: Conceptualization, S.G.-Z.; Writing—Original Draft Preparation, S.G.-Z., M.G.-A., A.F.-Q., I.E. and J.T.; Writing—Review and Editing, M.P.P., S.G.-Z., I.E. and J.T.; Project Administration, M.P.P.; Funding Acquisition, M.P.P. and A.F.-Q. All authors have read and agreed to the published version of the manuscript.

Funding: This study was supported by Instituto de Salud Carlos III (CIBERobn) under Grant CB12/03/30007, Basque Government under Grant PA20/04 and the University of the Basque Country under Grant GIU18-173.

Conflicts of Interest: The authors declare no conflict of interest.

References

1. Winkel-Shirley, B. Biosynthesis of flavonoids and effects of stress. *Curr. Opin. Plant. Biol.* **2002**, *5*, 218–223. [CrossRef]
2. Treutter, D. Significance of flavonoids in plant resistance and enhancement of their biosynthesis. *Plant Biol.* **2005**, *7*, 581–591. [CrossRef] [PubMed]
3. Sirotkin, A.V.; Harrath, A.H. Phytoestrogens and their effects. *Eur. J. Pharmacol.* **2014**, *741*, 230–236. [CrossRef] [PubMed]
4. Wang, X.; Dykens, J.A.; Perez, E.; Liu, R.; Yang, S.; Covey, D.F.; Simpkins, J.W. Neuroprotective effects of 17beta-estradiol and nonfeminizing estrogens against H_2O_2 toxicity in human neuroblastoma SK-N-SH cells. *Mol. Pharmacol.* **2006**, *70*, 395–404. [CrossRef]
5. Endoh, H.; Sasaki, H.; Maruyama, K.; Takeyama, K.; Waga, I.; Shimizu, T.; Kato, S.; Kawashima, H. Rapid activation of MAP kinase by estrogen in the bone cell line. *Biochem. Biophys. Res. Commun.* **1997**, *235*, 99–102. [CrossRef] [PubMed]
6. Lu, D.; Giguère, V. Requirement of Ras-dependent pathways for activation of the transforming growth factor beta3 promoter by estradiol. *Endocrinology* **2001**, *142*, 751–759. [CrossRef]
7. Sotoca, A.M.; van den Berg, H.; Vervoort, J.; van der Saag, P.; Ström, A.; Gustafsson, J.A.; Rietjens, I.; Murk, A.J. Influence of cellular ERalpha/ERbeta ratio on the ERalpha-agonist induced proliferation of human T47D breast cancer cells. *Toxicol. Sci.* **2008**, *105*, 303–311. [CrossRef]
8. Messina, M.J. Legumes and soybeans: Overview of their nutritional profiles and health effects. *Am. J. Clin. Nutr.* **1999**, *70*, 439S–450S. [CrossRef]

9. Zaheer, K.; Humayoun Akhtar, M. An updated review of dietary isoflavones: Nutrition, processing, bioavailability and impacts on human health. *Crit. Rev. Food Sci. Nutr.* **2017**, *57*, 1280–1293. [CrossRef]
10. Křížová, L.; Dadáková, K.; Kašparovská, J.; Kašparovský, T. Isoflavones. *Molecules* **2019**, *24*, 1076. [CrossRef]
11. Bustamante-Rangel, M.; Delgado-Zamarreño, M.M.; Perez-Martin, L.; Rodriguez-Gonzalo, E.; Dominguez-Alvarez, J. Analysis of isoflavones in Foods. *Compr. Rev. Food Sci. Food Saf.* **2018**, *17*, 391–411. [CrossRef]
12. Rossi, M.; Amaretti, A.; Roncaglia, R.; Leonardi, A.; Raimondi, S. Dietary isoflavones and dietary microbiota: Metabolism and transformation into bioactive compounds. In *Isoflavones Biosynthesis, Occurence and Health Effects*; Thompson, M.J., Ed.; Nova Science Publishers: Hauppauge, NY, USA, 2010; pp. 137–161.
13. Setchell, K.D.; Brown, N.M.; Lydeking-Olsen, E. The clinical importance of the metabolite equol-a clue to the effectiveness of soy and its isoflavones. *J. Nutr.* **2002**, *132*, 3577–3584. [CrossRef] [PubMed]
14. Collison, M.W. Determination of total soy isoflavones in dietary supplements, supplement ingredients, and soy foods by high-performance liquid chromatography with ultraviolet detection: Collaborative study. *J. AOAC Int.* **2008**, *91*, 489–500. [CrossRef] [PubMed]
15. Kurzer, M.S.; Xu, X. Dietary phytoestrogens. *Annu. Rev. Nutr.* **1997**, *17*, 353–381. [CrossRef] [PubMed]
16. Saloniemi, H.; Wähälä, K.; Nykänen-Kurki, P.; Kallela, K.; Saastamoinen, I. Phytoestrogen content and estrogenic effect of legume fodder. *Proc. Soc. Exp. Biol. Med.* **1995**, *208*, 13–17. [CrossRef] [PubMed]
17. Rizzo, G.; Baroni, L. Soy, Soy Foods and Their Role in Vegetarian Diets. *Nutrients* **2018**, *10*, 43. [CrossRef]
18. Bhathena, S.J.; Velasquez, M.T. Beneficial role of dietary phytoestrogens in obesity and diabetes. *Am. J. Clin. Nutr.* **2002**, *76*, 1191–1201. [CrossRef]
19. U.S. Department of Agriculture. Database for the Isoflavone Content of Selected Foods, Release 2.0. Available online: https://data.nal.usda.gov/dataset/usda-database-isoflavone-content-selected-foods-release-20 (accessed on 21 September 2020).
20. Messina, M.; Nagata, C.; Wu, A.H. Estimated Asian adult soy protein and isoflavone intakes. *Nutr. Cancer* **2006**, *55*, 1–12. [CrossRef]
21. Jun, S.; Shin, S.; Joung, H. Estimation of dietary flavonoid intake and major food sources of Korean adults. *Br. J. Nutr.* **2016**, *115*, 480–489. [CrossRef]
22. van Erp-Baart, M.A.; Brants, H.A.; Kiely, M.; Mulligan, A.; Turrini, A.; Sermoneta, C.; Kilkkinen, A.; Valsta, L.M. Isoflavone intake in four different European countries: The VENUS approach. *Br. J. Nutr.* **2003**, *89* (Suppl. 1), S25–S30. [CrossRef]
23. van der Schouw, Y.T.; Kreijkamp-Kaspers, S.; Peeters, P.H.; Keinan-Boker, L.; Rimm, E.B.; Grobbee, D.E. Prospective study on usual dietary phytoestrogen intake and cardiovascular disease risk in Western women. *Circulation* **2005**, *111*, 465–471. [CrossRef] [PubMed]
24. Valsta, L.M.; Kilkkinen, A.; Mazur, W.; Nurmi, T.; Lampi, A.M.; Ovaskainen, M.L.; Korhonen, T.; Adlercreutz, H.; Pietinen, P. Phyto-oestrogen database of foods and average intake in Finland. *Br. J. Nutr.* **2003**, *89* (Suppl. 1), S31–S38. [CrossRef]
25. Lund, S.; Holman, G.D.; Schmitz, O.; Pedersen, O. Contraction stimulates translocation of glucose transporter GLUT4 in skeletal muscle through a mechanism distinct from that of insulin. *Proc. Natl. Acad. Sci. USA* **1995**, *92*, 5817–5821. [CrossRef] [PubMed]
26. Chang, H.H.; Robinson, A.R.; Common, R.H. Excretion of radioactive diadzein and equol as monosulfates and disulfates in the urine of the laying hen. *Can. J. Biochem.* **1975**, *53*, 223–230. [CrossRef] [PubMed]
27. Heinonen, S.; Wähälä, K.; Adlercreutz, H. Identification of isoflavone metabolites dihydrodaidzein, dihydrogenistein, 6′-OH-O-dma, and cis-4-OH-equol in human urine by gas chromatography-mass spectroscopy using authentic reference compounds. *Anal. Biochem.* **1999**, *274*, 211–219. [CrossRef] [PubMed]
28. Barnes, S. The biochemistry, chemistry and physiology of the isoflavones in soybeans and their food products. *Lymphat. Res. Biol.* **2010**, *8*, 89–98. [CrossRef]
29. Day, A.J.; Cañada, F.J.; Díaz, J.C.; Kroon, P.A.; Mclauchlan, R.; Faulds, C.B.; Plumb, G.W.; Morgan, M.R.; Williamson, G. Dietary flavonoid and isoflavone glycosides are hydrolysed by the lactase site of lactase phlorizin hydrolase. *FEBS Lett.* **2000**, *468*, 166–170. [CrossRef]
30. Cheynier, V.; Sarni-Manchado, P.; Quideau, S. (Eds.) *Recent Advances in Polyphenol Research*; Wiley-Blackwell: Hoboken, NJ, USA, 2012; Volume 3.

31. Gee, J.M.; DuPont, M.S.; Day, A.J.; Plumb, G.W.; Williamson, G.; Johnson, I.T. Intestinal transport of quercetin glycosides in rats involves both deglycosylation and interaction with the hexose transport pathway. *J. Nutr.* **2000**, *130*, 2765–2771. [CrossRef]
32. Tamura, A.; Shiomi, T.; Hachiya, S.; Shigematsu, N.; Hara, H. Low activities of intestinal lactase suppress the early phase absorption of soy isoflavones in Japanese adults. *Clin. Nutr.* **2008**, *27*, 248–253. [CrossRef]
33. Sfakianos, J.; Coward, L.; Kirk, M.; Barnes, S. Intestinal uptake and biliary excretion of the isoflavone genistein in rats. *J. Nutr.* **1997**, *127*, 1260–1268. [CrossRef]
34. Ronis, M.J.; Little, J.M.; Barone, G.W.; Chen, G.; Radominska-Pandya, A.; Badger, T.M. Sulfation of the isoflavones genistein and daidzein in human and rat liver and gastrointestinal tract. *J. Med. Food* **2006**, *9*, 348–355. [CrossRef] [PubMed]
35. Hosoda, K.; Furuta, T.; Yokokawa, A.; Ishii, K. Identification and quantification of daidzein-7-glucuronide-4′-sulfate, genistein-7-glucuronide-4′-sulfate and genistein-4′,7-diglucuronide as major metabolites in human plasma after administration of kinako. *Anal. Bioanal. Chem.* **2010**, *397*, 1563–1572. [CrossRef] [PubMed]
36. Ko, K.P.; Kim, C.S.; Ahn, Y.; Park, S.J.; Kim, Y.J.; Park, J.K.; Lim, Y.K.; Yoo, K.Y.; Kim, S.S. Plasma isoflavone concentration is associated with decreased risk of type 2 diabetes in Korean women but not men: Results from the Korean Genome and Epidemiology Study. *Diabetologia* **2015**, *58*, 726–735. [CrossRef] [PubMed]
37. Gaya, P.; Medina, M.; Sánchez-Jiménez, A.; Landete, J.M. Phytoestrogen Metabolism by Adult Human Gut Microbiota. *Molecules* **2016**, *21*, 1034. [CrossRef]
38. Atkinson, C.; Frankenfeld, C.L.; Lampe, J.W. Gut bacterial metabolism of the soy isoflavone daidzein: Exploring the relevance to human health. *Exp. Biol. Med.* **2005**, *230*, 155–170. [CrossRef]
39. Ingram, D.; Sanders, K.; Kolybaba, M.; Lopez, D. Case-control study of phyto-oestrogens and breast cancer. *Lancet* **1997**, *350*, 990–994. [CrossRef]
40. Batterham, T.J.; Shutt, D.A.; Braden, A.W.H.; Tweeddale, H.J. Metabolism of intraruminally administered [4-^{14}C]formononetic and [4-^{14}C]biochanin a in sheep. *Aust. J. Agric. Res.* **1971**, *22*, 131–138. [CrossRef]
41. Murota, K.; Nakamura, Y.; Uehara, M. Flavonoid metabolism: The interaction of metabolites and gut microbiota. *Biosci. Biotechnol. Biochem.* **2018**, *82*, 600–610. [CrossRef]
42. Mace, T.A.; Ware, M.B.; King, S.A.; Loftus, S.; Farren, M.R.; McMichael, E.; Scoville, S.; Geraghty, C.; Young, G.; Carson, W.E.; et al. Soy isoflavones and their metabolites modulate cytokine-induced natural killer cell function. *Sci. Rep.* **2019**, *9*, 5068. [CrossRef]
43. Adlercreutz, H.; Markkanen, H.; Watanabe, S. Plasma concentrations of phyto-oestrogens in Japanese men. *Lancet* **1993**, *342*, 1209–1210. [CrossRef]
44. Lu, L.J.; Anderson, K.E. Sex and long-term soy diets affect the metabolism and excretion of soy isoflavones in humans. *Am. J. Clin. Nutr.* **1998**, *68*, 1500S–1504S. [CrossRef] [PubMed]
45. Kelly, G.E.; Nelson, C.; Waring, M.A.; Joannou, G.E.; Reeder, A.Y. Metabolites of dietary (soya) isoflavones in human urine. *Clin. Chim. Acta* **1993**, *223*, 9–22. [CrossRef]
46. Watanabe, S.; Yamaguchi, M.; Sobue, T.; Takahashi, T.; Miura, T.; Arai, Y.; Mazur, W.; Wähälä, K.; Adlercreutz, H. Pharmacokinetics of soybean isoflavones in plasma, urine and feces of men after ingestion of 60 g baked soybean powder (kinako). *J. Nutr.* **1998**, *128*, 1710–1715. [CrossRef] [PubMed]
47. Karr, S.C.; Lampe, J.W.; Hutchins, A.M.; Slavin, J.L. Urinary isoflavonoid excretion in humans is dose dependent at low to moderate levels of soy-protein consumption. *Am. J. Clin. Nutr.* **1997**, *66*, 46–51. [CrossRef] [PubMed]
48. Lambert, M.N.T.; Hu, L.M.; Jeppesen, P.B. A systematic review and meta-analysis of the effects of isoflavone formulations against estrogen-deficient bone resorption in peri- and postmenopausal women. *Am. J. Clin. Nutr.* **2017**, *106*, 801–811. [CrossRef]
49. Akhlaghi, M.; Ghasemi Nasab, M.; Riasatian, M.; Sadeghi, F. Soy isoflavones prevent bone resorption and loss, a systematic review and meta-analysis of randomized controlled trials. *Crit. Rev. Food Sci. Nutr.* **2020**, *60*, 2327–2341. [CrossRef]
50. Sansai, K.; Na Takuathung, M.; Khatsri, R.; Teekachunhatean, S.; Hanprasertpong, N.; Koonrungsesomboon, N. Effects of isoflavone interventions on bone mineral density in postmenopausal women: A systematic review and meta-analysis of randomized controlled trials. *Osteoporos. Int.* **2020**, *31*, 1853–1864. [CrossRef]

51. Abdi, F.; Alimoradi, Z.; Haqi, P.; Mahdizad, F. Effects of phytoestrogens on bone mineral density during the menopause transition: A systematic review of randomized, controlled trials. *Climacteric* **2016**, *19*, 535–545. [CrossRef]
52. Perna, S.; Peroni, G.; Miccono, A.; Riva, A.; Morazzoni, P.; Allegrini, P.; Preda, S.; Baldiraghi, V.; Guido, D.; Rondanelli, M. Multidimensional Effects of Soy Isoflavone by Food or Supplements in Menopause Women: A Systematic Review and Bibliometric Analysis. *Nat. Prod. Commun.* **2016**, *11*, 1733–1740. [CrossRef]
53. Chen, L.R.; Ko, N.Y.; Chen, K.H. Isoflavone Supplements for Menopausal Women: A Systematic Review. *Nutrients* **2019**, *11*, 2649. [CrossRef]
54. Arjmandi, B.H.; Smith, B.J. Soy isoflavones' osteoprotective role in postmenopausal women: Mechanism of action. *J. Nutr. Biochem.* **2002**, *13*, 130–137. [CrossRef]
55. Chadha, R.; Bhalla, Y.; Jain, A.; Chadha, K.; Karan, M. Dietary Soy Isoflavone: A Mechanistic Insight. *Nat. Prod. Commun.* **2017**, *12*, 627–634. [CrossRef] [PubMed]
56. Kim, Y.; Je, Y. Flavonoid intake and mortality from cardiovascular disease and all causes: A meta-analysis of prospective cohort studies. *Clin. Nutr. ESPEN* **2017**, *20*, 68–77. [CrossRef]
57. Yan, Z.; Zhang, X.; Li, C.; Jiao, S.; Dong, W. Association between consumption of soy and risk of cardiovascular disease: A meta-analysis of observational studies. *Eur. J. Prev. Cardiol.* **2017**, *24*, 735–747. [CrossRef] [PubMed]
58. Abshirini, M.; Omidian, M.; Kord-Varkaneh, H. Effect of soy protein containing isoflavones on endothelial and vascular function in postmenopausal women: A systematic review and meta-analysis of randomized controlled trials. *Menopause* **2020**, *12*, 1425–1433. [CrossRef] [PubMed]
59. Man, B.; Cui, C.; Zhang, X.; Sugiyama, D.; Barinas-Mitchell, E.; Sekikawa, A. The effect of soy isoflavones on arterial stiffness: A systematic review and meta-analysis of randomized controlled trials. *Eur. J. Nutr.* **2020**. [CrossRef] [PubMed]
60. Chalvon-Demersay, T.; Azzout-Marniche, D.; Arfsten, J.; Egli, L.; Gaudichon, C.; Karagounis, L.G.; Tomé, D. A Systematic Review of the Effects of Plant Compared with Animal Protein Sources on Features of Metabolic Syndrome. *J. Nutr.* **2017**, *147*, 281–292. [CrossRef]
61. Rienks, J.; Barbaresko, J.; Nöthlings, U. Association of isoflavone biomarkers with risk of chronic disease and mortality: A systematic review and meta-analysis of observational studies. *Nutr. Rev.* **2017**, *75*, 616–641. [CrossRef]
62. Nachvak, S.M.; Moradi, S.; Anjom-Shoae, J.; Rahmani, J.; Nasiri, M.; Maleki, V.; Sadeghi, O. Soy, Soy Isoflavones, and Protein Intake in Relation to Mortality from All Causes, Cancers, and Cardiovascular Diseases: A Systematic Review and Dose-Response Meta-Analysis of Prospective Cohort Studies. *J. Acad. Nutr. Diet.* **2019**, *119*, 1483–1500.e1417. [CrossRef]
63. Simental-Mendía, L.E.; Gotto, A.M.; Atkin, S.L.; Banach, M.; Pirro, M.; Sahebkar, A. Effect of soy isoflavone supplementation on plasma lipoprotein(a) concentrations: A meta-analysis. *J. Clin. Lipidol.* **2018**, *12*, 16–24. [CrossRef]
64. Gruber, C.J.; Tschugguel, W.; Schneeberger, C.; Huber, J.C. Production and actions of estrogens. *N. Engl. J. Med.* **2002**, *346*, 340–352. [CrossRef] [PubMed]
65. Xu, B.; Chibber, R.; Ruggiero, D.; Kohner, E.; Ritter, J.; Ferro, A.; Ruggerio, D. Impairment of vascular endothelial nitric oxide synthase activity by advanced glycation end products. *FASEB J.* **2003**, *17*, 1289–1291. [PubMed]
66. Vlachopoulos, C.; Aznaouridis, K.; Stefanadis, C. Prediction of cardiovascular events and all-cause mortality with arterial stiffness: A systematic review and meta-analysis. *J. Am. Coll. Cardiol.* **2010**, *55*, 1318–1327. [CrossRef] [PubMed]
67. Santo, A.S.; Santo, A.M.; Browne, R.W.; Burton, H.; Leddy, J.J.; Horvath, S.M.; Horvath, P.J. Postprandial lipemia detects the effect of soy protein on cardiovascular disease risk compared with the fasting lipid profile. *Lipids* **2010**, *45*, 1127–1138. [CrossRef] [PubMed]
68. Micek, A.; Godos, J.; Brzostek, T.; Gniadek, A.; Favari, C.; Mena, P.; Libra, M.; Del Rio, D.; Galvano, F.; Grosso, G. Dietary phytoestrogens and biomarkers of their intake in relation to cancer survival and recurrence: A comprehensive systematic review with meta-analysis. *Nutr. Rev.* **2020**, *79*, 42–65. [CrossRef] [PubMed]
69. Qiu, S.; Jiang, C. Soy and isoflavones consumption and breast cancer survival and recurrence: A systematic review and meta-analysis. *Eur. J. Nutr.* **2019**, *58*, 3079–3090. [CrossRef]

70. Zhao, T.T.; Jin, F.; Li, J.G.; Xu, Y.Y.; Dong, H.T.; Liu, Q.; Xing, P.; Zhu, G.L.; Xu, H.; Miao, Z.F. Dietary isoflavones or isoflavone-rich food intake and breast cancer risk: A meta-analysis of prospective cohort studies. *Clin. Nutr.* **2019**, *38*, 136–145. [CrossRef]
71. Zhong, X.S.; Ge, J.; Chen, S.W.; Xiong, Y.Q.; Ma, S.J.; Chen, Q. Association between Dietary Isoflavones in Soy and Legumes and Endometrial Cancer: A Systematic Review and Meta-Analysis. *J. Acad. Nutr. Diet.* **2018**, *118*, 637–651. [CrossRef]
72. Liu, J.; Yuan, F.; Gao, J.; Shan, B.; Ren, Y.; Wang, H.; Gao, Y. Oral isoflavone supplementation on endometrial thickness: A meta-analysis of randomized placebo-controlled trials. *Oncotarget* **2016**, *7*, 17369–17379. [CrossRef]
73. Grosso, G.; Godos, J.; Lamuela-Raventos, R.; Ray, S.; Micek, A.; Pajak, A.; Sciacca, S.; D'Orazio, N.; Del Rio, D.; Galvano, F. A comprehensive meta-analysis on dietary flavonoid and lignan intake and cancer risk: Level of evidence and limitations. *Mol. Nutr. Food Res.* **2017**, *61*, 1600930. [CrossRef]
74. Hua, X.; Yu, L.; You, R.; Yang, Y.; Liao, J.; Chen, D. Association among Dietary Flavonoids, Flavonoid Subclasses and Ovarian Cancer Risk: A Meta-Analysis. *PLoS ONE* **2016**, *11*, e0151134. [CrossRef] [PubMed]
75. Perez-Cornago, A.; Appleby, P.N.; Boeing, H.; Gil, L.; Kyrø, C.; Ricceri, F.; Murphy, N.; Trichopoulou, A.; Tsilidis, K.K.; Khaw, K.T.; et al. Circulating isoflavone and lignan concentrations and prostate cancer risk: A meta-analysis of individual participant data from seven prospective studies including 2,828 cases and 5,593 controls. *Int. J. Cancer* **2018**, *143*, 2677–2686. [CrossRef] [PubMed]
76. Applegate, C.C.; Rowles, J.L.; Ranard, K.M.; Jeon, S.; Erdman, J.W. Soy Consumption and the Risk of Prostate Cancer: An Updated Systematic Review and Meta-Analysis. *Nutrients* **2018**, *10*, 40. [CrossRef] [PubMed]
77. Zhang, Q.; Feng, H.; Qluwakemi, B.; Wang, J.; Yao, S.; Cheng, G.; Xu, H.; Qiu, H.; Zhu, L.; Yuan, M. Phytoestrogens and risk of prostate cancer: An updated meta-analysis of epidemiologic studies. *Int. J. Food Sci. Nutr.* **2017**, *68*, 28–42. [CrossRef] [PubMed]
78. He, X.; Sun, L.M. Dietary intake of flavonoid subclasses and risk of colorectal cancer: Evidence from population studies. *Oncotarget* **2016**, *7*, 26617–26627. [CrossRef]
79. Yu, Y.; Jing, X.; Li, H.; Zhao, X.; Wang, D. Soy isoflavone consumption and colorectal cancer risk: A systematic review and meta-analysis. *Sci. Rep.* **2016**, *6*, 25939. [CrossRef]
80. Jiang, R.; Botma, A.; Rudolph, A.; Hüsing, A.; Chang-Claude, J. Phyto-oestrogens and colorectal cancer risk: A systematic review and dose-response meta-analysis of observational studies. *Br. J. Nutr.* **2016**, *116*, 2115–2128. [CrossRef]
81. You, J.; Sun, Y.; Bo, Y.; Zhu, Y.; Duan, D.; Cui, H.; Lu, Q. The association between dietary isoflavones intake and gastric cancer risk: A meta-analysis of epidemiological studies. *BMC Public Health* **2018**, *18*, 510. [CrossRef]
82. Dong, J.Y.; Qin, L.Q. Soy isoflavones consumption and risk of breast cancer incidence or recurrence: A meta-analysis of prospective studies. *Breast Cancer Res. Treat.* **2011**, *125*, 315–323. [CrossRef]
83. Zamora-Ros, R.; Knaze, V.; Rothwell, J.A.; Hémon, B.; Moskal, A.; Overvad, K.; Tjønneland, A.; Kyrø, C.; Fagherazzi, G.; Boutron-Ruault, M.C.; et al. Dietary polyphenol intake in Europe: The European Prospective Investigation into Cancer and Nutrition (EPIC) study. *Eur. J. Nutr.* **2016**, *55*, 1359–1375. [CrossRef]
84. Bardin, A.; Boulle, N.; Lazennec, G.; Vignon, F.; Pujol, P. Loss of ERbeta expression as a common step in estrogen-dependent tumor progression. *Endocr. Relat. Cancer* **2004**, *11*, 537–551. [CrossRef] [PubMed]
85. Daily, J.W.; Ko, B.S.; Ryuk, J.; Liu, M.; Zhang, W.; Park, S. Equol Decreases Hot Flashes in Postmenopausal Women: A Systematic Review and Meta-Analysis of Randomized Clinical Trials. *J. Med. Food* **2019**, *22*, 127–139. [CrossRef] [PubMed]
86. Chen, M.N.; Lin, C.C.; Liu, C.F. Efficacy of phytoestrogens for menopausal symptoms: A meta-analysis and systematic review. *Climacteric* **2015**, *18*, 260–269. [CrossRef] [PubMed]
87. Li, L.; Lv, Y.; Xu, L.; Zheng, Q. Quantitative efficacy of soy isoflavones on menopausal hot flashes. *Br. J. Clin. Pharmacol.* **2015**, *79*, 593–604. [CrossRef]
88. Li, L.; Xu, L.; Wu, J.; Dong, L.; Zhao, S.; Zheng, Q. Comparative efficacy of nonhormonal drugs on menopausal hot flashes. *Eur. J. Clin. Pharmacol.* **2016**, *72*, 1051–1058. [CrossRef]
89. Sarri, G.; Pedder, H.; Dias, S.; Guo, Y.; Lumsden, M.A. Vasomotor symptoms resulting from natural menopause: A systematic review and network meta-analysis of treatment effects from the National Institute for Health and Care Excellence guideline on menopause. *BJOG* **2017**, *124*, 1514–1523. [CrossRef]

90. Franco, O.H.; Chowdhury, R.; Troup, J.; Voortman, T.; Kunutsor, S.; Kavousi, M.; Oliver-Williams, C.; Muka, T. Use of Plant-Based Therapies and Menopausal Symptoms: A Systematic Review and Meta-analysis. *JAMA* **2016**, *315*, 2554–2563. [CrossRef]
91. Hairi, H.A.; Shuid, A.N.; Ibrahim, N.; Jamal, J.A.; Mohamed, N.; Mohamed, I.N. The Effects and Action Mechanisms of Phytoestrogens on Vasomotor Symptoms During Menopausal Transition: Thermoregulatory Mechanism. *Curr. Drug Targets* **2019**, *20*, 192–200. [CrossRef]
92. Magee, P.J. Is equol production beneficial to health? *Proc. Nutr. Soc.* **2011**, *70*, 10–18. [CrossRef]
93. EFSA Pannel on Food Additives and Nutrient Sources Added to Food (ANS). Risk assessment for peri- and post-menopausal women taking food su pplements containing isola ted isoflavones. *EFSA J.* **2015**, *13*, 4246. [CrossRef]
94. Duffy, C.; Perez, K.; Partridge, A. Implications of phytoestrogen intake for breast cancer. *CA Cancer J. Clin.* **2007**, *57*, 260–277. [CrossRef] [PubMed]
95. Schmidt, M.; Arjomand-Wölkart, K.; Birkhäuser, M.H.; Genazzani, A.R.; Gruber, D.M.; Huber, J.; Kölbl, H.; Kreft, S.; Leodolter, S.; Linsberger, D.; et al. Consensus: Soy isoflavones as a first-line approach to the treatment of menopausal vasomotor complaints. *Gynecol. Endocrinol.* **2016**, *32*, 427–430. [CrossRef] [PubMed]
96. Matulka, R.A.; Matsuura, I.; Uesugi, T.; Ueno, T.; Burdock, G. Developmental and Reproductive Effects of SE5-OH: An Equol-Rich Soy-Based Ingredient. *J. Toxicol.* **2009**, *2009*, 307618. [CrossRef] [PubMed]

Publisher's Note: MDPI stays neutral with regard to jurisdictional claims in published maps and institutional affiliations.

Article

The Acute and Chronic Cognitive Effects of a Sage Extract: A Randomized, Placebo Controlled Study in Healthy Humans

Emma L. Wightman [1,2,*], Philippa A. Jackson [1], Bethany Spittlehouse [1], Thomas Heffernan [1], Damien Guillemet [3] and David O. Kennedy [1]

1 Brain Performance and Nutrition Research Centre, Northumbria University, Newcastle Upon Tyne NE1 8ST, UK; philippa.jackson@northumbria.ac.uk (P.A.J.); b.spittlehouse@northumbria.ac.uk (B.S.); tom.heffernan@northumbria.ac.uk (T.H.); david.kennedy@northumbria.ac.uk (D.O.K.)

2 NUTRAN, Northumbria University, Newcastle Upon Tyne NE1 8ST, UK

3 Nexira SAS, 129 Chemin de Croisset-CS 94151, 76723 Rouen, France; d.guillemet@nexira.com

* Correspondence: emma.l.wightman@northumbria.ac.uk; Tel.: +44-191-227-3725

Abstract: The sage (*Salvia*) plant contains a host of terpenes and phenolics which interact with mechanisms pertinent to brain function and improve aspects of cognitive performance. However, previous studies in humans have looked at these phytochemicals in isolation and following acute consumption only. A preclinical in vivo study in rodents, however, has demonstrated improved cognitive outcomes following 2-week consumption of Cognivia^{TM}, a proprietary extract of both *Salvia officinalis* polyphenols and *Salvia lavandulaefolia* terpenoids, suggesting that a combination of phytochemicals from sage might be more efficacious over a longer period of time. The current study investigated the impact of this sage combination on cognitive functions in humans with acute and chronic outcomes. Participants (n = 94, 25 M, 69 F, 30–60 years old) took part in this randomised, double-blind, placebo-controlled, parallel groups design where a comprehensive array of cognitions were assessed 120- and 240-min post-dose acutely and following 29-day supplementation with either 600 mg of the sage combination or placebo. A consistent, significant benefit of the sage combination was observed throughout working memory and accuracy task outcome measures (specifically on the Corsi Blocks, Numeric Working Memory, and Name to Face Recall tasks) both acutely (i.e., changes within day 1 and day 29) and chronically (i.e., changes between day 1 to day 29). These results fall slightly outside of those reported previously with single *Salvia* administration, and therefore, a follow-up study with the single and combined extracts is required to confirm how these effects differ within the same cohort.

Keywords: sage; *Salvia officinalis*; *Salvia lavandulaefolia*; polyphenols; cognition

Citation: Wightman, E.L.; Jackson, P.A.; Spittlehouse, B.; Heffernan, T.; Guillemet, D.; Kennedy, D.O. The Acute and Chronic Cognitive Effects of a Sage Extract: A Randomized, Placebo Controlled Study in Healthy Humans. *Nutrients* **2021**, *13*, 218. https://doi.org/10.3390/nu13010218

Received: 20 November 2020
Accepted: 11 January 2021
Published: 14 January 2021

1. Introduction

The Nepetoideae subfamily of the Lamiaceae family of plants, which provides most of our culinary herbs and many essential oils, is a particularly rich source of plants that use volatile terpenes in symbiotic ecological roles. Members include psychoactive herbs such as rosemary, lemon-balm, peppermint and sage, and this group typically synthesise mono- and sesquiterpenes such as 1,8-cineole, α-pinene, camphor, geraniol, geranial, borneol, camphene and β-caryophyllene. These plants also express high levels of phenolics including, in all cases, rosmarinic acid and its derivatives, alongside other phenolic acids and polyphenols [1].

Extracts from this family of plants often share common (but variable) mechanisms of action relevant to the brain, for instance, inhibition of acetylcholinesterase and binding allosterically to gamma-Aminobutyric acid ($GABA^A$), nicotinic and muscarinic receptors [1]. These effects are most likely attributable to the terpene content outlined above [2–6]. The phenolic constituents have also demonstrated effects pertinent to brain function, e.g., increases in brain-derived neurotrophic factor (BDNF), and antioxidant and anti-inflammatory activities in neurons (see [7] for a review). More importantly, a recent trial by our own lab demonstrated the more robust effects (in this case, acutely on energy and mood in humans) which can be achieved when combining phenolics (coffee berry, apple catechins and blueberry anthocyanins) and terpenes (sage and ginseng) into a single supplement [8].

The psychoactive effects of consuming single doses of extracts and dried leaf of *Salvia officinalis* and *Salvia lavandulaefolia* (common and Spanish sage, respectively), containing the above-defined full spectrum of phytochemicals, also extend to cognitive function. (It is interesting to note here that, while there are some similarities in the phytochemical composition of sage species, differences also exist. For instance, *salvia officinalis* and *lavandulaefolia* can contain similar percentages of 1,8-cineole but significantly different levels of camphor (22.2% versus 19.6% 1,8-cineole, respectively, and 33.6% versus 15.5% camphor, respectively) [3,9]. With regards to phenolic composition, the two species seem broadly in line although *salvia officinalis* appears to synthesise salvianolic acid derivatives of rosmarinic acid where *salvia lavandulaefolia* does not [1].) Here, vigilance and memory in both younger and older adults seem to be most sensitive to improvements [4,5]. Several studies have also assessed the effects of a single dose of essential oils composed solely of the volatile terpenes present in *Salvia lavandulaefolia* plant material. In the first of these studies, healthy young participants took single oral doses of 50 and 100 µL of encapsulated *salvia officinalis* dried leaf [4] and exhibited improvements in memory (immediate and delayed recall of words) within the first 2.5 h post-dose. Similar mnemonic effects were subsequently confirmed following single doses of 25 and 50 µL of the same essential oil along with improved performance on a mental arithmetic task and improved levels of subjective alertness, calmness and contentment [10]. Most recently, the psychoactive properties of *Salvia lavandulaefolia* were confirmed in healthy young adults who consumed single doses of essential oil that exclusively contained monoterpenes. Specifically, this oil provided a high concentration of 1,8-cineol and presented a particularly potent acetylcholinesterase inhibitory profile. Within the first four hours, single doses improved memory and attention task performance, increased alertness and reduced mental fatigue during extended performance of difficult tasks [11].

Taken together, these human randomised controlled trials demonstrate robust and consistent acute cognitive effects of *Salvia* when administered as a single species (either *officinalis* or *lavandulaefolia*). However, what is not known is whether chronic effects could also be elicited nor whether, when the species are co-administered, these effects pervade and/or synergistic mechanisms of action produce different effects entirely. This question was partly answered recently by a preclinical in vivo rodent study where a combination of essential oil from *Salvia lavandulaefolia* and a leaf extract from *Salvia officinalis* (Cogniva™) was investigated [12]. Here, the combination was compared to each species alone, and a control, which demonstrated a significant acute effect on memory in the Y-maze after all single doses of *Salvia* and a significant chronic (following 2 weeks of treatment) effect on visuospatial memory in the Morris water maze following the combination only. Biochemical and histological investigation after the end of the administration period revealed that sage stimulated the expression of calcium–calmodulin-dependent protein kinase II (CaMKII), a mechanism that has been proposed to regulate the biochemical neuronal processes supporting working memory, learning and interpretation [13–16].

Taken together, the original research question addressed by this study is whether this *Salvia* and phenolic combination can produce similarly positive effects on memory in humans and how it might mediate persistent attentional and memory demands across various media, in particular, in an attempt to mirror real-world demands.

2. Methods

2.1. Study Design and Participants

The study aimed to analyse a final data set of $n = 90$ ($n = 45$ per condition) healthy adults from this randomised (simple), double-blind, placebo-controlled, parallel groups design. A secondary aim was to balance the age of participants across 3 age categories: $n = 30$ between 30–40 years, $n = 30$ between 41–50 years and $n = 30$ between 51–60 years. This was, in large part, to help balance the age of participants across this broad range, and it was hoped that, in doing so, analyses could take into account potential age differences (However, as it transpired, no interpretable interactions between treatment × age were revealed in the analyses (likely because the analysis simply was not powered to include this), and therefore, it was decided that, in order to maintain clarity with so many other outcomes, the reported analyses omit this factor.) Participants were recruited from the local area via social media, internal email for Northumbria University staff and students, and advertisements in the local newspaper. The actual number recruited, to allow for any loss pre-analysis, was $n = 94$ (69 female and 25 male, mean age 43.9 years (SD 8.6 years), 88 right-handed and 6 left-handed, mean years in education 16.7 years (SD 3.6 years) and mean BMI 25.4 (SD 3.5)). The number of participants in the placebo condition was 49, and 45 participants consumed the active intervention. $n = 36$ were aged 30–40 years, $n = 32$ were between 41–50 years and $n = 26$ were between 51–60 years. See Figure 1 for participant disposition diagram and Table 1 for demographic breakdown for each treatment condition.

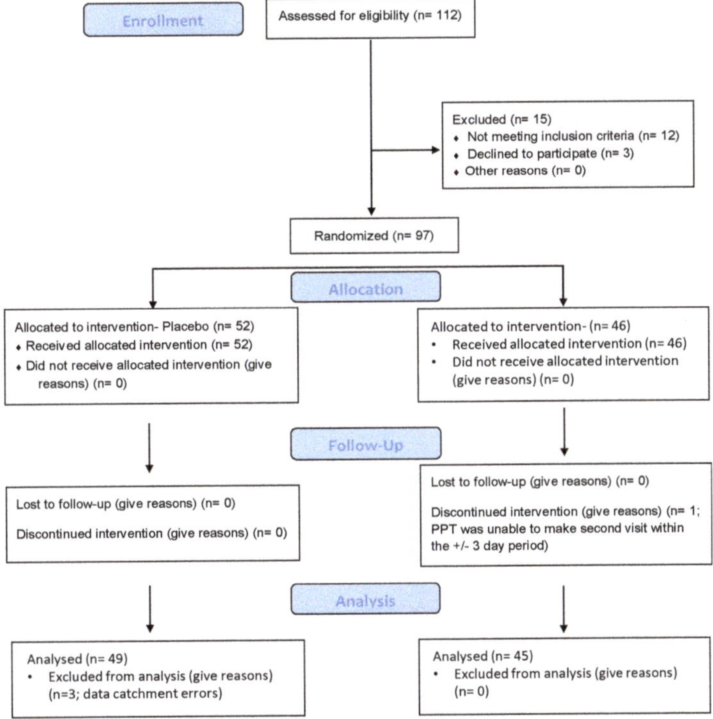

Figure 1. Participant disposition diagram. From the 112 potential volunteers enrolled, 97 were randomised and allocated to receive either treatment A (placebo; $n = 52$) or treatment B (sage; $n = 46$). With 1 participant lost to follow up in the latter group and 3 excluded from analysis due to data catchment errors in the former group, the final total of data sets available for analysis were $n = 49$ in the placebo condition and $n = 45$ in the sage condition.

Participants were excluded if they had any pre-existing medical condition/illness which would impact taking part in the study (this was necessarily vague to encompass any unforeseen issues on a case-by-case basis); were currently taking prescription medications which would contraindicate with the study (i.e., which might impact the outcome measures directly themselves); had high blood pressure (systolic over 159 mm Hg or diastolic over 99 mm Hg); had a BMI outside of the range 18.5–30 kg/m^2 (the exclusion of those in the underweight and obese BMI ranges was an attempt to mitigate against any potentially excluding health complaints associated with these extreme ends of adiposity); were pregnant, seeking to become pregnant or lactating; had learning and/or behavioural difficulties such as dyslexia or ADHD; had a visual impairment that could not be corrected with glasses or contact lenses (including colour-blindness); smoked; consumed excessive amounts of caffeine (>500 mg per day); had food intolerances/sensitivities; had taken antibiotics, prebiotics or probiotics (including drinks, e.g., Yakult® or Actimel®) within the past 8 weeks; were currently participating in other clinical or nutrition intervention studies or had in the past 4 weeks; had been diagnosed with/undergoing treatment for alcohol or drug abuse in the last 12 months; had been diagnosed with/undergoing treatment for a psychiatric disorder in the last 12 months; suffered from frequent migraines that require medication (more than or equal to 1 per month); had sleep disturbances (including night-shift work) and/or were taking sleep aid medication; or had active infections.

Table 1. Participant demographics for the placebo and sage conditions.

Measure	Placebo	Sage	Total
n	49	45	94
Average years of age (SD)	44.7 (8.4)	43 (8.9)	43.9 (8.6)
Male/Female	11/38	13/32	25/69
Left-/Right-handed	3/46	3/42	6/88
Average years in education (SD)	17 (3.5)	16.5 (3.6)	16.7 (3.6)
Average Body Mass Index (SD)	24.8 (3.3)	26.1 (3.6)	25.4 (3.5)

It outlines the demographic composition of participants as assessed at screening/training in both the placebo and sage conditions as well as an average across both groups. No significant differences between treatment groups were observed on these demographic features.

This study was conducted according to the guidelines laid down in the Declaration of Helsinki 1975, and all procedures were approved by the department of Psychology (Northumbria University) staff ethics committee (code: 8970). Written informed consent was obtained from all participants.

2.2. Treatments

Participants consumed either placebo or 600 mg Cognivia™ (a proprietorial supplement (with more detail available in [12]) which is available to buy over the counter) every day for 29 +/− 3 days (This was the permitted range for those who could not return on day 29, but in actuality, nobody was under 29 days). The selected dose contained 400 mg of aqueous extract from *Salvia officinalis* leaves characterized for its content in polyphenols (as rosmarinic acid, apigenin glucosides, luteolin glucosides and others). The remaining 200 mg contained 50 μL of *Salvia lavandulaefolia* essential oil characterized for its content in terpenoids (as eucalyptol, camphor, α- and β-pinene, and others) and encapsulated with gum acacia. The posology of both active substances was selected in accordance with descriptions of the most effective dosages described in the clinical acute studies introduced above (i.e., 50 μL of *Salvia lavandulaefolia* essential oil and extract of *Salvia officinalis* with a ratio equivalent to 2.25 g of dried leaves). The encapsulated powder of essential oils with gum acacia has been proposed to facilitate posology and observance compared to the liquid form and to protect terpenoids from evaporation and oxidation. Acacia gum has a long history of use with terpenes and essential oil protection [17], and the encapsulation has not compromised the nootropic activities of *Salvia lavandulaefolia* essential oil in a previous preclinical study [12]. Treatment was in the form of blue capsules (both active and placebo) and was dispensed from identical white bottles; participants took their first and last lab-based doses from these bottles and self-supplemented at home in the interim. Two compliance measures were used to determine adherence to the treatment regimen; the primary measure was a capsule count of returned treatment, and the second was reference to a treatment diary in which the participant noted the time of their treatment consumption each day.

2.3. *Cognitive, Mood and Blood Pressure (BP) Assessment*

Computerised Mental Performance Assessment System (COMPASS):

This testing system delivers a bespoke collection of tasks, with fully randomised parallel versions of each task delivered at each assessment for each individual. The battery has been in use within Northumbria University for over 10 years; is now commercially available for other research organisations (www.cognitivetesting.com); and is currently in use within a number of UK, US, New Zealand and Australian Universities, companies and research organisations. Each battery is entirely self-contained, e.g., the stimuli presented at the start (i.e., the pictures and words) are recalled at the end and any subsequent batteries present a novel selection of stimuli. All tasks are briefly summarised below, and full descriptions can be found elsewhere [18].

2.3.1. Picture Presentation

Fifteen colour photographic images of everyday objects (such as a telephone, car or cup) were presented sequentially on screen for the participant to remember at the rate of 1 every 3 s, with a stimulus duration of 1 s.

2.3.2. Face Presentation

Twelve passport-style photographic images of people, containing a first and last name underneath, were presented sequentially in a random order to participants. Stimulus duration was 3 s, with a 1-s interstimulus duration.

2.3.3. Word Presentation

Fifteen words were presented sequentially on screen for participants to remember. Stimulus duration and interstimulus time were both 1 s.

2.3.4. Immediate Word Recall

The participant was allowed 60 s to write down as many of the words that were just presented as possible. The task was scored for number correct and errors.

2.3.5. Numeric Working Memory

Five digits (1–9) were presented sequentially for the participant to remember. This was followed by a series of 30 probe digits (15 targets and 15 distractors), and the participant indicated whether it had been in the original series by a simple "yes" or "no" key press. The task consisted of 3 separate trials. Accuracy and mean reaction time for correct responses were recorded.

2.3.6. Choice Reaction Time

Participants responded with a left or right key press corresponding to the direction of the arrows appearing on screen. The randomly varying interstimulus interval was between 1 and 3 s for a total of 50 stimuli. Accuracy and mean reaction time for correct responses were recorded.

2.3.7. Corsi Blocks Task

A set number of blocks, from a maximum of 9, changed colour from blue to red in a randomly generated sequence and, once finished, participants were instructed to repeat the sequence by clicking on the blocks using the mouse and cursor. The task was repeated five times at each level of difficulty, from 4 upwards, until the participant could no longer correctly recall the sequence. The task was scored for "span score", calculated by averaging the level of the last three correctly completed trials.

2.3.8. Cognitive Demand Battery (CDB)

The CDB comprises repeated (in this case, 3 repetitions) of the Serial 3s subtraction task (2 min), Serial 7s subtraction task (2 min) and Rapid Visual Information Processing (RVIP, 5 min) task. For the serial subtraction tasks, participants subtracted either 3 or 7 consecutively from a randomly generated number between 800 and 999 for the duration of the task, entering their responses on the keyboard's linear number pad. These tasks were scored for the number of correct subtractions and the number of errors. For the RVIP task, participants monitored a continuous series of single digits (1–9 at a rate of 100 per minute) for targets of three consecutive odd or three consecutive even digits (8 per minute). The task was scored for number of correctly identified target strings and average reaction time for correct detections.

2.3.9. Peg and Ball Task

Participants were presented with 2 configurations of 3 coloured balls (blue, green and red) on 3 pegs that each hold 3 balls. Participants had to rearrange the balls, moving one ball at a time, from the starting configuration so that they matched the position of the balls in the goal configuration. Each trial (of 5) generated scores for planning times prior to moving, time to completion and errors.

2.3.10. Delayed Word Recall

The participant was again given 60 s to write down as many of the words presented previously as possible. Total number of correct responses and errors were recorded.

2.3.11. Delayed Word Recognition

The original 15 words, plus 15 distractor words, were presented one at a time in a random order with participants responding "yes" or "no" as to whether they were originally presented. Accuracy and mean reaction time for correct responses were recorded.

2.3.12. Delayed Picture Recognition

The original 15 pictures plus 15 distractor pictures were presented one at a time in a random order with participants responding "yes" or "no" as to whether they were originally presented. Accuracy and mean reaction time for correct responses were recorded.

2.3.13. Name-to-Face Recall

The twelve original photographs presented at outset were presented on the screen, one at a time with a list of 4 different first names and 4 different last names underneath. Participants had to choose the first and last name that was originally presented with the photograph. The numbers of correct responses for first and last names were recorded and collapsed to give an overall score for this task.

2.3.14. Global Cognitive Measures

Almost all of the above cognitive tasks produce multiple outcome measures, e.g., a measure of accuracy, speed and error in performing the individual task. This allows one to investigate performance further by combining those same measures across all appropriate tasks. From this, global cognitive measures can then be derived: accuracy of attention (comprised of accuracy in relation to the choice reaction time and rapid visual information processing tasks), speed of attention (comprised of the reaction time performance in relation to the choice reaction time and rapid visual information processing tasks), working memory (comprised of accuracy in relation to the numeric working memory task and span score on the Corsi Blocks tasks), speed of memory (comprised of reaction time performance in relation to the numeric working memory, picture recognition and word recognition tasks) and episodic memory (comprised of accuracy in relation to the immediate and delayed word recall tasks, name-to-face recall, and picture and word recognition).

Figure 2 outlines the individual tasks used, the order in which they were presented and the approximate timings as well as the primary cognitive domain of each task (left-hand-side of the diagram). The figure also depicts the global cognitive domains (right-hand-side) and where the data was sourced from the individual cognitive tasks to derive these global scores.

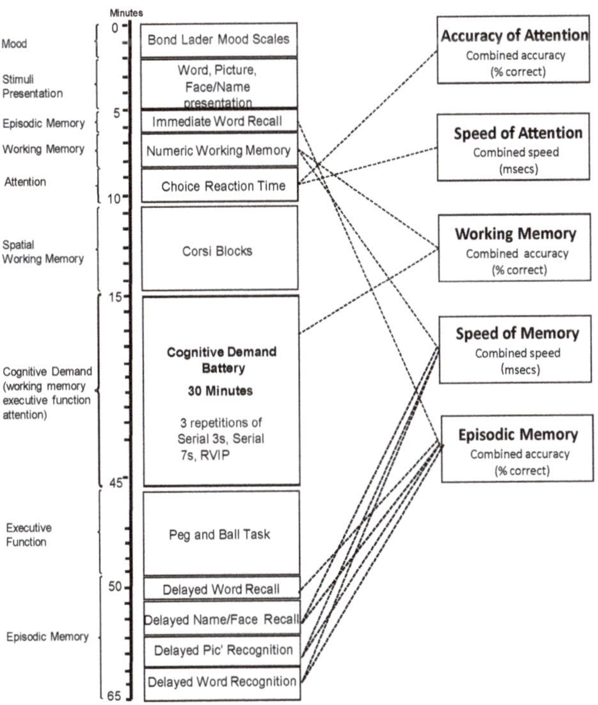

Figure 2. The running order of the individual cognitive assessments. The tasks are shown in order of completion with approximate timings. On the left, the "cognitive domain" assessed by the task is shown, and the boxes to the right show global measures into which data from several tasks can be collapsed.

2.3.15. Prospective Remembering Video Procedure (PRVP)

The PRVP task, a location-learning task, has proven sensitive to memory failings during impaired states, e.g., during binge drinking [19]. Less is known about its sensitivity to normal memory degradation over time and whether it can detect attenuation of this in response to nutritional supplementation. As such, its use here is somewhat exploratory. The task required participants to encode a list of 18 locations and their matched action within 60 s. These location–action pairs then unfolded during a 10-min video clip of a walk down a shopping street, e.g., "At Thorntons, buy a bag of toffees.", during which participants must note down as many pairs as possible, along with how many pushchairs they saw during the clip (acting as a distractor and not analysed). The performance of participants for their first completion of this task (i.e., day 25) was scored as previously described [19], i.e., 1 point for each correctly remembered location–action (with a potential total of 18 points), as this is the original procedure for completing the task. However, we also wanted to measure whether memory degradation of these locations–actions was affected by treatment and so, on day 29, participants completed this task again but without seeing the list of locations–actions first. Here, participants were awarded 1 point for each location and action (i.e., with a potential total of 36 points) as it was anticipated that remembering would be significantly more challenging when relying on encoding from 4 days ago, and the original scoring method may lead to floor effects.

2.3.16. Mood

To assess mood, the current study used both the Bond–Lader [20] mood scales (completed at the beginning of each task battery repetition) and the State Trait Anxiety Inventory (STAI) [21]. Both state and trait anxiety were measured during the screening/training visit to provide a baseline measure of mood. Subsequently, during each testing session, only state was assessed as trait mood should be stable across this relatively short period.

2.3.17. Blood Pressure

Sitting blood pressure and heart rate readings were collected using a Boso Medicus Prestige blood pressure monitor with the subject's arm supported at the level of the heart and with their feet flat on the floor. Readings were taken following completion of the baseline tasks and again following completion of the post-dose tasks.

2.4. Procedure

Testing took place between August 2018–April 2019 at Northumbria University, UK, within a suite of testing facilities with participants visually isolated from each other. Participants attended the laboratory on 4 separate occasions: an introductory visit between 1 and 14 days before the first day of treatment, two testing days (day 1 and day 29) and an interim visit (day 25).

The introductory visit to the laboratory was comprised of briefing on the requirements of the study, obtaining informed consent, health screening, completing the Caffeine Consumption Questionnaire (CCQ) and State-Trait Anxiety Inventory (STAI) trait subscale, training on the cognitive and mood measures, and collecting demographic data.

For the two ensuing laboratory-based testing sessions (day 1 and day 29), participants attended the laboratory before 8.00 a.m. after having consumed a standardised breakfast of cereal and/or toast at home no later than an hour before arrival. They must have refrained from alcohol for 24 h and caffeine for 18 h. On arrival, on each day, participants completed the State-Trait Anxiety Inventory (STAI) state subscale and the computerised cognitive assessment (as per Figure 2), followed by measurements of blood pressure and heart rate. Immediately following this, they consumed their treatment for that day. Two further cognitive assessments (plus blood pressure and heart rate) identical to the pre-dose assessment commenced at 120 (approximately 11:00 a.m.) and 240 (approximately 02:00 p.m.) minutes post-dose; the latter was taken in order to take advantage of the natural decline in performance during the day. Participants were offered a standardised lunch at approximately 12:10 p.m. in order to remove the potential confound of hunger for the final cognitive assessment in the afternoon. Lunch comprised 1 cheese sandwich (Hovis soft white bread 2 slices, with 186 kcal, 1.4 g fat, 2.8 g sugar and 7 g protein; Sainsbury's British Medium Grated Cheddar Cheese at 30 g, with 127 kcal, 10.5 g fat, <0.5 g sugar and 7.6 g protein; and Lurpack slightly salted spread at approximately 10 g, with 72 kcal, 8 g fat, <0.1 g sugar and <0.1 g protein), 1 packet of ready salted flavour crisps (Walkers 25-g bag, with 132 kcal, 8 g fat, 0.1 g sugar and 1.5 g protein) and 1 pot of custard (Ambrosia 125-g pot, with 124 kcal, 3.5 g fat, 14.3 g sugar and 3.6 g protein). This lunch was optional (as long as non/consumption of components was the same for both visits) to avoid the potentially more disruptive effects of eating items which were unpalatable to participants.

Additionally, on day 25, participants came into the lab and completed the PRVP task as described above. At baseline on day 29, participants were prompted to recall the locations and actions from this PRVP task again but without the prompt of the video.

Figures 3 and 4 depict the laboratory-based testing session timeline and chronic study overview, respectively.

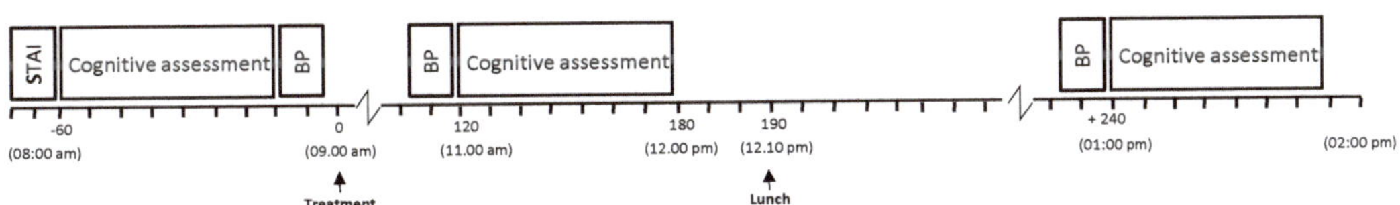

Figure 3. Testing session timeline on both the acute and chronic visits. Participants completed the state subscale of the State Trait Anxiety Inventory (STAI) followed by a full cognitive assessment pre-dose and at 120 and 240 min post-dose. Blood pressure and heart rate were recorded after each cognitive assessment. Lunch was provided at approximately 12:10 p.m.

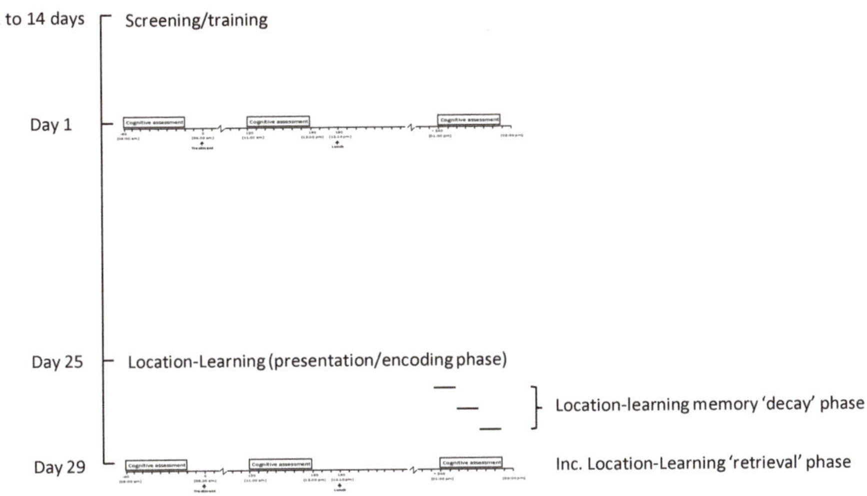

Figure 4. Overall trial diagram. The diagram includes screening/training and the main acute (day 1) and chronic (day 29) lab visit sessions. Participants returned to the lab on day 25 to first complete the location-learning task. The 3 intervening days between this and day 29 when they again completed the location-learning task acted as a "decay" phase for these encoded memories.

2.5. Statistics

An *A Priori* G*Power [22] calculation determined that, to achieve a medium effect size (Cohen's f = 0.15) with a minimum power of 0.8 utilizing the following analysis plan, a sample size of $n = 86$ would be required, which we rounded up to $n = 90$.

All of the below analyses were conducted with IBM SPSS Statistics 25 and were first investigated for normality and baseline differences between treatment groups. Unless reported, no baseline differences were detected. The confidence intervals for all analyses are set at 95% and, if post hoc analyses were required, these were student's t-tests.

The COMPASS, Bond–Lader and blood pressure data were then analysed for acute effects within day 1 and day 29 as well as pure chronic effects within day 29:

1. Acute effects within day 1 and 29

Here, the post-dose performance for each day was changed from its own baseline and two-way, repeated measures ANOVAs were conducted with "treatment" as a fixed factor:

2. Pure chronic effects within day 29

Here, pre-dose and the two post-dose assessments on day 29 were changed from the day 1 baseline and two-way, repeated measures ANOVAs were conducted using "treatment" as a fixed factor.

The PRVP data was analysed via four methods:

1. Day 25 data: A univariate ANOVA with "treatment" as a fixed factor was used to analyse performance following 25 days (±3 days) of treatment. One full mark was awarded for correctly remembering the location with its action (this is defined as the "original scoring").

2. Day 29 data (original scoring): See 1.

3. Day 29 data (lenient scoring): It was considered that the original scoring described above may be too conservative to utilize on performance captured 4 days (±3 days) after having encoded the location/action list, and therefore, a further analysis applying more lenient scoring was also undertaken. Here, 1 mark was awarded for all correctly remembered locations and all correctly remembered actions and a univariate ANOVA with treatment as a fixed factor compared this score between treatments.

4. "Decay" score: the score on day 29 was subtracted from the day 25 performance (original scoring method was used for both) to create a "decay" score, defining how much memory had degraded across the 4 days (±3 days), and a univariate ANOVA with treatment as a fixed factor compared this decay score between treatments.

The STAI mood data were analysed via two approaches. Firstly, the trait anxiety scores (only one value for each participant collected pre-dose) were analysed via univariate ANOVA, with treatment as a fixed factor, to determine whether the two groups had any intrinsic differences in anxiety. Secondly, to determine the effects of treatment on state anxiety scores, change scores (day 29 minus day 1) were compared between treatments via repeated measures ANOVA with treatment as a between-subjects factor.

3. Results

3.1. Compliance, Treatment Guess and Adverse Events

Mean compliance was 101%, with compliances ranging 83–121%.

A chi-square analysis showed that participants were able to subjectively detect that they were in the active condition (69% guessed correctly): $X(1) = 5.79$, $p = 0.02$.

Over the course of the study, 44 adverse events presented which could possibly be related to the study treatment (10 placebo and 34 sage treatments). These comprised:

- Muscular/bodily pain/injury: 9
- Cold/flu symptoms: 5
- Headache/migraine: 41 (individual reports from 14 participants)

All participants reported thinking that their adverse events were *not* related to supplement use, and symptoms in all cases resolved during the course of the study. All adverse events were reported as "mild" or "moderate" apart from 2 participants in the sage condition who reported headaches as severe. However, one of them also reported flu/cold-like symptoms.

3.2. Blood Pressure

No pure chronic effects of treatment on blood pressure were observed, but an acute interaction between "treatment × repetition" on heart rate was observed on day 1: $F(1,93) = 4.15$, $p = 0.04$. However, post hoc interrogation revealed no significant differences at either post-dose time point.

3.3. Prospective Remembering Video Procedure (PRVP)

No significant effects were observed on any of the four analysis approaches.

3.4. Mood

No significant effects were observed on state or trait anxiety as indexed by the STAI. With regards to the Bond–Lader mood scales, the only effect involving treatment was a single acute trend towards significance for "treatment × repetition" on day 29 for contentment, $F(1,92) = 3.21$, $p = 0.08$, but no significant effects at any repetition was observed in the post hoc comparisons.

3.5. COMPASS Tasks

Because the COMPASS task outcomes comprise both acute and chronic effects and they are more abundant than the above outcomes, they are separated here into first acute and then chronic effects for clarity. (See Table 2 for cognitive task scores in comparison with baseline and Table 3 (which also includes trends towards significance although these are reported more fully in Supplementary Materials) for a full summary of the cognitive effects of treatment).

Table 2. Cognitive task scores in comparison with baseline.

	Treatment Condition	Day 1						Day 29					
		Baseline		Post-Dose 1		Post-Dose 2		Pre-Dose		Post-Dose 1		Post-Dose 2	
				Difference from Baseline		Difference from Baseline		Difference from Baseline		Difference from Baseline		Difference from Baseline	
		Mean	SD	Mean	SD	Mean	SD	Mean	SD	Mean	SD	Mean	SD
Numeric Working Memory Accuracy	Placebo	95.29	4.66	0.64	4.13	0.67	5.74	1.33	4.77	−0.09	4	0.67	4.71
	600 mg Sage	95.51	4.04	0.35	3.94	0.88	3.87	0.08	3.92	1.44	4.38	0.43	3.88
Corsi Blocks Span	Placebo	6.07	0.82	−0.13	0.8	−0.16	0.67	−0.04	0.74	−0.04	0.72	−0.27	0.98
	600 mg Sage	5.67	1	0.18	0.96	0.12	0.9	−0.16	1.15	0.3	1.03	0.26	1.07
Name-to-face-Recall Accuracy	Placebo	65.42	16.86	−9.1	13.14	−8.83	17.59	−2.5	15.77	−6.75	14.7	−9.67	16.78
	600 mg Sage	63.35	18.63	−4.26	19.26	−5.97	17.58	4.17	19.06	−0.19	19.15	−4.17	17.66

It shows the scores on COMPASS tasks which are implicated in significant differences between groups. Day 1 baseline values denote raw means, and subsequent columns denote means and standard deviations (SD)s that changed from this baseline.

Table 3. Summary of cognitive effects of treatment.

Task	Outcome Measure	Acute		Pure Chronic
		Day 1	Day 29	
Corsi Blocks	Span Score	treatment $p = 0.04$	treatment $p = 0.002$	treatment × repetition $p = 0.001$
Numeric Working Memory	Accuracy	*n/A*	treatment $p = 0.03$ and treatment × repetition $p = 0.04$	treatment × repetition $p = 0.01$
Peg and Ball	Thinking Time	treatment × repetition $p = 0.08$	N/A	N/A
Name-To-Face Recall	Reaction Time	N/A	treatment $p = 0.06$	treatment × repetition $p = 0.07$
	Accuracy	N/A	N/A	treatment $p = 0.03$
Serial 3 Subtractions	Errors	N/A	treatment × repetition $p = 0.06$	treatment × repetition $p = 0.06$
	Total	treatment × repetition $p = 0.07$	N/A	N/A
	Accuracy	treatment × repetition $p = 0.07$	N/A	N/A
Serial 7 Subtractions	Accuracy	treatment × repetition $p = 0.08$	N/A	N/A
Working Memory	Accuracy	treatment $p = 0.006$	treatment $p = 0.07$	treatment $p = 0.07$ and treatment × repetition $p = 0.03$
Overall Accuracy	Accuracy	treatment $p = 0.09$	treatment $p = 0.09$	treatment $p = 0.08$

It shows the outcome variables from COMPASS which evinced significant effects involving treatment as a factor. For completeness, this table also depicts the trends towards significance (latter in italics) involving treatment as a factor. These were not included in the above textural report of the results for brevity, but the authors believe that they do contribute to the full picture of results and so the full report of trends towards significance can be found in the online Supplementary Materials. Outcomes defined within a thick black border denote the global cognitive domains. N/A = not applicable (i.e., no significant effect containing "treatment" was observed here).

3.5.1. Acute Effects

Numeric working memory accuracy evinced a significant acute main effect of "treatment" in favour of sage within day 29; F(1,92) = 4.87, $p = 0.03$. A significant "treatment × repetition" in the same direction was also observed on day 29, F(1,92) = 4.21, $p = 0.04$, and post-hoc ANOVAs revealed that this was influenced by post-dose repetition 1, F(1,93) = 8.4, $p = 0.005$, as no significant effect was observed at post-dose repetition 2, F(1,93) = 0.50, $p = 0.48$. On the **Corsi blocks span score**, a significant baseline difference was detected on day 1, where placebo participants had a significantly higher span score (mean 6.06) than those in the sage condition (mean 5.65), F(1, 88) = 4.49, $p = 0.04$. A significant acute effect of "treatment" was then observed on day 1, F(1, 92) = 4.2, $p = 0.04$ (where span score was higher for sage compared to the placebo) and on day 29, F(1, 92) = 10.58, $p = 0.002$. Again, span score was higher for sage versus placebo. (See Figure 5).

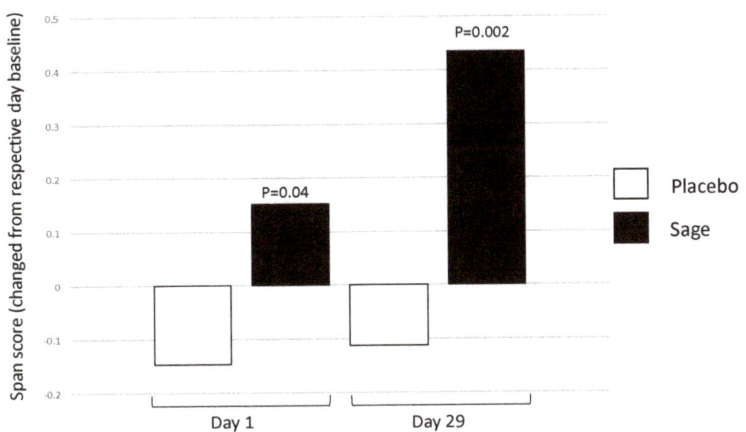

Figure 5. Acute effect of treatment on Corsi blocks span score on day 1 and day 29: a significant main acute effect of treatment was observed on both day 1 ($p = 0.04$) and day 29 ($p = 0.002$) when post-dose performance was compared to the pre-dose performance on the same day.

3.5.2. Chronic Effects

Numeric working memory accuracy evinced a significant pure chronic interaction between "treatment × repetition" on day 29 in favour of sage, F(2, 184) = 4.49, $p = 0.01$, with post-hoc ANOVAs revealing no significant effects at pre-dose, F(1,93) = 0.80, $p = 0.37$; a trend towards significance at post-dose repetition 1, F(1,93) = 3.1, $p = 0.08$; and no significant difference at post-dose repetition 2, F(1,93) = 0.07, $p = 0.79$. (See Figure 6 for acute effect of "treatment" on day 29 and Figure 7 for acute and pure chronic interactions between "treatment × repetition" on day 29). On **Corsi blocks span score**, a significant pure chronic interaction between "treatment × repetition" was observed on day 29, F(2,184) = 7.15, $p = 0.001$. Post hoc ANOVAs revealed that, whilst there was no effect at the pre-dose time point, F(1,93) = 0.36, $p = 0.55$, sage was trending towards better performance at post-dose repetition 1, F(1,93) = 3.4, $p = 0.07$, and this reached significance at post-dose repetition 2, F(1,93) = 6.2, $p = 0.02$. (See Figure 8). **On name-to-face recall accuracy,** a significant pure chronic main effect of "treatment" was observed on day 29, F (1,92) = 4.98, $p = 0.03$, where accuracy was better for sage compared to the placebo. (See Figure 9).

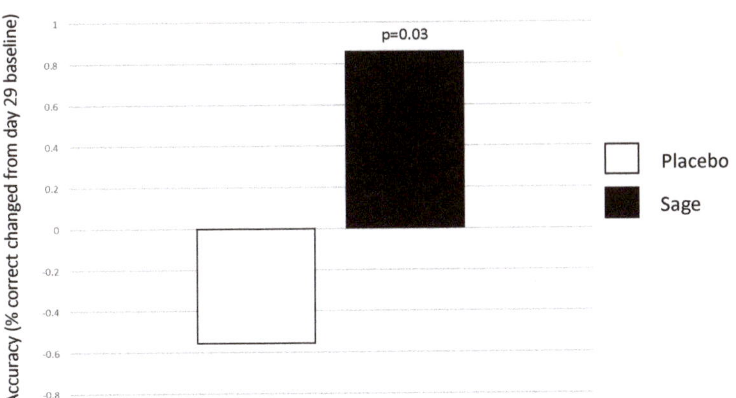

Figure 6. Acute main effect of treatment on day 29 on numeric working memory accuracy: a significant acute main effect of treatment was observed on day 29 ($p = 0.03$). Here, percentage change from baseline (day 29 pre-dose) accuracy was higher in the sage condition than the placebo.

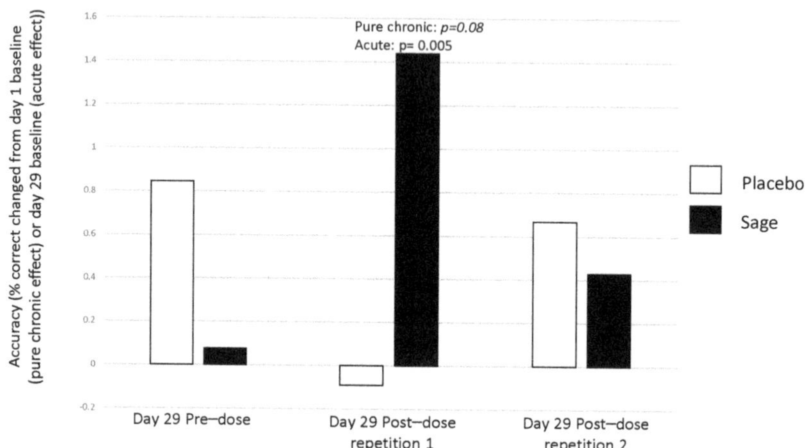

Figure 7. Acute (day 29) and pure chronic interaction between treatment × repetition on numeric working memory accuracy: both an acute interaction ($p = 0.04$) and pure chronic interaction ($p = 0.01$) between treatment × repetition was observed on day 29. Both revealed that this effect took place at post-dose repetition 1; the acute effect compares this post-dose repetition to the day 29 pre-dose performance ($p = 0.005$) and the pure chronic effect compares this post-dose repetition to the day 1 baseline ($p = 0.08$).

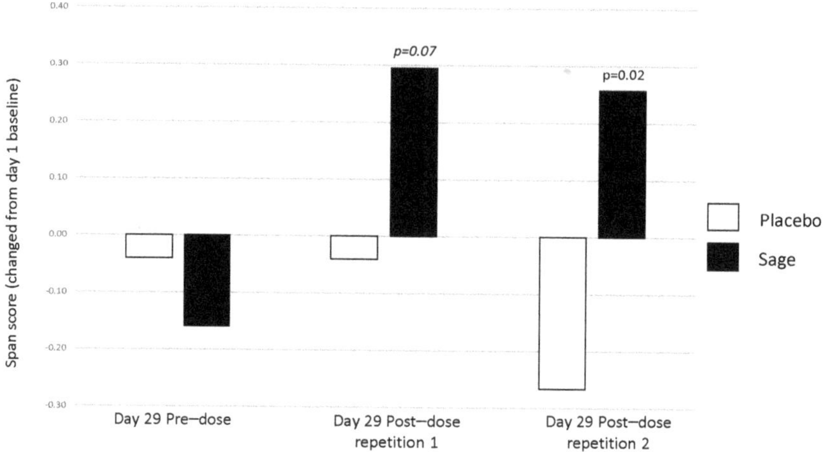

Figure 8. Pure chronic interaction between treatment × repetition on Corsi blocks span score: a significant pure chronic interaction between treatment × repetition ($p = 0.001$) was observed, and here, when compared to day 1 baseline performance, the post-dose repetition 1 span score trended ($p = 0.07$) towards being significantly better in the sage condition than the placebo and was significantly better at post-dose repetition 2 ($p = 0.02$).

Of the 5 global cognitive domains that can be derived from combining COMPASS task outcome measures, 2 evinced significant effects pertaining to "treatment".

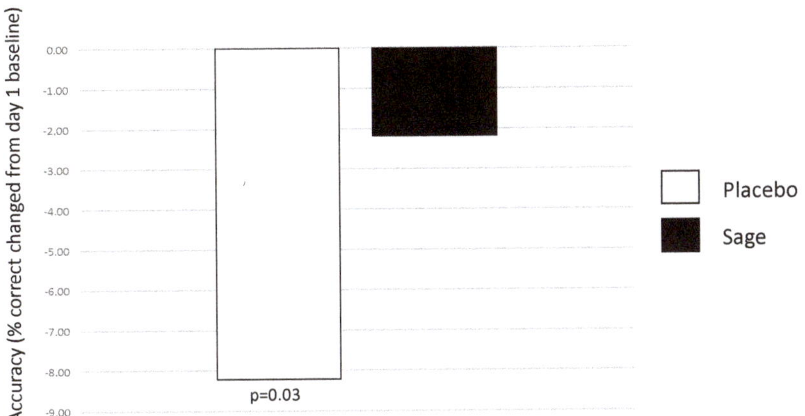

Figure 9. Pure chronic main effect of treatment on day 29 for name-to-face recall accuracy: a significant acute main effect of treatment ($p = 0.03$) was observed on day 29 where accuracy was significantly lower (as compared to pre-dose performance on day 1) in the placebo condition as compared to the sage condition.

3.5.3. Working Memory "Accuracy"

A significant acute effect of treatment was seen on day 1, $F(1,93) = 7.8$, $p = 0.006$, and a trend towards significance was seen on day 29, $F(1,93) = 3.49$, $p = 0.07$. On both days, accuracy was reduced in the placebo condition and sage attenuated this. (See Figure 10).

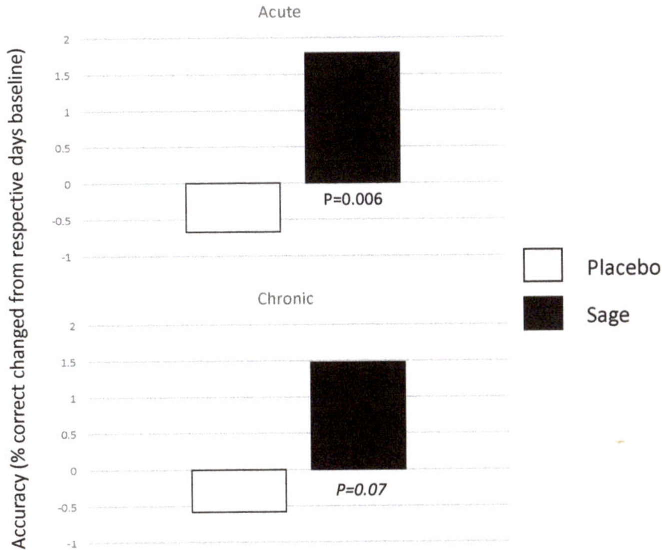

Figure 10. Acute effect of treatment on working memory accuracy on day 1 (top) and day 29 (bottom): the effects of treatment were observed acutely on day 1 ($p = 0.006$), and this was trending towards significance for day 29 also ($p = 0.07$). In both cases, accuracy was higher (as compared to their own days' pre-dose performance) in the sage condition as compared to the placebo condition.

A pure chronic trend towards significance was observed for "treatment" on day 29, $F(1,93) = 3.50$, $p = 0.07$, and a significant interaction between "treatment × repetition" was also seen, $F(1,93) = 3.7$, $p = 0.03$. Here, accuracy reduced from baseline in the placebo condition at both post-dose repetition 1 and post-dose repetition 2, whereas it was increased in the sage group. (See Figure 11).

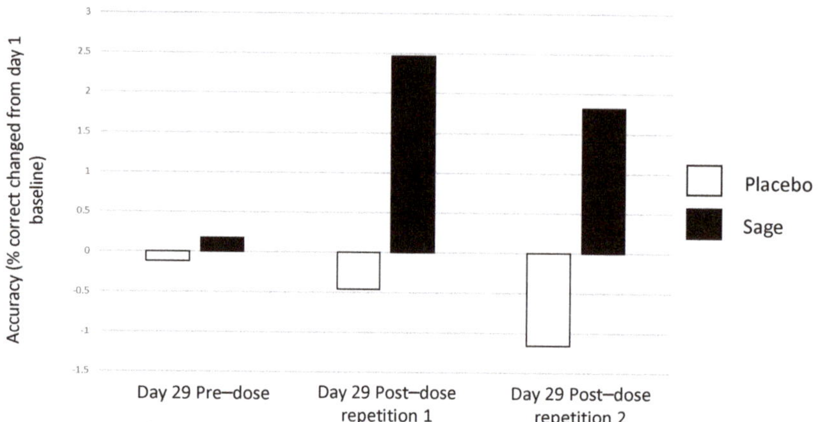

Figure 11. Pure chronic interaction between treatment × repetition on working memory accuracy on day 29: a significant pure chronic interaction between treatment × repetition ($p = 0.03$) was observed on day 29, where accuracy was significantly better (as compared to baseline performance on day 1) in the sage condition at both post-dose repetition 1 and 2 as compared to the placebo condition.

3.5.4. Overall "Accuracy"

Acutely, a trend towards significance was observed for "treatment", F(1,93) = 3.0, $p = 0.09$, and here, we see a similar pattern to the above working memory accuracy findings; on both days, accuracy was reduced in the placebo participants, and this was attenuated in the sage participants. (See Figure 12).

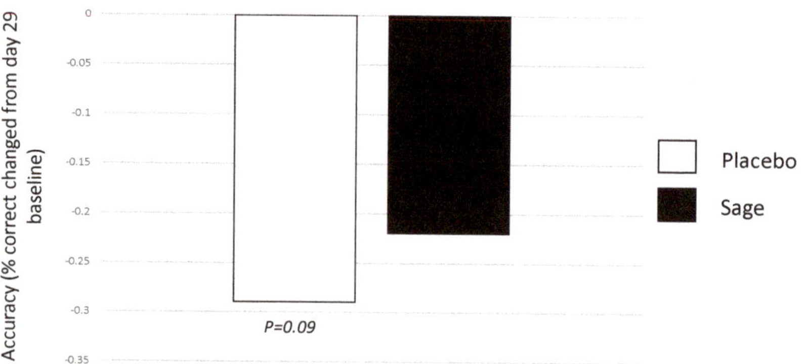

Figure 12. Acute effect of treatment on day 29 for overall accuracy: a trend towards a significant main effect of treatment ($p = 0.09$) was observed for overall accuracy on day 29, where accuracy was poorer (as compared to pre-dose performance on the same day, day 29) in the placebo condition as compared to the sage condition.

This acute effect on day 29 was supported by a trend towards significance for a pure chronic effect of "treatment" on day 29 in favour of sage, F(1,93) = 3.0, $p = 0.08$. (See Figure 13).

Due to space constraints, tables depicting means and standard deviations for all trial outcomes can be found in the online Supplementary Materials. Here, you can also find the statistical analysis tables for all significant and non-significant outcomes.

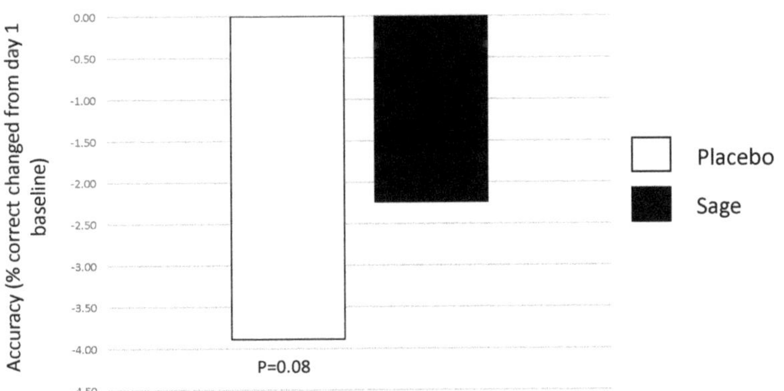

Figure 13. Pure chronic effect of treatment on day 29 for overall accuracy: a trend towards a significant pure chronic effect of treatment ($p = 0.08$) was observed, where accuracy was significantly poorer (as compared to baseline performance on day 1) in the placebo condition as compared to the sage condition.

4. Discussion

The results here demonstrate some clear, additive benefits of sage to individual task performance as well as an attenuation of natural declines in performance; with the effects clearly isolated to the accuracy and working memory performance cognitive domains. The most convincing effects of sage are seen on day 29 and the existence of pure chronic effects on these measures suggests that these effects are the result of a cumulative effect of sage consumed over 29 days.

As an example, span score on the Corsi blocks task was significantly better in the sage condition acutely on day 1 ($p = 0.04$) and more so on day 29 ($p = 0.002$) and the pure chronic effect here also ($p = 0.001$) reinforces that this is likely due to a cumulative effect of sage over 29 days. Improved accuracy on the numeric working memory task was also seen on day 29 following sage ("treatment", $p = 0.03$, and "treatment × repetition", $p = 0.04$), and these findings were reinforced by pure chronic effects on day 29 too ($p = 0.01$). One outcome measure evinced only a pure chronic effect on day 29 in response to sage, accuracy on the name-to-face recall task ($p = 0.03$), and without an acute effect within day 29. We would argue that this supports the use of the analysis plan used here, where both acute effects (comparing post-treatment performance to pre-treatment performance on that same day) and pure chronic effects (comparing post-treatment performance on day 29 to pre-treatment performance on day 1) are investigated. This approach disentangles these effects, where taking treatment on day 29 does not have a significant acute effect in itself and is only revealed when analysis views this in relation to being the 29th dose of the intervention.

Taken together, it is clear that the benefits of sage are focused on the accuracy of performance (and on tasks with working memory as their key cognitive domain) in particular and so it is not surprising that, when the global cognitive domains were analysed, it was the accuracy of working memory factor which yielded a significant acute result on day 1 ($p = 0.006$) and a pure chronic effect on day 29 ($p = 0.03$). It is worth noting here that some of the above effects were significant main effects of treatment and/or interactions between treatment and repetition, and regarding the latter, it was always the case that one, not both, of the post-dose repetitions evinced significant differences between treatments. This likely speaks to the impact of individual variability in any number of factors, e.g., attentional focus and/or the pharmacokinetics of nutritional interventions. This justifies the use of protocols like that employed here, where cognition is assessed over relatively long periods of time and repeated multiple times. This not only amplifies the power of analysing these outcomes but also extends the window of opportunity for cognitive assessment, capturing effects in those who achieve plasma levels much sooner or later than others or during a window of the day where responsiveness to the intervention is higher.

Future investigation into this potential pharmacokinetic variability would be insightful and would likely confirm that variable effects at different time points is the result of fluctuating plasma levels of the intervention and/or time of day effects.

One key aim of this study was to mimic the persistent attentional and memory demands elicited in everyday life across various media, and so, it is particularly interesting that some of the above benefits of sage can be viewed as an attenuation of naturally depleting performance. Both span score on the Corsi blocks task and accuracy on the working memory accuracy global cognitive domain were depleted in the placebo condition (i.e., performance reduced from pre- to post-dose), but sage was able to prevent this depletion from being as severe. Why the performance is so depleted in the placebo group is an interesting point. A recent review [23] demonstrated that both time of day (namely later in the day) and lack of movement were two key predictors of poorer cognition, alongside mood and motivation. As this trial required participants to be sedentary for large periods of time during cognitive task completion and the fact that this continued into the afternoon might simply suggest that sitting for so long was the cause. Whilst this does represent many real-world scenarios, e.g., where we may need to sit and focus on a task for large portions of the day, it would be interesting to see if sage could outperform the placebo in situations where participants were more active.

Historically, the cognitive effects of sage seemed to be isolated to improved memory (specifically recall) and attention alongside increased alertness, calmness, contentment and reduced mental fatigue [2,4,5,10,11], which obviously presents a partial deviation from the results seen here. Here, we saw no changes in mood, and whilst memory does seem to be the prevailing cognitive domain affected, here, it is working memory rather than recall, as seen in the aforementioned original trials conducted by this lab. When interrogating the differences/commonalities between these historical trials and the current trial, it is important to note that the cognitive tasks used were delivered via a platform similar to COMPASS (and so the tasks and their completion requirements were very close) and, in some trials, the testing time was comparable. The age of participants in the current trial were also investigated in these previous trials; albeit no one trial covered the whole range used here. As such, the only notable deviances which might explain the different results reported here are the increased power ($n = 94$ here compared to Ns ranging from the 20 s to 30 s previously) and that the original trials utilized *Salvia officinalis/Salvia lavandulaefolia* in isolation in the form of an essential oil compared to the use of a *Salvia* combination and the addition of the naturally co-occurring polyphenols in a dried-leaf form used here.

The polyphenols identified in *Salvia officinalis* and *Salvia lavandulaefolia* include rosmarinic acid, methyl carnosate, caffeic acid, luteolin 7–0-glucoside, luteolin, apigenin and hispidulin. The former also synthesizes salvianolic acid derivatives of rosmarinic acid. These polyphenols are present alongside terpenes like α- and β-pinene, 1,8-cineole, camphor, geraniol, borneol and camphene (see [1] for review), and therefore, the individual and symbiotic effects of individual phytochemicals cannot be attributed to the effects seen here. One might consider this attribution unimportant anyway considering that, when consuming this specific intervention, similar products and indeed dietary sage, it would be as part of a phytochemical cocktail and not isolated compounds. This may explain why no reports exist on the effects of the abovementioned individual phenolic acid and terpene constituents on memory function in humans.

The hypothesis of differential effects in response to this altered intervention is strengthened by the preclinical trial described in the Introduction section which was conducted in rodent models using the combination of sage used here (Cognivia™) [12]. This study demonstrated a synergistic effect of the combination in comparison with a *Salvia officinalis* aqueous extract or *Salvia lavandulaefolia* essential oil alone, and these findings were particularly apparent following 2 weeks of intervention. Moreover, the effects of this combination in rodents mirrored those seen here in humans, focused as they were on learning and spatial memory functions as observed in the Morris water maze.

This recent preclinical trial might shed some light on why this combination would preferentially target working memory and accuracy performance, specifically its finding that calcium/calmodulin-dependent protein kinase II (CaMKII) expression was increased. CaMKII is a key enzyme in all brain regions, recruited around synapses, and is implicated in several neurotransmitter metabolic pathways, including serotonin, with relevant activity in postsynaptic signal propagation (ion channel modulation and calcium homeostasis), synaptogenesis and synaptic plasticity. In particular, CaMKII-induced synaptic strengthening leads to long-term polarisation (via actin remodelling), a major biochemical pathway supporting working memory and cognitive processes such as learning, reasoning and interpretation [13–16].

As supplementary points, it is worth noting that the effects of treatment were observed neither on mood (which is likely explained by the abovementioned difference in the active ingredients between this and historic trials) nor on the PRVP (video) task. The absence of effects on the video task may be as a result of this task hitherto proving sensitivity only to the effects of insults to prospective memory rather than any additive benefits to performance from some intervention. This does not mean that this task or the domain of prospective memory should not be taken forward in future intervention trials but rather that its sensitivity may be more subtle than other tasks/tools that have been long-validated in this area.

Secondly, this study did not restrict or assess the diet of participants prior to or during the study. The aim of this trial was to recruit a random cohort of participants and, with little to no research to indicate that particular dietary approaches would impact the effects of a salvia intervention, the inclusion of any kind of dietary measurement/control never entered considerations for the methodology employed here. Without question, assessing the interaction between diet and other lifestyle factors on all interventions/supplements is an interesting point. However, due to the incalculable variability within and across participants when it comes to diet and lifestyle, if this is not going to be a primary outcome measure of a trial design (with the cohort size to sustain it—one which we certainly did not have here), then a crude assessment of diet and/or a small sample size likely would not provide any great insight and would be more likely to confuse interpretation of other outcome measures.

Thirdly, the decision to supplement for 29 days was somewhat arbitrary in the face of no evidence on the cumulative effects of *Salvia* in humans; as such, a month was deemed an appropriate and relatively easily achievable (confirmed by the mean compliance of 101%) supplementation period for this initial investigation. Practically, chronic supplementation studies must allow a window of time during which participants can return for the final visit, and here, +/− 3 days was decided on as an appropriate range as this would incorporate when the return visit fell on the weekend. Ultimately, however, 87 of the 94 (93%) analysed participants were supplemented between 28–30 days, and again, as no data suggests that supplementary effects of sage would differ outside of this range for the remaining 7% of participants, one cannot say whether this would have impacted the results.

Finally, it should not go unmentioned that participants were able to subjectively detect that they had been in the active intervention group (i.e., 69% correctly guessed this when asked at their final visit), and it would be remiss to think that this could not have impacted the results via participant expectancy effects. Some of this awareness was due the perception of a herbal taste during reflux, but some was based on an identification of a change in mental ability. It could be argued that the ability to subjectively detect this in oneself is a positive outcome, especially as it coincided with statistically detectable improvements in mental ability and speaks to the potency of this extract in being able to induce such changes. Whilst very little can be done to remove the subjective perception of increased mental ability (really, the only option here would be to incorporate an active control into the design), certainly, future studies could improve on the formulation or encapsulation method of the intervention to conceal the herbal taste.

In conclusion, we have observed a consistent significant benefit of a sage combination intervention (Cognivia™) in healthy adult humans on working memory and accuracy of performance cognitive domains. This significant activity was observed both acutely (after just 2 h following consumption) and chronically (after 29 days of administration). The pattern and magnitude of significance points towards an increase in product efficacy over the administration period and, taken together, suggests that future trials should focus on disentangling the working and spatial memory effects of this intervention in humans with an extended timeframe of perhaps several months. Validating the CaMKII mechanism in humans would also be advantageous.

Supplementary Materials: The following are available online at https://www.mdpi.com/2072-6643/13/1/218/s1, Table S1: Results and statistical tables.

Author Contributions: E.L.W. designed the trial design with support from D.O.K., D.G. and T.H. B.S. collected the data supervised by P.A.J., E.L.W. analysed the data and wrote the initial draft of the manuscript, and all authors contributed to revisions on drafts. All authors have read and agree to the published version of the manuscript.

Funding: The study was sponsored by Nexira SAS.

Institutional Review Board Statement: The study was conducted according to the guidelines of the Declaration of Helsinki, and approved by the department of Psychology (Northumbria University) staff ethics committee (code: 8970) on 19/03/2018.

Informed Consent Statement: All subjects gave their informed consent for inclusion before they participated in the study.

Data Availability Statement: The data presented in this study are available on request from the corresponding author. The data are not publicly available as they pertain to a proprietorial product.

Acknowledgments: All of the authors (E.L.W., P.A.J., B.S., T.H., D.G. and D.O.K.) were actively involved in the planning of the research described herein and in writing the paper. All authors contributed to and reviewed the final publication.

Conflicts of Interest: D.G. is affiliated with Nexira SAS.

Clinical Trials Number: NCT03695003.

Abbreviations

BP	Blood Pressure
PRVP	Prospective Remembering Video Procedure
BL	Bond–Lader
STAI	State Trait Anxiety Index
COMPASS	Computerised Mental Performance Assessment System

References

1. Kennedy, D.O. *Plants and the Human Brain*; Oxford University Press: New York, NY, USA, 2014.
2. Savelev, S.; Okello, E.; Perry, N.; Wilkins, R.; Perry, E. Synergistic and antagonistic interactions of anticholinesterase terpenoids in Salvia lavandulaefolia essential oil. *Pharmacol. Biochem. Behav.* **2003**, *75*, 661–668. [CrossRef]
3. Porres-Martínez, M.; González-Burgos, E.; Carretero, M.E.; Gómez-Serranillos, M.P. Major selected monoterpenes α-pinene and 1,8-cineole found inSalvia lavandulifolia(Spanish sage) essential oil as regulators of cellular redox balance. *Pharm. Biol.* **2014**, *53*, 921–929. [CrossRef] [PubMed]
4. Kennedy, D.O.; Pace, S.; Haskell, C.F.; Okello, E.J.; Milne, A.; Scholey, A.B. Effects of Cholinesterase Inhibiting Sage (Salvia officinalis) on Mood, Anxiety and Performance on a Psychological Stressor Battery. *Neuropsychopharmacology* **2005**, *31*, 845–852. [CrossRef] [PubMed]
5. Scholey, A.; Tildesley, N.T.J.; Ballard, C.G.; Wesnes, K.A.; Tasker, A.; Perry, E.K.; Kennedy, D.O. An extract of Salvia (sage) with anticholinesterase properties improves memory and attention in healthy older volunteers. *Psychopharmacology* **2008**, *198*, 127–139. [CrossRef] [PubMed]
6. Kennedy, D.; Dodd, F.L.; Robertson, B.C.; Okello, E.J.; Reay, J.L.; Scholey, A.B.; Haskell, C.F. Monoterpenoid extract of sage (Salvia lavandulaefolia) with cholinesterase inhibiting properties improves cognitive performance and mood in healthy adults. *J. Psychopharmacol.* **2010**, *25*, 1088–1100. [CrossRef] [PubMed]
7. Lopresti, A.L. Salvia (Sage): A Review of its Potential Cognitive-Enhancing and Protective Effects. *Drugs R D* **2017**, *17*, 53–64. [CrossRef] [PubMed]

8. Jackson, P.; Wightman, E.; Veasey, R.; Forster, J.; Khan, J.; Saunders, C.; Mitchell, E.S.; Haskell-Ramsay, C.; Kennedy, D. A Randomized, Crossover Study of the Acute Cognitive and Cerebral Blood Flow Effects of Phenolic, Nitrate and Botanical Beverages in Young, Healthy Humans. *Nutrients* **2020**, *12*, 2254. [CrossRef] [PubMed]
9. Kammoun El Euch, S.; Hassine, D.B.; Cazaux, S.; Bouzouita, N.; Bouajila, J. Salvia officinalis essential oil: Chemical analysis and evaluation of antienzymatic and antioxidant bioactivities. *S. Afr. J. Bot.* **2019**, *120*, 253–260. [CrossRef]
10. Tildesley, N.T.J.; Kennedy, D.O.; Perry, E.K.; Ballard, C.G.; Savelev, S.; Wesnes, K.A.; Scholey, A. Salvia lavandulaefolia (Spanish Sage) enhances memory in healthy young volunteers. *Pharmacol. Biochem. Behav.* **2003**, *75*, 669–674. [CrossRef]
11. Tildesley, N.T.J.; Kennedy, D.O.; Perry, E.K.; Ballard, C.G.; Wesnes, K.A.; Scholey, A. Positive modulation of mood and cognitive performance following administration of acute doses of Salvia lavandulaefolia essential oil to healthy young volunteers. *Physiol. Behav.* **2005**, *83*, 699–709. [CrossRef] [PubMed]
12. Dinel, A.-L.; Lucas, C.; Guillemet, D.; Layé, S.; Pallet, V.; Joffre, C. Chronic Supplementation with a Mix of *Salvia officinalis* and *Salvia lavandulaefolia* Improves Morris Water Maze Learning in Normal Adult C57Bl/6J Mice. *Nutrients* **2020**, *12*, 1777. [CrossRef] [PubMed]
13. Yamauchi, T. Neuronal Ca^{2+}/Calmodulin-Dependent Protein Kinase II—Discovery, Progress in a Quarter of a Century, and Perspective: Implication for Learning and Memory. *Biol. Pharm. Bull.* **2005**, *28*, 1342–1354. [CrossRef] [PubMed]
14. Shonesy, B.C.; Jalan-Sakrikar, N.; Cavener, V.S.; Colbran, R.J. CaMKII: A molecular substrate for synaptic plasticity and memory. In *Progress in Molecular Biology and Translational Science*; Academic Press: Cambridge, MA, USA, 2014; Volume 122, pp. 61–87.
15. Rossetti, T. Erasure and Occlusion of a Behavioral Memory by Mutant Forms of CaMKII Supports the Role of CaMKII in Memory Maintenance. Ph.D. Thesis, Brandeis University, Waltham, MA, USA, 2017.
16. Fraize, N.; Hamieh, A.M.; Joseph, M.A.; Touret, M.; Parmentier, R.; Salin, P.A.; Malleret, G. Differential changes in hippocampal CaMKII and GluA1 activity after memory training involving different levels of adaptive forgetting. *Learn. Mem.* **2017**, *24*, 86–94. [CrossRef]
17. Charve, J.; Reineccius, G.A. Encapsulation Performance of Proteins and Traditional Materials for Spray Dried Flavors. *J. Agric. Food Chem.* **2009**, *57*, 2486–2492. [CrossRef] [PubMed]
18. Kennedy, D.; Wightman, E.; Forster, J.; Khan, J.; Jones, M.B.; Jackson, P. Cognitive and Mood Effects of a Nutrient Enriched Breakfast Bar in Healthy Adults: A Randomised, Double-Blind, Placebo-Controlled, Parallel Groups Study. *Nutrients* **2017**, *9*, 1332. [CrossRef] [PubMed]
19. Heffernan, T.M.; Clark, R.; Bartholomew, J.; Ling, J.; Stephens, S.; Stephens, R. Does binge drinking in teenagers affect their everyday prospective memory? *Drug Alcohol. Depend.* **2010**, *109*, 73–78. [CrossRef] [PubMed]
20. Bond, A.; Lader, M. Use of Analog Scales in Rating Subjective Feelings. *Br. J. Med. Psychol.* **1974**, *47*, 211–218. [CrossRef]
21. Speilberger, C.D.; Gorsuch, R.L.; Lushene, R.E. *The State Trait Anxiety Inventory Manual*; Consulting Psychologists Press: Palo Alto, CA, USA, 1969.
22. Erdfelder, E.; Faul, F.; Buchner, A. GPOWER: A general power analysis program. *Behav. Res. Methods Instrum. Comput.* **1996**, *28*, 1–11. [CrossRef]
23. Weizenbaum, E.; Torous, J.; Fulford, D. Cognition in Context: Understanding the Everyday Predictors of Cognitive Performance in a New Era of Measurement. *JMIR mHealth uHealth* **2020**, *8*, e14328. [CrossRef] [PubMed]

Article

Effects of 12 Weeks *Cosmos caudatus* Supplement among Older Adults with Mild Cognitive Impairment: A Randomized, Double-Blind and Placebo-Controlled Trial

Yee Xing You [1], Suzana Shahar [1,*], Nor Fadilah Rajab [2], Hasnah Haron [3], Hanis Mastura Yahya [3], Mazlyfarina Mohamad [4], Normah Che Din [5] and Mohamad Yusof Maskat [6]

1 Dietetics Programme and Centre for Healthy Aging and Wellness (H-Care), Faculty of Health Sciences, Universiti Kebangsaan Malaysia, Jalan Raja Muda Abdul Aziz, Kuala Lumpur 50300, Malaysia; yeexing@gmail.com
2 Biomedical Science Programme and Centre for Healthy Aging and Wellness (H-Care), Faculty of Health Sciences, Universiti Kebangsaan Malaysia, Jalan Raja Muda Abdul Aziz, Kuala Lumpur 50300, Malaysia; nfadilah@ukm.edu.my
3 Nutritional Sciences Programme and Centre for Healthy Aging and Wellness (H-Care), Faculty of Health Sciences, Universiti Kebangsaan Malaysia, Jalan Raja Muda Abdul Aziz, Kuala Lumpur 50300, Malaysia; hasnaharon@ukm.edu.my (H.H.); hanis.yahya@ukm.edu.my (H.M.Y.)
4 Diagnostic Imaging and Radiotherapy Programme and Centre for Diagnostic, Therapeutic and Investigative Studies, Faculty of Health Sciences, Universiti Kebangsaan Malaysia, Jalan Raja Muda Abdul Aziz, Kuala Lumpur 50300, Malaysia; mazlyfarina@ukm.edu.my
5 Health Psychology Programme, Centre of Rehabilitation and Special Needs, Faculty of Health Sciences, Universiti Kebangsaan Malaysia, Jalan Raja Muda Abdul Aziz, Kuala Lumpur 50300, Malaysia; normahcd@ukm.edu.my
6 Department of Food Sciences, Faculty of Science and Technology, Universiti Kebangsaan Malaysia, UKM, Bangi 43600, Malaysia; yusofm@ukm.edu.my
* Correspondence: suzana.shahar@ukm.edu.my; Tel.: +60-3-9289-7651

Citation: You, Y.X.; Shahar, S.; Rajab, N.F.; Haron, H.; Yahya, H.M.; Mohamad, M.; Din, N.C.; Maskat, M.Y. Effects of 12 Weeks *Cosmos caudatus* Supplement among Older Adults with Mild Cognitive Impairment: A Randomized, Double-Blind and Placebo-Controlled Trial. *Nutrients* **2021**, *13*, 434. https://doi.org/10.3390/nu13020434

Academic Editor: Tsuyoshi Chiba
Received: 7 January 2021
Accepted: 24 January 2021
Published: 29 January 2021

Abstract: *Cosmos caudatus* (CC) contains high flavonoids and might be beneficial in neuroprotection. It has the potential to prevent neurodegenerative diseases. Therefore, we aimed to investigate the effects of 12 weeks of *Cosmos caudatus* supplement on cognitive function, mood status, blood biochemical profiles and biomarkers among older adults with mild cognitive impairment (MCI) through a double-blind, placebo-controlled trial. The subjects were randomized into CC supplement ($n = 24$) and placebo group ($n = 24$). Each of them consumed one capsule of CC supplement (250 mg of CC/capsule) or placebo (500 mg maltodextrin/capsule) twice daily for 12 weeks. Cognitive function and mood status were assessed at baseline, 6th week, and 12th week using validated neuropsychological tests. Blood biochemical profiles and biomarkers were measured at baseline and 12th week. Two-way mixed analysis of variance (ANOVA) analysis showed significant improvements in mini mental state examination (MMSE) (partial $\eta^2 = 0.150$, $p = 0.049$), tension (partial $\eta^2 = 0.191$, $p = 0.018$), total mood disturbance (partial $\eta^2 = 0.171$, $p = 0.028$) and malondialdehyde (MDA) (partial $\eta^2 = 0.097$, $p = 0.047$) following CC supplementation. In conclusion, 12 weeks CC supplementation potentially improved global cognition, tension, total mood disturbance, and oxidative stress among older adults with MCI. Larger sample size and longer period of intervention with incorporation of metabolomic approach should be conducted to further investigate the underlying mechanism of CC supplementation in neuroprotection.

Keywords: *Cosmos caudatus*; cognitive function; mood; biomarkers; flavonoids

1. Introduction

Aging has become a global issue. The number of people over 60 years old is expected to increase from 507.95 million in 2015 to 1293.7 million by 2050 [1]. Population aging has been fastest in South-East Asia including Malaysia. The percentage of the population aged

65 years and above almost doubled from 6% in 1990 to 11% in 2019 in South-East Asia [2]. In Malaysia, a similar scenario can be observed in which the aging population has risen from 5.6% to 10.7% of the total population between the year 1991 to 2020 [3,4]. The recent World Health Organization report, dementia has been diagnosed among 50 million people globally, with approximately 60% living in low- and middle-income countries. The total number of people with dementia is estimated to reach 82 million in 2030 and 152 million in 2050 [5]. Particularly in Malaysia, the prevalence of dementia in Malaysia is estimated at 0.126% and 0.454% in 2020 and 2050, respectively [3] and the prevalence of mild cognitive impairment is 16% [6].

Mild cognitive impairment (MCI) is an etiologically heterogeneous syndrome characterized by memory performance below the age norm and represents a transitional state between normal aging and dementia disorders. Globally, efforts have been taken to explore neuroprotective effects of nutraceutical products among the aging population aiming for the prevention of neurodegenerative diseases such as dementia through food-based recommendations [7–10]. Asian in particular is rich in traditional herbs and vegetables potentially use as nutraceutical. Several studies have proven beneficial effects of nutraceutical products on the cognitive function of older adults. These included indigenous herbs rich in polyphenols and antioxidants such as *Persicaria minor* aqueous extract supplementation [11], *Ginkgo biloba* [12,13] and *Centella asiatica* [14]. In particular, for example, an Asian herbs, *Persicaria minor* supplementation for 24 weeks promoted both attention and mood among older adults with mild cognitive impairment [11,15]. Association between dietary nutrients and cognitive function among older adults have been reported elsewhere [16–19]. Nevertheless, the evidence from randomized controlled trials is limited to justify the recommendation of nutraceutical products are useful to prevent neurodegeneration and cognitive impairment [5,19]. Therefore, more efforts should be conducted to determine the efficacy of the nutraceutical products.

A traditional vegetable, *Cosmos caudatus* (CC) or locally known as *ulam raja* (Kings of *ulam*) is an annual plant in the genus *Cosmos* and it is widely distributed in South East Asia including Indonesia, Thailand, and Malaysia [20]. It showed the highest total phenolic content as compared to nine other common Malaysian traditional vegetables [21]. CC exhibits strong antioxidant properties and may have the ability to prevent neurodegenerative diseases such as dementia [20,22,23]. One of its beneficial based on the medicinal properties are as antidiabetic, anti-obesity, antimicrobial, antihypertensive, and anticancer. However, it is yet to be determined on its potential in dementia prevention [20]. On the other hand, the phytochemicals of CC are known for their free radical scavenging properties that may help to reduce the oxidative stress and lipid peroxidation in neuronal membrane [22,24]. Flavonoid such as quercetin which can be found in CC which potentially improved memory deficits and cognitive impairment in a mice model [25]. In fact, CC also contains other flavonoids such as catechin, quercetin, proanthocyanidin, and rutin which could pass across the blood–brain barrier and might be able to protect against cognitive decline [20]. However, none of the studies examined the effects of CC on cognitive function, mood status, biochemical profiles, as well as biomarkers among older adults with mild cognitive impairment (MCI).

To address the research gap, a randomized, double-blind, placebo-controlled clinical trial was conducted to investigate the effects of 12 weeks of CC supplement on cognitive function, mood status, blood biochemical profiles and biomarkers among older adults with MCI. We hypothesized that 12 weeks of CC supplement has the ability to improve cognitive function, mood status, and blood biomarkers among older adults with MCI.

2. Materials and Methods

2.1. Study Design and Subject's Selection

A 12-week randomized, double blind, placebo-controlled trial was designed to investigate the effects of CC supplementation on brain function including cognitive function, mood status and biochemical indices among older adults with MCI. The study protocol was approved by the Medical Research Ethics Committee Universiti Kebangsaan Malaysia with code NN-2019-137 and written informed consent was obtained from all the subjects prior to data collection. This study was also registered under the International Standard Randomized Controlled Trial Number (ISRCTN) Registry (ISRCTN16793907) and conducted in accordance with Good Clinical Practice Guidelines and the ethical principles of the Declaration of Helsinki 1964.

Calculation for sample size was determined by using the Randomized Controlled Trials formula proposed by Chan (2003) [26]. Regarding the study of Gschwind et al. (2017), the mean difference of cognitive test score between supplement and control groups (12) with pooled standard deviation (17) were substituted into the formula [13]. The calculated sample size was 24 per group after consideration of 30% drop out rate. The sample size calculation as shown below:

$$n1 = n2 = (C/\delta^2) + 2$$
$$n = C/(\mu 2 - \mu 1/\sigma)^2 + 2$$
$$n = 7.9/(12/17.4)^2 + 2$$
$$n = 18$$

where n = sample size

$\delta = \mu 2 - \mu 1/\sigma$

$\mu 1 - \mu 2$ = mean difference between the intervention and control group after the intervention [13]

σ = the standard deviation of the intervention group [13]

C = constant: 7.9 (80% power and 95% confidence interval)

With consideration that the dropout rate of subjects is 30%, the sample size would be 18 + 6 = 24 subjects per arm.

This study involved older adults with mild cognitive impairment aged between 60 to 75 years living in the Klang Valley, a central of Malaysia. Subjects were screened and recruited via poster advertisements on social media such as WhatsApp and Facebook. Open health check-ups related to the screening were held at the Centre for Healthy Ageing and Wellness (H-Care) to ensure willingness of the potential subjects to participate into the study. Subjects were allocated based on a simple randomization method using a computer-generated software according to gender run by the investigator with no involvement in the clinical trial. All the study fieldworkers and subjects were kept blinded to the group assignment, study product distribution, and trial findings. The supplement manufacturer unblinded the study label after the data analysis was completed. The inclusion criteria of the study is older adults aged 60 to 75 years old with MCI based on Petersen (2014) criteria [27] and body mass index must be within 20–30 kg/m^2. Older adults with self-reported neurodegenerative diseases, smokers, regular consumption of traditional herbs or nutraceutical products for the past 6 months, depressive symptoms (score > 5 in Geriatric Depression Scale), contract serious comorbidities such as renal and kidney failure (based on blood analysis report), undergo hormone therapy and consumption of warfarin medication were excluded from the study.

Figure 1 shows the study flow chart. During screening, a total of 200 older adults aged 60 to 75 years were enrolled and assessed for eligibility. A total of 152 older adults were excluded due to several reasons such as not meeting the inclusion criteria (n = 121), declined to participate (n = 29) and other personal reasons (n = 2). Subjects who met the inclusion and exclusion criteria were then randomized into placebo and CC supplement groups, with 24 subjects per arm. On the first follow up at the 6th week, there was one dropout from the CC supplement group due to the subject being uninterested in continuing with the study and failure to answer calls; however, there were no dropouts in the placebo group. In conclusion, only 23 subjects of the CC supplement group and 24 placebo subjects completed the study at the end of the 12-week study (n = 47).

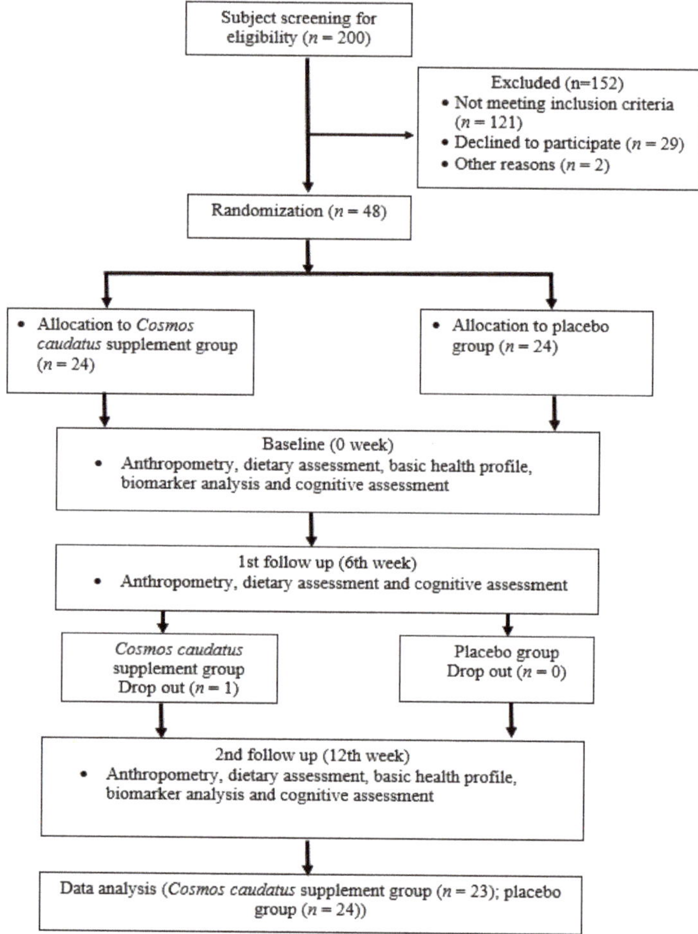

Figure 1. Consort study flow chart.

2.2. Study Product and Dosage

A finished product in the form of a capsule containing 250 mg of CC powder and 250 mg of maltodextrin was produced. This product was developed by the Institute Bioproduct Development (IBD), Universiti Teknologi Malaysia. On a daily basis, two capsules of the supplement were taken by the subjects either during breakfast, lunch or dinner. Each capsule can be orally taken during alternate main mealtime. On the other hand, the placebo used in this study was a 500 mg of maltodextrin. Both capsules were sensory identical and packed in the bottles labelled as either A or B by the manufacturer. We referred to the available toxicity study by Mohamed et al. (2013) which was tested among Wistar rats for acute oral toxicity and it was safe to consume up to 500 mg/kg body weight [28]. In an in vivo animal study examining the antioxidant effects of CC supplement among 30 mice (divided into 5 different doses), findings showed that 100 mg/kg of CC aqueous extract have the potential to increase superoxide dismutase (SOD) and catalase (antioxidant biomarkers) levels after 21 days of treatment [29]. Our dosage (500 mg of CC/day) also exceeded the in vivo animal therapeutic dosage as we hypothesised that it might give the same antioxidant effect in humans. The heavy metals and microbial analysis were conducted and it was reported that composition of heavy metals and microbial was below the toxicity level. The nurition composition of CC supplement and placebo is shown in Table 1. Compliance was assessed by performing capsule count at the end of the 6th and 12th week. The compliance rate of this study was 90.4%.

Table 1. Nutrient composition of *Cosmos caudatus* (CC) supplement and placebo (per 100 g).

Nutrients	CC Supplement	Placebo
Energy (kcal)	284 ± 4	376 ± 3
Carbohydrate (g)	47.35 ± 0.92	94.05 ± 0.64
Protein (g)	20.30 ± 0.42	0
Fat (g)	1.40 ± 0.14	0
Ash content (g)	26.70 ± 1.27	0.05 ± 0.07
Moisture content (g)	4.25 ± 0.49	5.85 ± 0.50
Vitamin A Retinol (mg)	0	0
Vitamin C (mg)	100.50 ± 2.12	0
Vitamin E Alpha-Tocopherol (mg)	2.10 ± 0.42	0
Calcium (mg)	2255.00 ± 7.07	9.06 ± 0.54
Iron (mg)	7.69 ± 0.16	0.62 ± 0.13
Potassium (mg)	9610 ± 466.69	15.00 ± 0.71
Sodium (mg)	71.60 ± 5.52	69.15 ± 2.62
Zinc (mg)	2.77 ± 0.52	0.70 ± 0.14
Total Dietary Fibre (g)	7.80 ± 2.97	1.8 ± 0.14
Total phenolic content (mg Gallic acid equivalent)	1482 ± 101	3.56 ± 0.10
DPPH (mmol Trolox equivalent)	330.86 ± 9.48	4.57 ± 2.01
FRAP (mmol Trolox equivalent)	393.57 ± 9.78	1.45 ± 1.17
Quercetin (%*w/w*)	0.9	NA
Quercitrin (%*w/w*)	1.0	NA

The data are nutritional information of CC supplement and placebo per 100 g, where subjects consumed two capsules per day (500 mg/day) for the intervention. Abbreviation: DPPH: 2,2-diphenyl-1-picrylhydrazyl; FRAP: Ferric Reducing Antioxidant Power.

2.3. Study Procedures

Data collection was conducted three times starting from baseline, 6th week and 12th week of supplementation. The data that was taken during that period included details such as subjects' sociodemographic information, anthropometric measurements (body mass index), vital signs (pulse rate and blood pressure), dietary intake information using validated dietary history questionnaire (DHQ) [30] and neuropsychological tests for every visit. In addition, the blood biochemical profiles and biomarkers were measured during baseline and the 12th week only. A total of five trained fieldworkers participated in the data collection procedures. Those five fieldworkers were assigned to collect the same subjects' information and conduct the respective cognitive tests from baseline to the 12th week of the intervention to avoid inconsistent results. The data collection was one-to-one and face to face interview. Each neuropsychological tests took 15 to 20 min to complete. Same sets of validated neuropsychological tests were used throughout the study. No pre-testing of the questionnaire was conducted as validated questionnaires were used for all parameters. The outcomes of the study was shown in Table 2.

Table 2. Outcomes of the study.

Outcomes	Neuropsychological Assessment
Primary outcomes	
MMSE	Global cognitive function
Digit Span	Attention and working memory
RAVLT (immediate and delayed recall)	Verbal immediate memory
VR (immediate and delayed recall)	Visuo–spatial function
Digit symbol substitution	Psychomotor speed
POMS	Tension, depression, anger, fatigue, esteem-related effect, vigor, and confusion
Secondary outcomes	
Biomarkers	BDNF, MDA, iNOS, COX-2, SOD, GSH
Biochemical profiles	Fasting blood sugar, lipid profile, liver function test, renal function test

Abbreviation: BDNF: Brain derived neurotrophic factor; COX-2: Cyclooxygenase-2; GSH: Glutathione; iNOS: Inducible nitric oxide synthase; MDA: Malondialdehyde; MMSE: Mini-mental State Examination; POMS: Profile of Mood State; RAVLT: Rey Auditory Verbal Learning Test; SOD: Superoxide dismutase; VR: Visual reproduction.

2.3.1. Cognitive Function Assessment

A series of neuropsychological tests such as Mini-Mental State of Examination [31], Digit Span [32], Rey Auditory Verbal Learning Test [33], Digit Symbol [32], and Visual Reproduction [32] were utilized to assess global cognitive function, working memory, psychomotor speed and visual-spatial memory of the subjects. Their mood status was assessed using the validated Profile of Mood State (POMS) questionnaire [34].

2.3.2. Blood Biochemical Profile and Blood Biomarkers Tests

At baseline and during the 12th week follow up session, subjects were asked to fast overnight for at least 10 h for blood sample collecting purpose. Peripheral venous blood samples were drawn by a trained phlebotomist. A total of 20 mL blood was collected into tubes and immediately stored in an ice box for delivery. All the basic biochemical profile analysis such as fasting blood sugar, lipid profile, liver function test and renal profile were analysed at the medical laboratory Pathlab Malaysia Sdn Bhd, Selangor, Malaysia. The serum was centrifuged and stored under −80 °C for one month before biomarker analysis was carried out using commercial ELISA kits at Bioserasi and toxicology laboratory at the Faculty of Health Sciences, Universiti Kebangsaan Malaysia, Kuala Lumpur. The oxidative stress biomarkers (malondialdehyde, MDA), inflammatory biomarkers (inducible nitric oxide synthase, iNOS and cyclooxygenase-2, COX-2), antioxidant biomarkers (superoxide dismutase, SOD and glutathione, GSH) and brain derived neurotrophic factor (BDNF) were measured in this study using commercial ELISA kits (Elabscience, Houstan, TX, USA). This ELISA kit uses Competitive-ELISA as a method. The standard concentrations of was stated in the kits' manual. The optical density (OD) was measured spectrophotometrically at a wavelength of 450 nm. The concentrations of all the biomarkers in the duplicate samples were determined by comparing the OD of the samples to the standard curve.

2.4. Statistical Analysis

Normality of data was analysed by using the Shapiro-Wilk test and significant value $p > 0.05$ which indicates normal distribution. The mean differences of the baseline data between the supplement and placebo group were analysed using the Independent-t test for continuous parameters. On the other hand, Chi-Square was selected for the analysis of categorical variables such as gender or race. Furthermore, the interaction effects were analyzed using a two-way repeated measure analysis of variance (ANOVA) that was adjusted for age, education years, body mass index, physical activity, MMSE, energy intake, vitamin A and C after the Bonferroni adjustment. The covariates and confounding factors for the repeated measure analysis were selected from the variables which may have contributed possible confounding effects on the outcome of measurements. Besides that, the percentage of mean change for each interval (baseline to 6th week and baseline to

12th week) was calculated and presented as a line chart. Percentage of mean change for each subject in both the groups was calculated using the formula; [(score at 6th or 12th week-score of baseline)/score of baseline] × 100%. The mean difference of the percentage mean change in both the groups was analyzed using the Independent-*t*-test.

3. Results

3.1. Subjects' Characteristics

Table 3 shows a total of 48 subjects involved in this study with mean age of 65.11 ± 4.05 years. These subjects were randomized into CC supplement group (65.83 ± 4.35 years old) and placebo group (64.42 ± 3.71 years old). Majority of the subjects were women (66.7%), Malay (60.4%), married (77.1%), received secondary education (61.7%) and with a mean household income of RM1991.57 ± 844.94 (USD 475 ± 201). Based on the subjects' self-reported medical condition, 35.4% of the subjects diagnosed with hypertension, 22.9%, 31.3% and 4.2% of them diagnosed with diabetes, hyperlipidemia and other diseases such as gout and gastroesophageal reflux disease, respectively. The sociodemographic and self-reported medical condition were not statistically significant between both groups at baseline ($p > 0.05$).

Table 3. Baseline sociodemography information and self-reported medical condition between CC supplement and placebo group subjects [presented as mean ± standard deviation or *n*(%)].

Parameter	CC Supplement (*n* = 24)	Placebo (*n* = 24)	Total (*n* = 48)	*p*-Value
Age [1]	65.83 ± 4.35	64.42 ± 3.71	65.11 ± 4.05	0.237
Gender [2]				0.917
Male	8 (33.33)	8 (33.3)	16 (33.3)	
Female	16 (66.7)	16 (66.7)	32 (66.7)	
Ethnicity [2]				0.483
Malay	16 (66.7)	13 (54.2)	29 (60.4)	
Chinese	6 (25.0)	10 (41.7)	16 (33.3)	
Indian	2 (8.3)	1 (4.2)	3 (6.3)	
Formal education (years) [1]	11.39 ± 2.39	10.17 ± 3.20	10.77 ± 2.87	0.145
Education level [2]				0.184
Primary school	1 (4.2)	3 (12.5)	4 (8.3)	
Secondary school	13 (54.2)	17 (70.8)	30 (62.5)	
Diploma/Certificate	9 (37.5)	3 (12.5)	12 (25.0)	
Degree	1 (4.2)	1 (4.2)	2 (4.2)	
Marital status [2]				0.123
Single	1 (4.2)	3 (12.5)	4 (8.3)	
Married	22 (91.7)	15 (62.5)	37 (77.1)	
Divorce	0 (0)	2 (8.3)	2 (4.2)	
Widow/widower	1 (4.2)	4 (16.7)	5 (10.4)	
Household income (RM) [1]	2021.83 ± 904.81	1962.58 ± 801.86	1991.57 ± 844.94	0.813
Hypertension [2]				0.159
Yes	6 (25.0)	11 (45.8)	17 (35.4)	
No	18 (75.0)	13 (54.2)	31 (64.6)	
Diabetes [2]				0.671
Yes	6 (25.0)	5 (20.8)	11 (22.9)	
No	18 (75.0)	19 (79.2)	37 (77.1)	
Hyperlipidaemia [2]				0.587
Yes	7 (29.2)	8 (33.3)	15 (31.3)	
No	17 (70.8)	16 (66.7)	33 (68.7)	
Others [2]				0.975
Yes	1 (4.2)	1 (4.2)	2 (4.2)	
No	23 (95.8)	23 (95.8)	46 (95.8)	
Physical activity [2]				0.591
Everyday	1 (4.2)	0 (0)	1 (2.1)	
3–5 times per week	7 (29.2)	4 (16.7)	11 (22.9)	
1–2 times per week	8 (33.3)	11 (45.8)	19 (39.6)	
None	8 (33.3)	9 (37.5)	17 (35.4)	
Body mass index (kg/m^2)	25.67 ± 3.02	25.72 ± 2.29	25.70 ± 2.64	0.952

[1] Independent-*t* test, not significant at $p > 0.05$; [2] Cross tabs Chi-square test, not significant at $p > 0.05$; N/A: Not applicable.

3.2. Cognitive Function and Mood Status

Table 4 demonstrates the intervention effects after controlling for confounding factors on cognitive functions between CC supplement group and placebo group. Mini-Mental State Examination (MMSE) showed significant intervention effect ($p = 0.049$, partial $\eta^2 = 0.150$, power = 0.586). Both groups showed increment in MMSE mean scores, however, CC supplement group had significant percentage of mean change on the 6th and 12th week of intervention as compared to the placebo group ($p < 0.05$) using the independent-*t* test (Figure 2). Although there were no significant treatment × time effects from the Digit Span test among the CC supplement and placebo groups, independent-*t* test showed that the percentage mean change in the Digit Span test was significantly higher in the CC supplement group (20.6%) than the placebo group (8.32%) on the 6th week of intervention ($p < 0.05$).

Table 4. Intervention effect of cognitive function and mood state from baseline to 12th week.

	CC Supplement (n = 23)	Placebo (n = 24)	Treatment × Time Effect		
			p	Partial Eta Squared	Power
Mini-mental State Examination (MMSE)					
Baseline	27.09 ± 1.38	26.58 ± 1.35			
6th week	28.30 ± 0.70	26.92 ± 1.02	0.049 *	0.150	0.586
12th week	28.91 ± 0.95	27.38 ± 1.28			
Digit Span					
Baseline	8.39 ± 1.23	7.92 ± 1.28		0.040	0.173
6th week	10.09 ± 1.41	8.58 ± 2.34	0.466		
12th week	10.91 ± 1.70	9.50 ± 2.59			
RAVLT (Immediate Recall)					
Baseline	6.30 ± 1.02	6.29 ± 1.12	0.058	0.143	0.560
6th week	10.22 ± 1.28	8.96 ± 1.76			
12th week	11.26 ± 2.18	10.75 ± 2.29			
RAVLT (Delayed recall)					
Baseline	5.74 ± 0.92	5.63 ± 1.24	0.070	0.068	0.529
6th week	9.39 ± 1.37	7.83 ± 1.49			
12th week	9.96 ± 2.65	9.58 ± 3.02			
Digit Symbol					
Baseline	8.87 ± 1.71	8.00 ± 2.63	0.264	0.069	0.278
6th week	10.04 ± 2.01	8.88 ± 2.59			
12th week	10.48 ± 1.97	9.88 ± 2.89			
Visual Reproduction (Immediate recall)					
Baseline	32.22 ± 4.62	29.50 ± 6.45	0.212	0.080	0.320
6th week	34.83 ± 3.30	29.54 ± 7.25			
12th week	35.13 ± 3.94	30.88 ± 7.06			
Visual reproduction (delayed recall)					
Baseline	31.61 ± 6.06	27.83 ± 7.98	0.242	0.074	0.295
6th week	35.13 ± 4.35	28.88 ± 7.85			
12th week	35.30 ± 3.40	29.88 ± 8.73			
Tension					
Baseline	5.09 ± 3.46	6.67 ± 3.95	0.018 *	0.191	0.733
6th week	3.91 ± 2.04	6.13 ± 2.53			
12th week	3.22 ± 1.65	5.46 ± 2.25			
Anger					
Baseline	1.47 ± 0.92	2.58 ± 1.16	0.139	0.099	0.401
6th week	1.13 ± 1.58	3.21 ± 1.83			
12th week	1.09 ± 0.64	3.08 ± 2.39			
Fatigue					
Baseline	3.65 ± 2.53	5.17 ± 2.35	0.811	0.011	0.081
6th week	4.17 ± 1.87	5.33 ± 2.91			
12th week	4.00 ± 1.57	5.54 ± 1.25			

Table 4. *Cont.*

Depression					
Baseline	1.70 ± 0.36	3.00 ± 1.22	0.921	0.004	0.062
6th week	1.39 ± 0.37	2.38 ± 2.20			
12th week	1.34 ± 0.77	2.17 ± 1.93			
Esteem-related effect					
Baseline	16.74 ± 2.88	14.54 ± 3.06	0.149	0.095	0.387
6th week	17.30 ± 1.69	15.21 ± 3.46			
12th week	18.35 ± 1.34	15.00 ±2.83			
Vigor					
Baseline	12.83 ± 3.39	10.25 ± 3.72	0.050	0.243	0.862
6th week	13.30 ± 1.55	10.29 ± 2.16			
12th week	14.35 ± 2.17	9.00 ± 2.11			
Confusion					
Baseline	2.74 ± 1.15	4.13 ± 1.84	0.983	0.001	0.052
6th week	3.22 ± 2.15	3.88 ± 2.86			
12th week	2.00 ± 1.45	2.83 ± 1.34			
Total mood disturbance					
Baseline	85.09 ± 10.58	96.75 ± 12.74	0.028 *	0.171	0.672
6th week	83.22 ± 6.18	95.42 ± 12.04			
12th week	78.96 ± 5.06	95.08 ± 9.08			

* Significance at $p < 0.05$. Controlled for age, body mass index, physical activity, energy intake, vitamin A and C.

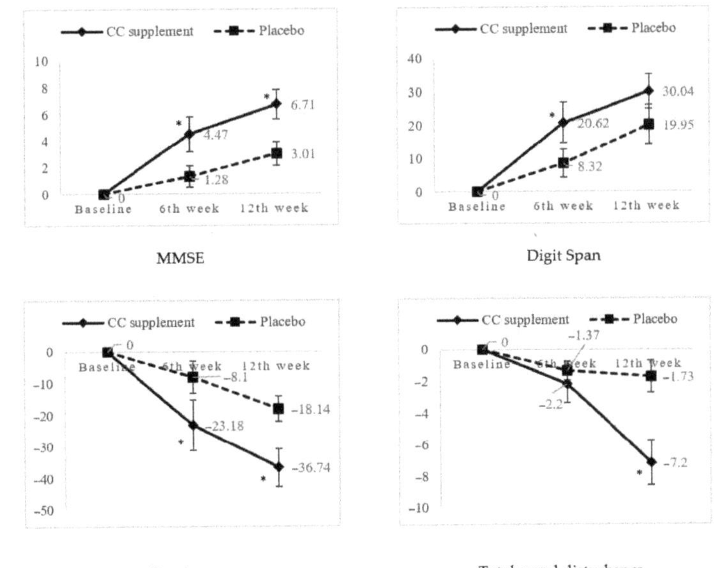

Figure 2. Percentage mean change of significant cognitive and mood parameters from baseline to 6th week and 12th week follow-ups. * Significant at $p < 0.05$ using independent t-test.

A significant intervention effect was observed in tension ($p = 0.018$, partial $\eta^2 = 0.191$, power = 0.733) and total mood disturbance ($p = 0.028$, partial $\eta^2 = 0.171$, power = 0.672). The percentage of mean change for tension and total mood disturbance showed statistically significant improvement in CC supplement as compared to placebo groups at 6th and 12th week (Figure 2).

3.3. Biochemical Profiles and Biomarker Analysis

All blood biochemical profiles did not show any treatment × time effects, and there was also no significant difference in percentage mean change between the CC supplement and placebo groups at any time points ($p > 0.05$) which is shown in Supplementary Materials. Subsequently, malondialdehyde (MDA) showed significant intervention effects after 12 weeks of intervention ($p = 0.047$, partial $\eta^2 = 0.097$, power = 0.516) (Table 5). The mean concentration of MDA decreased from 350.14 ± 161.08 to 313.32 ± 153.35 ng/mL (−10.52%) in the CC supplement group. Independent-*t* test also showed the percentage mean change in MDA was significantly decreased from baseline to the 12th week in CC supplement as compared to placebo group ($p < 0.05$). Although there was no significant group, time and intervention effects in glutathione (GSH) between the CC supplement and placebo group, the percentage mean change in GSH was significantly higher in the CC supplement group (+8.7%) as compared to the placebo group (+1.04%) ($p < 0.05$) (Figure 3).

Table 5. Intervention effect of biomarkers from baseline to 12th week.

	CC Supplement (n = 23)	Placebo (n = 24)	Treatment × Time Effect		
			p	Partial Eta Squared	Power
Brain derived neurotrophic factor (BDNF) (pg/mL)					
Baseline	180.80 ± 86.42	180.09 ± 95.78	0.884	0.001	0.052
12th week	194.74 ± 137.49	171.22 ± 113.87			
Inducible nitric oxide synthase (iNOS) (pg/mL)					
Baseline	220.38 ± 66.56	218.99 ± 74.66	0.994	0.000	0.050
12th week	216.34 ± 81.39	219.64 ± 80.67			
Cyclooxygenase-2 (COX-2) (ng/mL)					
Baseline	0.86 ± 0.42	1.08 ± 0.59	0.867	0.001	0.053
12th week	0.78 ± 0.43	1.15 ± 0.70			
Superoxide dismutase (SOD) (pg/mL)					
Baseline	49.13 ± 3.81	49.46 ± 2.64	0.577	0.008	0.085
12th week	51.63 ± 2.04	51.22 ± 3.50			
Malondialdehyde (MDA) (ng/mL)					
Baseline	350.14 ±161.08	358.21 ± 172.59	0.047 *	0.097	0.516
12th week	313.32 ± 153.35	398.94 ± 181.85			
Glutathione (GSH) (mM)					
Baseline	0.46 ± 0.09	0.48 ± 0.05	0.111	0.064	0.356
12th week	0.50 ± 0.06	0.48 ± 0.07			

* Significance at $p < 0.05$. Controlled for age, physical activity, BMI, energy intake, vitamin A and C.

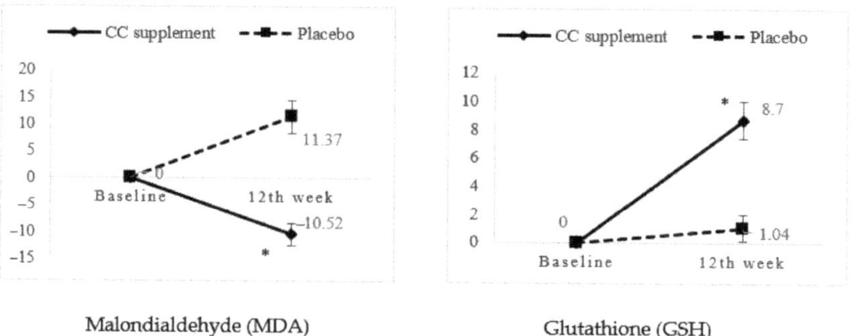

Figure 3. Percentage mean change of significant biomarkers from baseline to 6th week and 12th week follow-ups. * Significant at $p < 0.05$ using independent *t*-test.

4. Discussion

Our study has successfully found that 12 weeks of CC supplementation has the ability to improve global cognitive function as assessed using MMSE. Both CC supplement and placebo groups have improved in these cognitive domains, however, the percentage mean change of CC supplement group was significantly higher than the placebo group. This phenomenon might due to the placebo effect, where placebo subjects expected that the supplements, they received were effective [35]. We are cautious while interpreting the significant result, it showed small effect size, however, it should be recruited more samples to improve the significance of the result in future study. We reasoned the significant improvement as CC supplement contains high flavonoids content such as quercetin, catechin, epicatechin and proanthocyanidins which have a potential involvement in neuroprotection pathway [20,36]. The bioactive compounds such as quercetin and quercitrin of this supplement were higher than the previous proven traditional vegetable *Persicaria minor* supplement in neuroprotection among older adults with MCI [11]. However, the flavonoid contents (excluding quercetin and quercitrin) in CC supplement remain unknown.

In addition, we believe that the cognitive improvement might also cause by the possible synergistic effects between antioxidants and flavonoids activities in CC supplement. It has been reported that certain flavonoids could cross the blood–brain barrier (BBB) in vitro [37–39]. The past studies relating to in vivo and in vitro have indicated that flavonoids in herbal extracts that are similar with the CC supplement are competent enough to be absorbed following an oral administration, passing through the blood–brain barrier, and the absorption would lead to different impacts on the central nervous system [38]. Flavonoids such as quercetin might exert antioxidant properties and biochemical impacts with regards to many oxidative stress-induced ailments for the elderly [40,41]. As per emerging evidence, flavonoids have been reported to offer protection against degeneration and neural injuries in dementia and Alzheimer's diseases [42,43]. However, the types of flavonoids in CC supplement (excluding quercetin) and the bioavailability of CC's flavonoids in human body is yet to be investigated in future.

Several pathways have been recommended by means of which quercetin and quercitrin in CC supplement might impact cognitive functioning. For instance, in vitro research indicates that quercetin is an effective antioxidant and might safeguard the neuronal cells from neurotoxicity related to oxidative stress [44]. Furthermore, past in vitro studies have pointed out that quercetin is an adenosine antagonist, as it might improve the cognitive functioning and reduce physical and cognitive fatigues [45]. In addition, the quercitrin was found to be involved in the dilution of brain oxidative stress, clampdown of inflammation, and the improvement of neurotransmitter dysfunction [44,46,47]. Flavonoids such as quercetin can deter the mitogen-activated protein kinase (MAPK) signaling pathway by inhibiting the expression of inflammatory protein and cytokine generation [48]. The flavonoids (excluding quercetin and quercitrin) might be beneficial in cognitive functioning, it should be identified their mechanism of action in future.

Additionally, our study shows that there is an improvement in tension and total mood disturbance in CC supplement group as compared to the placebo group. We reasoned that the positive effect of mood might be due to the presence of quercetin and quercitrin in CC supplement. This findings are supported by Yahya et al. (2017) and Lau et al. (2020) in which both studies have reported that the structural formula pertaining to *Polygonum minus* aqueous extract possessing two flavonoids (quercitrin and quercetin) have anti-anxiety and anti-depressant characteristics [11,15]. Likewise, Udani (2013) noted an important enhancement in depression, tension, and anger of POMS following the supplementation of *Superulam*, a mixture of *ulam* extracts which contains CC supplement [49]. When *Nigella sativa* L., also called as black cumin, that also contains quercetin, was supplemented for four weeks, the subject's mood and anxiety would improve and decrease respectively due to the herbs' potent anticholinesterase and antioxidant activities [50]. We also reasoned the promising results on mood might be because of the possible synergistic effects between the antioxidants and flavonoids in CC supplement. As stated by Daramola (2018), when natural antioxidant in plants is combined with other antioxidants, it could result in additive and synergistic impacts [51]. However, the underlying mechanisms of CC supplementation on mood is yet to be discovered.

Interestingly, there is a significant decrease in malondialdehyde (MDA) levels in CC supplement group as compared to the placebo group after 12 weeks of supplementation. This phenomenon might be due to the presence of antioxidants and flavonoids in CC supplement are believed to be crucial in stopping the creation of free radicals as it is able to curb few lethal actions of reactive oxygen species (ROS) on DNA, lipids, and proteins [52,53]. The flavonoids in CC supplement also possess the abilities to avert DNA oxidative damage, satiate lipid peroxidation, and rummage reactive oxygen species (ROS) such as hydrogen peroxide, superoxide, and hydroxyl radicals [54]. In addition, we also reasoned that antioxidants and flavonoids in CC supplement has potential to reduce the autoxidation pathway by preventing the formation of free radicals. The flavonoids in CC supplement such as quercetin, catechin, and epicatechin might also possibly aid the species that tend to initiate peroxidation, quench superoxide radicals, and disrupt the autoxidative chain reaction, and inhibit the formation of peroxides [20,55,56].

On the other hand, we have also found that there is a significant increase in the percentage of mean change of serum glutathione (GSH) levels in CC supplement group as compared to the placebo group after 12 weeks of supplementation. Glutathione is an antioxidant that protects against cell damage caused by ROS, including peroxides, lipid peroxides, heavy metals and free radicals [57–59]. This could be due to the role of flavonoids in CC that assist the increment of glutathione level in which further scavenged ROS through enzymatic as well as non-enzymatic reactions. The mechanism possibly via the bioavailability of glutathione or glutathione precursors presence that help to boost glutathione levels in the human body [60]. However, the exact amount of glutathione in CC supplement and the bioavailability of glutathione in human body are yet to be investigated.

The possible reasons for our insignificant results in other biochemical profiles and biomarkers may be due to the low bioavailability of flavonoids that reduces the antioxidant and anti-inflammatory capacities, as well as the short duration of the study to improve these biomarkers outcomes. Furthermore, we also did not investigate on the bioavailability of flavonoids from CC supplement in human models. Undoubtedly, there are also other possible risk factors that could affect the biochemical profiles and biomarkers such as stress, social activity involvement, dietary practices, and physical activity that should be considered and measured in future studies [61–64].

The scientific evidence from this study has the potential to trigger health promotions of CC consumption that could also consequently lead to the population's healthier lifestyle, improved quality of life, a further reduced risk of contracting neurodegenerative diseases and lowering healthcare costs related to the disease burden [65]. Furthermore, it would also create a "Knowledge-Economy", of which knowledge would lead to plantations and agricultural industry related to CC plant, thus increase the income of the nation, and further reduce poverty.

The primary strength of our study is the study design of randomized, double-blind, placebo-controlled trial to investigate the effects of CC supplement on cognitive status, biomarkers, health parameters and mood status. This study also focused on older adults with mild cognitive impairment (MCI) who have higher risk of contracting Alzheimer's disease. The main limitation of this study the sample size is small and the duration of the study is considered short as some of the neuropsychological tests and biochemical profile's results were not significantly improved. The mood status that was assessed using self-reported questionnaire is less sensitive for neurological disorders, such as cognitive impairment individuals [66]. The flavonoids other than quercetin and quercitrin were not quantified in this supplement. Future studies would benefit from using a larger and more diverse sample size to further explore the efficacy of CC supplements to generate further statistically significant results on cognitive function, biochemical profiles and mood status in a longer period (i.e., 6 months or 12 months). The full bioactive compound profiles should be identified and quantified using liquid chromatography–mass spectrometry method. The metabolomic approach is also recommended to further determine the efficacy of CC to gauge a better understanding of its cognitive decline preventive properties among older adults with MCI [67,68]. Simultaneously, the serum quercetin and other potential flavonoids are suggested to be included in future studies to have a better comprehension on the bioavailability of flavonoids in human body after consuming the CC supplements.

5. Conclusions

Cosmos caudatus supplement might potentially reduce lipid peroxidation using malondialdehyde biomarkers and increase serum glutathione level. It might also have the ability to improve global cognitive function using neuropsychological test, reduce tension and total mood disturbance among older adults with MCI. The results must be considered preliminary until such effects can be studied in a longer clinical trial and larger sample size to elucidate the neuroprotective effects of CC supplement. The CC supplement reportedly has no harmful effects based on the vital signs, liver function test and renal profile, and there are no serious adverse events were reported after 12 weeks.

6. Patents

The patent has been filed with the code: PI2020005611.

Supplementary Materials: The following is available online at https://www.mdpi.com/2072-6643/13/2/434/s1, Table S1: Baseline blood biochemical profiles between *Cosmos caudatus* supplement group and placebo group subjects.

Author Contributions: Conceptualization, Y.X.Y. and S.S.; methodology, Y.X.Y.; software, Y.X.Y. and M.M.; validation, N.F.R., H.H., M.M., and S.S.; formal analysis, Y.X.Y.; investigation, S.S.; resources, S.S.; data curation, Y.X.Y.; writing—original draft preparation, Y.X.Y.; writing—review and editing, N.F.R., H.H., M.Y.M., H.M.Y., N.C.D., M.M., and S.S.; visualization, Y.X.Y.; supervision, N.F.R., H.H., M.Y.M., H.M.Y., N.C.D., M.M., and S.S.; project administration, Y.X.Y.; funding acquisition, S.S. All authors have read and agreed to the published version of the manuscript.

Funding: This research was funded by the Fundamental Research Grant Scheme (FGRS) provided by the Ministry of Education Malaysia (FRGS/1/2019/SKK02/UKM/01/1) and *Dana Impak Perdana* (DIP-2019-029) funded by the National University of Malaysia.

Institutional Review Board Statement: The study was conducted according to the guidelines of the Declaration of Helsinki, and approved by the Medical Research Ethics Committee Universiti Kebangsaan Malaysia with code NN-2019-137 on 9 October 2019.

Informed Consent Statement: Informed consent was obtained from all subjects involved in the study prior to data collection.

Data Availability Statement: The datasets of this study available from the corresponding author on reasonable request.

Acknowledgments: We would like to express our gratitude to all fieldworkers, HCARE staffs, local authorities, enumerators and participants for their involvement in this study.

Conflicts of Interest: The authors declare no conflict of interest.

References

1. United Nations. *World Populaton Ageing*; United Nations: New York, NY, USA, 2015.
2. United Nations. *World Populaton Ageing*; United Nations: New York, NY, USA, 2019.
3. Tey, N.P.; Siraj, S.B.; Kamaruzzaman, S.B.; Chin, A.V.; Tan, M.P.; Sinnappan, G.S.; Muller, A.M. Aging in Multi-ethnic Malaysia. *Gerontologist* **2016**, *56*, 603–609. [CrossRef] [PubMed]
4. Department of Statistics Malaysia. *Current Population Estimates Malaysia 2018–2019*; Department of Statistics: Putrajaya, Malaysia, 2019.
5. World Health Organization. *Risk Reduction of Cognitive Decline and Dementia: WHO Guidelines*; World Health Organization: Geneva, Switzerland, 2019.

6. Vanoh, D.; Shahar, S.; Din, N.C.; Omar, A.; Vyrn, C.A.; Razali, R.; Ibrahim, R.; Hamid, T.A. Predictors of poor cognitive status among older Malaysian adults: Baseline findings from the LRGS TUA cohort study. *Aging Clin. Exp. Res.* **2017**, *29*, 173–182. [CrossRef] [PubMed]
7. Boccardi, V.; Baroni, M.; Mangialasche, F.; Mecocci, P. Vitamin E family: Role in the pathogenesis and treatment of Alzheimer's disease. *Alzheimer's Dement. Transl. Res. Clin. Interv.* **2016**, *2*. [CrossRef]
8. Mostafavi, S.-A.; Hosseini, S. Chapter 7—Foods and Dietary Supplements in the Prevention and Treatment of Neurodegenerative Diseases in Older Adults. In *Foods and Dietary Supplements in the Prevention and Treatment of Disease in Older Adults*; Watson, R.R., Ed.; Academic Press: San Diego, CA, USA, 2015; pp. 63–67. [CrossRef]
9. Nuzzo, D.; Amato, A.; Picone, P.; Terzo, S.; Galizzi, G.; Bonina, F.P.; Mulè, F.; Di Carlo, M. A Natural Dietary Supplement with a Combination of Nutrients Prevents Neurodegeneration Induced by a High Fat Diet in Mice. *Nutrients* **2018**, *10*, 1130. [CrossRef] [PubMed]
10. Rosenberg, A.; Mangialasche, F.; Ngandu, T.; Solomon, A.; Kivipelto, M. Multidomain Interventions to Prevent Cognitive Impairment, Alzheimer's Disease, and Dementia: From FINGER to World-Wide FINGERS. *J. Prev. Alzheimer's Dis.* **2020**, *7*, 29–36. [CrossRef]
11. Lau, H.; Shahar, S.; Mohamad, M.; Rajab, N.F.; Yahya, H.M.; Din, N.C.; Hamid, H.A. The effects of six months Persicaria minor extract supplement among older adults with mild cognitive impairment: A double-blinded, randomized, and placebo-controlled trial. *BMC Complement. Med. Ther.* **2020**, *20*, 315. [CrossRef]
12. Chen, L.-E.; Wu, F.; Zhao, A.; Ge, H.; Zhan, H. Protection Efficacy of the Extract ofGinkgo biloba against the Learning and Memory Damage of Rats under Repeated High Sustained +Gz Exposure. *Evid. Based Complement. Altern. Med.* **2016**, *2016*, 6320586. [CrossRef]
13. Gschwind, Y.J.; Bridenbaugh, S.A.; Reinhard, S.; Granacher, U.; Monsch, A.U.; Kressig, R.W. Ginkgo biloba special extract LI 1370 improves dual-task walking in patients with MCI: A randomised, double-blind, placebo-controlled exploratory study. *Aging Clin. Exp. Res.* **2017**, *29*, 609–619. [CrossRef]
14. Puttarak, P.; Dilokthornsakul, P.; Saokaew, S.; Dhippayom, T.; Kongkaew, C.; Sruamsiri, R.; Chuthaputti, A.; Chaiyakunapruk, N. Effects of *Centella asiatica* (L.) Urb. on cognitive function and mood related outcomes: A Systematic Review and Meta-analysis. *Sci. Rep.* **2017**, *7*, 10646. [CrossRef]
15. Yahya, H.; Shahar, S.; Nur Arina Ismail, S.; Aziz, A.; Normah, C.D.; Hakim, B.N. Mood, Cognitive Function and Quality of Life Improvements in Middle Aged Women Following Supplementation with Polygonum minus Extract. *Sains Malays.* **2017**, *46*, 245–254. [CrossRef]
16. Hooshmand, B.; Mangialasche, F.; Kalpouzos, G.; Solomon, A.; Kåreholt, I.; Smith, D.; Refsum, H.; Wang, R.; Muehlmann, M.; Birgit, E.-W.; et al. Association of Vitamin B12, Folate, and Sulfur Amino Acids with Brain Magnetic Resonance Imaging Measures in Older Adults: A Longitudinal Population-Based Study. *JAMA Psychiatry* **2016**, *73*. [CrossRef] [PubMed]
17. Mangialasche, F.; Solomon, A.; Kåreholt, I.; Hooshmand, B.; Cecchetti, R.; Fratiglioni, L.; Soininen, H.; Laatikainen, T.; Mecocci, P.; Kivipelto, M. Serum levels of vitamin E forms and risk of cognitive impairment in a Finnish cohort of older adults. *Exp. Gerontol.* **2013**, *48*. [CrossRef] [PubMed]
18. Lau, H.; Shahar, S.; Mohamad, M.; Rajab, N.F.; Yahya, H.M.; Din, N.C.; Hamid, H.A. Relationships between dietary nutrients intake and lipid levels with functional MRI dorsolateral prefrontal cortex activation. *Clin. Interv. Aging* **2018**, *14*, 43–51. [CrossRef] [PubMed]
19. Scarmeas, N.; Anastasiou, C.A.; Yannakoulia, M. Nutrition and prevention of cognitive impairment. *Lancet Neurol.* **2018**, *17*, 1006–1015. [CrossRef]
20. Moshawih, S.; Cheema, M.; Ahmad, Z.; Zakaria, Z.A.; Nazrul Hakim, M. A Comprehensive Review on Cosmos caudatus (Ulam Raja): Pharmacology, Ethnopharmacology, and Phytochemistry. *Int. Res. J. Educ. Sci.* **2017**, *1*, 14–31.
21. Sumazian, Y.; Ahmad, S.; Hakiman, M.; Maziah, M. Antioxidant Activities, Flavonoids, Ascorbic Acid and Phenolic Content of Malaysian Vegetables. *J. Med. Plants Res.* **2010**, *4*, 881–890. [CrossRef]
22. You, Y.; Shahar, S.; Haron, H.; Yahya, H. More Ulam for Your Brain: A Review on the Potential Role of Ulam in Protecting Against Cognitive Decline. *Sains Malays.* **2018**, *47*, 2713–2729. [CrossRef]
23. You, Y.X.; Shahar, S.; Mohamad, M.; Yahya, H.M.; Haron, H.; Abdul Hamid, H. Does traditional asian vegetables (ulam) consumption correlate with brain activity using fMRI? A study among aging adults from low-income households. *J. Magn. Reson. Imaging* **2019**, *51*. [CrossRef]
24. Spencer, B.; Verma, I.; Desplats, P.; Morvinski, D.; Rockenstein, E.; Adame, A.; Masliah, E. A neuroprotective brain-penetrating endopeptidase fusion protein ameliorates Alzheimer disease pathology and restores neurogenesis. *J. Biol. Chem.* **2014**, *289*, 17917–17931. [CrossRef]
25. Lei, X.; Chao, H.; Zhang, Z.; Lv, J.; Li, S.; Wei, H.; Xue, R.; Li, F.; Li, Z. Neuroprotective effects of quercetin in a mouse model of brain ischemic/reperfusion injury via anti-apoptotic mechanisms based on the Akt pathway. *Mol. Med. Rep.* **2015**, *12*, 3688–3696. [CrossRef]
26. Chan, Y.H. Randomised controlled trials (RCTs)—Sample size: The magic number? *Singap. Med. J.* **2003**, *44*, 172–174.
27. Petersen, R.C.; Caracciolo, B.; Brayne, C.; Gauthier, S.; Jelic, V.; Fratiglioni, L. Mild cognitive impairment: A concept in evolution. *J. Intern. Med.* **2014**, *275*, 214–228. [CrossRef] [PubMed]

28. Mohamed, N.; Ehsan, S.Z.; Noor 'Adilah, K.; Lee, C.P.; Farhana, E.; Derick, P.; Soelaiman, I.; Nazrun, A.S.; Muhammad, N. Acute Toxicity Study of Cosmos caudatus on Biochemical Parameters in Male Rats. *Sains Malays.* **2013**, *42*, 1247–1251.
29. Abdullah, A.; Kaur Dhaliwal, K.; Nabillah, N.; Roslan, F.; Chai, H.; Lee, M.; Miszard Kalaiselvam, H.; Radman, Q.; Haji, M.; Saad, K.; et al. The Effects of Cosmos caudatus (Ulam Raja) on Detoxifying Enzymes in Extrahepatic Organs in Mice ARTICLE INFO ABSTRACT. *J. Appl. Pharm. Sci.* **2015**, *5*, 82–88. [CrossRef]
30. Shahar, S.; Earland, J.; Abdulrahman, S. Validation of a Dietary History Questionnaire against a 7-D Weighed Record for Estimating Nutrient Intake among Rural Elderly Malays. *Malays. J. Nutr.* **2000**, *6*, 33–44.
31. Ibrahim, N.M.; Shohaimi, S.; Chong, H.T.; Rahman, A.H.; Razali, R.; Esther, E.; Basri, H.B. Validation study of the Mini-Mental State Examination in a Malay-speaking elderly population in Malaysia. *Dement. Geriatr. Cogn. Disord.* **2009**, *27*, 247–253. [CrossRef]
32. Weshsler, D. *Wechsler Adult Intelligence Scale-III*; The Psychological Corporation San Antonio: San Antonio, TX, USA, 1997.
33. Jamaluddin, R.; Othman, Z.; Musa, K.I.; Alwi, M.N.M. Validation of The Malay Version of Auditory Verbal Learning Test (Mvavlt) Among Schizophrenia Patients In Hospital Universiti Sains Malaysia (Husm), Malaysia. *Asean J. Psychiatry* **2009**, *10*, 54–68.
34. McNair, D.M.; Lorr, M.; Droppleman, L.F. *Profile of Mood States*; Educational and Industrial Testing Service: San Diego, CA, USA, 1971.
35. Gupta, U.; Verma, M. Placebo in clinical trials. *Perspect. Clin. Res.* **2013**, *4*, 49–52. [CrossRef]
36. Ferri, P.; Angelino, D.; Gennari, L.; Benedetti, S.; Ambrogini, P.; Del Grande, P.; Ninfali, P. Enhancement of flavonoid ability to cross the blood-brain barrier of rats by co-administration with α-tocopherol. *Food Funct.* **2015**, *6*, 394–400. [CrossRef]
37. Youdim, K.; Shukitt-Hale, B.; Joseph, J. Flavonoids and the brain: Interactions at the blood-brain barrier and their physiological effects on the central nervous system. *Free Radic. Biol. Med.* **2005**, *37*, 1683–1693. [CrossRef]
38. Figueira, I.; Garcia, G.; Pimpão, R.C.; Terrasso, A.P.; Costa, I.; Almeida, A.F.; Tavares, L.; Pais, T.F.; Pinto, P.; Ventura, M.R.; et al. Polyphenols journey through blood-brain barrier towards neuronal protection. *Sci. Rep.* **2017**, *7*, 11456. [CrossRef] [PubMed]
39. Rendeiro, C.; Rhodes, J.S.; Spencer, J.P. The mechanisms of action of flavonoids in the brain: Direct versus indirect effects. *Neurochem. Int.* **2015**, *89*, 126–139. [CrossRef] [PubMed]
40. Jäger, K.A.; Saaby, L. Flavonoids and the CNS. *Molecules* **2011**, *16*, 1471–1485. [CrossRef] [PubMed]
41. Vauzour, D.; Vafeiadou, K.; Rodriguez-Mateos, A.; Rendeiro, C.; Spencer, J.P.E. The neuroprotective potential of flavonoids: A multiplicity of effects. *Genes Nutr.* **2008**, *3*, 115–126. [CrossRef]
42. Folch, J.; Ettcheto, M.; Petrov, D.; Abad, S.; Pedros, I.; Marin, M.; Olloquequi, J.; Camins, A. Review of the advances in treatment for Alzheimer disease: Strategies for combating beta-amyloid protein. *Neurologia* **2018**, *33*, 47–58. [CrossRef]
43. Jiang, N.; Doseff, A.I.; Grotewold, E. Flavones: From Biosynthesis to Health Benefits. *Plants* **2016**, *5*, 27. [CrossRef]
44. Costa, L.G.; Garrick, J.M.; Roquè, P.J.; Pellacani, C. Mechanisms of Neuroprotection by Quercetin: Counteracting Oxidative Stress and More. *Oxid. Med. Cell Longev* **2016**, *2016*, 2986796. [CrossRef]
45. Bigelman, K.; Chapman, D.; Freese, E.; Trilk, J.; Cureton, K. Effects of 6 Weeks of Quercetin Supplementation on Energy, Fatigue, and Sleep in ROTC Cadets. *Mil. Med.* **2011**, *176*, 565–572. [CrossRef]
46. Spencer, J.P.; Rice-Evans, C.; Williams, R.J. Modulation of pro-survival Akt/protein kinase B and ERK1/2 signaling cascades by quercetin and its in vivo metabolites underlie their action on neuronal viability. *J. Biol. Chem.* **2003**, *278*, 34783–34793. [CrossRef]
47. Ma, J.Q.; Luo, R.Z.; Jiang, H.X.; Liu, C.M. Quercitrin offers protection against brain injury in mice by inhibiting oxidative stress and inflammation. *Food Funct.* **2016**, *7*, 549–556. [CrossRef]
48. Yahfoufi, N.; Alsadi, N.; Jambi, M.; Matar, C. The Immunomodulatory and Anti-Inflammatory Role of Polyphenols. *Nutrients* **2018**, *10*, 1618. [CrossRef] [PubMed]
49. Udani, J.K. Effects of SuperUlam on Supporting Concentration and Mood: A Randomized, Double-Blind, Placebo-Controlled Crossover Study. *Evid. Based Complement. Altern. Med. Ecam* **2013**, *2013*, 238454. [CrossRef] [PubMed]
50. Sayeed, M.S.B.; Shams, T.; Fahim Hossain, S.; Rahman, M.R.; Mostofa, A.; Fahim Kadir, M.; Mahmood, S.; Asaduzzaman, M. Nigella sativa L. seeds modulate mood, anxiety and cognition in healthy adolescent males. *J. Ethnopharmacol.* **2014**, *152*, 156–162. [CrossRef] [PubMed]
51. Daramola, B. Preliminary investigation on antioxidant interactions between bioactive components of Solanum anguivi and Capsicum annuum. *J. Food Sci. Technol.* **2018**, *55*, 3827–3832. [CrossRef] [PubMed]
52. Chang, J.; Tse, C.S.; Leung, G.T.; Fung, A.W.; Hau, K.T.; Chiu, H.F.; Lam, L.C. Bias in discriminating very mild dementia for older adults with different levels of education in Hong Kong. *Int. Psychogeriatr.* **2014**, *26*, 995–1010. [CrossRef] [PubMed]
53. Suganthy, N.; Devi, K.P.; Nabavi, S.F.; Braidy, N.; Nabavi, S.M. Bioactive effects of quercetin in the central nervous system: Focusing on the mechanisms of actions. *Biomed. Pharmacother. Biomed. Pharmacother.* **2016**, *84*, 892–908. [CrossRef] [PubMed]
54. García-Blanco, A.; Baquero, M.; Vento, M.; Gil, E.; Bataller, L.; Cháfer-Pericás, C. Potential oxidative stress biomarkers of mild cognitive impairment due to Alzheimer disease. *J. Neurol. Sci.* **2017**, *373*, 295–302. [CrossRef]
55. Gaschler, M.M.; Stockwell, B.R. Lipid peroxidation in cell death. *Biochem. Biophys. Res. Commun.* **2017**, *482*, 419–425. [CrossRef]
56. Wojsiat, J.; Zoltowska, K.M.; Laskowska-Kaszub, K.; Wojda, U. Oxidant/Antioxidant Imbalance in Alzheimer's Disease: Therapeutic and Diagnostic Prospects. *Oxid. Med. Cell Longev.* **2018**, *2018*, 6435861. [CrossRef]
57. Pisoschi, A.M.; Pop, A.; Cimpeanu, C.; Predoi, G. Antioxidant Capacity Determination in Plants and Plant-Derived Products: A Review. *Oxid. Med. Cell Longev.* **2016**, *2016*, 9130976. [CrossRef]

58. Mandal, P.K.; Shukla, D.; Tripathi, M.; Ersland, L. Cognitive Improvement with Glutathione Supplement in Alzheimer's Disease: A Way Forward. *J. Alzheimer's Dis.* **2019**, *68*, 531–535. [CrossRef] [PubMed]
59. Youssef, P.; Chami, B.; Lim, J.; Middleton, T.; Sutherland, G.T.; Witting, P.K. Evidence supporting oxidative stress in a moderately affected area of the brain in Alzheimer's disease. *Sci. Rep.* **2018**, *8*, 11553. [CrossRef] [PubMed]
60. Winterbourn, C.C. Revisiting the reactions of superoxide with glutathione and other thiols. *Arch. Biochem. Biophys.* **2016**, *595*, 68–71. [CrossRef] [PubMed]
61. Desai, S.J.; Prickril, B.; Rasooly, A. Mechanisms of Phytonutrient Modulation of Cyclooxygenase-2 (COX-2) and Inflammation Related to Cancer. *Nutr. Cancer* **2018**, *70*, 350–375. [CrossRef] [PubMed]
62. Kelly, Á.M. Exercise-Induced Modulation of Neuroinflammation in Models of Alzheimer's Disease. *Brain Plast.* **2018**, *4*, 81–94. [CrossRef] [PubMed]
63. Morgan, A.; Kondev, V.; Bedse, G.; Baldi, R.; Marcus, D.; Patel, S. Cyclooxygenase-2 inhibition reduces anxiety-like behavior and normalizes enhanced amygdala glutamatergic transmission following chronic oral corticosterone treatment. *Neurobiol. Stress* **2019**, *11*, 100190. [CrossRef]
64. Newcombe, E.A.; Camats-Perna, J.; Silva, M.L.; Valmas, N.; Huat, T.J.; Medeiros, R. Inflammation: The link between comorbidities, genetics, and Alzheimer's disease. *J. Neuroinflamm.* **2018**, *15*, 276. [CrossRef]
65. Shahar, S.; Lau, H.; Puteh, S.E.W.; Amara, S.; Razak, N.A. Health, access and nutritional issues among low-income population in Malaysia: Introductory note. *BMC Public Health* **2019**, *19*, 552. [CrossRef]
66. Siwek, M.; Dudek, D.; Rybakowski, J.; Łojko, D.; Pawłowski, T.; Kiejna, A. Mood Disorder Questionnaire—Characteristic and indications. *Psychiatr. Pol.* **2009**, *43*, 287–299.
67. McGhie, T.K.; Rowan, D.D. Metabolomics for measuring phytochemicals, and assessing human and animal responses to phytochemicals, in food science. *Mol. Nutr. Food Res.* **2012**, *56*, 147–158. [CrossRef]
68. Beckonert, O.; Keun, H.C.; Ebbels, T.M.; Bundy, J.; Holmes, E.; Lindon, J.C.; Nicholson, J.K. Metabolic profiling, metabolomic and metabonomic procedures for NMR spectroscopy of urine, plasma, serum and tissue extracts. *Nat. Protoc.* **2007**, *2*, 2692–2703. [CrossRef] [PubMed]

MDPI
St. Alban-Anlage 66
4052 Basel
Switzerland
Tel. +41 61 683 77 34
Fax +41 61 302 89 18
www.mdpi.com

Nutrients Editorial Office
E-mail: nutrients@mdpi.com
www.mdpi.com/journal/nutrients

www.ingramcontent.com/pod-product-compliance
Lightning Source LLC
LaVergne TN
LVHW070923170726
843515LV00005B/850

* 9 7 8 3 0 3 6 5 6 1 1 9 6 *